THE PSEUDOSCIENCE WARS

IMMANUEL VELIKOVSKY (1895–1979), circa 1977. Source: IVP 145:2.

THE PSEUDOSCIENCE WARS

IMMANUEL VELIKOVSKY
and the Birth of the Modern Fringe

Michael D. Gordin

The University of Chicago Press CHICAGO AND LONDON

MICHAEL D. GORDIN is a professor in the Department of History at Princeton University. He is the author of *A Well-Ordered Thing: Dmitrii Mendeleev and the Shadow of the Periodic Table; Five Days in August: How World War II Became a Nuclear War;* and *Red Cloud at Dawn: Truman, Stalin, and the End of the Atomic Monopoly.*

The University of Chicago Press, Chicago 60637
The University of Chicago Press, Ltd., London
© 2012 by The University of Chicago
All rights reserved. Published 2012.
Printed in the United States of America

21 20 19 18 17 16 15 14 13 12   1 2 3 4 5

ISBN-13: 978-0-226-30442-7 (cloth)
ISBN-10: 0-226-30442-6 (cloth)
ISBN-13: 978-0-226-30443-4 (e-book)
ISBN-10: 0-226-30443-4 (e-book)

Library of Congress Cataloging-in-Publication Data

Gordin, Michael D.
   The pseudoscience wars: Immanuel Velikovsky and the birth of the modern fringe / Michael D. Gordin.
      pages; cm
   Includes bibliographical references and index.
   ISBN-13: 978-0-226-30442-7 (cloth: alkaline paper)
   ISBN-10: 0-226-30442-6 (cloth: alkaline paper)
   ISBN-13: 978-0-226-30443-4 (e-book)
   ISBN-10: 0-226-30443-4 (e-book)   1. Pseudoscience—History—
20th century.   2. Velikovsky, Immanuel, 1895–1979—Appreciation.
3. Velikovsky, Immanuel, 1895–1979—Influence.   4. Lysenko, Trofim
Denisovich, 1898–1976.   5. Creationism—History.   6. Eugenics—
History.   7. Science—History—20th century.   I. Title.
   Q172.5.P77G674 2012
   523.2—dc23   2012003653

♾ This paper meets the requirements of ANSI/NISO Z39.48-1992 (Permanence of Paper).

FOR ERIKA,
who was there at the beginning and the end

No probability, however seductive, can protect us from error; even if all parts of a problem seem to fit together like the pieces of a jigsaw puzzle, one has to remember that the probable need not necessarily be the truth, and the truth not always probable.

SIGMUND FREUD, *Moses and Monotheism*

Nonsense is nonsense, but the history of nonsense is a very important science.

attributed to SAUL LIEBERMAN

CONTENTS

ACKNOWLEDGMENTS

This book began in the archives, and my greatest debts are to those who made the research possible. At the Princeton University Library, I would like to thank Ben Primer and especially Don Skemer for taking the time to discuss with me how they came to acquire the Immanuel Velikovsky Papers. The entire staff at the Manuscripts Reading Room also provided a wealth of help at all stages of this project, as did the Article Express and Interlibrary Loan office, who once again managed to acquire for me publications that seemed to have sunk without a trace. I could never have written this book without them.

I would also like to thank the librarians at the American Philosophical Society Library, the Manuscripts Room of the New York Public Library, and Harvard University Archives for making their collections available to me. All of these institutions, as well as the Albert Einstein Papers Project, kindly granted me permission to reproduce materials from their holdings. My parents, Gila and Rafael Gordin, put me in touch with Rafael Vieser, formerly of the Hebrew University Library, and he kindly informed me about the Velikovsky collections there.

Many friends and colleagues have been exposed to a great deal of Velikovskiana over the last few years, and I extend thanks to all of them (you know who you are). Peter Brown, Nathaniel Comfort, Angela Creager, Yaacob Dweck, James Gilbert, Owen Gingerich, Matthew Jones, David Kaiser, Joshua Katz, Philip Kitcher, Michiko Kobayashi, George Laufenberg, Patrick McCray, Erika Milam, Bhavani Raman, Matthew Stanley, Robin Wasserman, Eric Weitz, and Nasser Zakariya read parts or all of the manuscript, and their comments have improved the result in innumerable ways. I presented various parts of this project at the University of Maryland–College Park, Caltech, Tel Aviv University, both the History of Science Program Seminar and the Shelby Cullom Davis Center for Historical Studies at Princeton University, Cornell University, and ETH-Zürich, and the probing questions I received guided many of the revisions. Finally, Nikolai Krementsov, Michael Ruse, and Michael Shermer evaluated this manuscript for the University of Chicago Press, and I very much appreciate both their positive and their critical readings of the text. I have incorporated everything I could. The remaining errors are, of course, entirely my own.

I owe a special debt of gratitude to two individuals involved in the events related here, from very different positions. Immanuel Velikovsky's daughter Dr. Ruth Velikovsky Sharon very graciously met with me early in my research and spoke to me about her father, and I appreciate both her time and her understanding of the nature of this historical project. She is also responsible for the generous donation of his papers to Princeton University, and I am tremendously grateful for that as well.

In addition, C. Leroy Ellenberger shared many valuable materials he has gathered on the various aspects of the Velikovsky debates and also read a draft of the manuscript. His comments have always proven helpful, even when I elected not to adopt them.

A version of chapter 3 was published as "How Lysenkoism Became Pseudoscience: Dobzhansky to Velikovsky," *Journal of the History of Biology* 45 (2012).

A great deal of labor besides the author's is involved in turning an idea into a book. Karen Merikangas Darling, my editor at the University of Chicago Press, has been fantastic at every stage. She read each chapter hot off the keyboard and helped to shape this book into its final form. It was a true pleasure working with her. Don Lamm and Christy Fletcher found this project a home, and Ryan Dahn assisted in obtaining permissions for the archival material. My thanks to all of them.

Last, but furthest from least, my gratitude to Erika Milam. When we met, this book did not even exist as an idea in my head, and thanks to her it is now here. She stopped it from taking me over entirely. Her thoughts, love, and support have meant everything. She can rest assured that it is now done.

Introduction: Bad Ideas

No one in the history of the world has ever self-identified as a pseudo-scientist. There is no person who wakes up in the morning and thinks to himself, "I'll just head into my pseudolaboratory and perform some pseudoexperiments to try to confirm my pseudotheories with pseudo-facts." As is surely obvious, "pseudoscience" is a term of abuse, an epithet attached to certain points of view to discredit those ideas, complemented by "pseudoscientist" to designate the practitioner. Just as no adherents of a religious doctrine ever really consider themselves "heretics," alleged pseudoscientists have a very specific understanding of their activities. To their minds, they are doing science, full stop. This does not mean they are necessarily correct—lots of people are mistaken about what they are actually doing—but it should give us pause to think a bit harder concerning what the word "pseudoscience" really does.

Does, not means. "Pseudoscience" is a term, I maintain, without real content, and yet the notion performs active work in the world, separating off certain doctrines from those deemed to be science proper. On the imagined scale that has excellent science at one end and then slides through good science, mediocre science (the vast majority of what is done), poor science, to bad science on the other end, it is *not* the case that pseudoscience lies somewhere on this continuum. It is off the grid altogether.[1] The process of demarcating science from non-science is a central and quite general aspect of all scientific activities, but pseudoscience attracts particular vehemence as compared to, say, non-science. Scientists rarely spend much energy arguing that the Catholic Church or Vietnamese literature is pseudoscience; they are just not science—and devotees of those domains are quite happy with that designation. Pseudoscience is different. This is a combative notion deployed to categorize (and, its users hope, weaken or eliminate) doctrines that are non-science but pretend to be, aspire to be, or are simply mistaken for scientific. The effect of this demarcation through use of the moniker "pseudoscience," when it works, is to preserve the accepted boundaries of knowledge from intrusion.[2] In the end, pseudoscience is a bad idea. I do not mean that one should not practice phrenology, astrology, or what have you, but that the very notion of "pseudoscience" lacks a core.

Although pseudoscience is a fairly common epithet, it is not exactly universal. Scientists do not just call anything they do not like "pseudoscience." They are perfectly happy to declare many of their peers' work to be "bad" or "substandard" science. "Pseudoscience" is used in a targeted way, at certain times, and against specific enemies. This implies that there is no unified pseudoscience; the various doctrines labeled "pseudosciences" over the last two centuries actually have very little in common with one another besides being hated by assorted scientists. Ever since the term was introduced into the English language—at roughly the same moment as the word "scientist,"[3] which is surely no accident, for how could you mimic a category that does not exist?—skirmishes over designating certain fields as pseudosciences have escalated and de-escalated along with the general perception of the threatened or secure status of science. The *Oxford English Dictionary* tells us the term, meaning "a spurious or pretended science," entered the English lexicon in 1796 to refer to "alchymy," and then popped up again around 1823 concerning blazonry (the interpretation of heraldic insignias), of all things.[4] Surely those two fields were not related then, as they are not now. We are faced with a variant of the classic story of three blind men encountering an elephant. One holds the tail, and thinks it is a piece of string; another grabs a leg, and thinks he is holding a tree; the third holds the trunk, and believes he grasps a snake. Only, in the case of pseudoscience, they really are holding a piece of string, a tree trunk, and a snake. There is no elephant.

What unifies the so-called pseudosciences is that scientists in various fields have chosen to ostracize them in this particular way (as opposed to declaring them incorrect scientific theories). It is a core argument of this book that individual scientists (as distinct from the monolithic "scientific community") designate a doctrine a "pseudoscience" only when they perceive themselves to be threatened—not necessarily by the ideas themselves, but by what those ideas represent about the authority of science, science's access to resources, or some other broader social trend.[5] If one is not threatened, there is no need to lash out at the perceived pseudoscience;

instead, one continues with one's work and happily ignores the cranks. This means that we can examine the history of debates over pseudosciences past in order to explore not what disqualified a particular doctrine (say, astrology[6]) from membership in the scientific club, but rather to understand *science* and what scientists thought about their standards, their position in society, and their future. Pseudoscience, as historian of science Mark Adams points out in an essay on the history of eugenics, is "less interesting as a mode of historical explanation than as an object of historical study; it is not part of the solution, but part of the problem."[7] Each use of pseudoscience is tied intimately to its historical context. If you want to know what science is or has been, show me the contemporary pseudoscience.[8]

This book examines a specific contentious period for the status of science—Cold War America, from the late 1940s to the late 1970s—by exploring a series of debates over what counted as real science. (I exclude the cases of so-called pseudomedicine or quackery, which is a vastly larger topic and quite amenable to a similar investigation.[9]) These "pseudoscience wars," as I call them, raised scientists' anxiety over the incursions of "pseudoscience" among their students and the public at large to a fever pitch. Before 1950, debates over pseudoscience ran hot, but they did not in general exhibit the character of conspiracy theorizing. During the pseudoscience wars, doctrines that were relegated kicking and screaming to the "fringe" began to respond by deploying new arguments against the establishment, claiming not just that mainstream science was incorrect or incomplete (as, for example, in the controversies over J. B. Rhine's parapsychology experiments in the 1930s), but that scientists were engaged in a conspiracy to suppress new knowledge. It was no accident that this transition unfolded in the early 1950s, when America was gripped in the frenzy of McCarthyist red-baiting and nationwide panic about conspiracies to undermine the West. Through the contingent juxtaposition of a new bout of disputes over the boundaries of science and this tense domestic Cold War context, features of the paranoid style of the moment became rooted into the discourse of the fringe, a pattern that has stuck with us long after the passing of anti-Communist hysteria.[10] Arguments from the fringes of science today carry some of the last vestiges of this particular moment of American history, fossilized in amber.

This transformation was large, but it began with a specific controversy over one work and its author, and the chapters below will follow both from 1950 to 1980 to show how this one controversy carried along other pseudoscience conflicts with it, as a mishmash of diverse doctrines began to gel, if not into a single pseudoscience (for there is no elephant), certainly into the coherent conflict of the pseudoscience wars. It started with a book and a man. The book was *Worlds in Collision*, and the man was Immanuel Velikovsky (1895–1979).

If anyone has ever been tarnished by the accusation of pseudoscientist, it is Immanuel Velikovsky, a Russian-born psychoanalyst who arrived on American shores in 1939 after several convoluted peregrinations. In many of the accounts of this man and his life, the word "pseudoscience" crops up.[11] He has been variously dubbed one of the "deans of modern pseudoscience,"[12] the "first grand wizard of the Universal Order of Mass Pseudo-Scholarship,"[13] "an almost perfect textbook example of the pseudo-scientist,"[14] one of the "triad of pseudoscience gurus" (along with L. Ron Hubbard and Charles Fort),[15] and "the very model of a crank."[16] These allegations were not uncontested, and his supporters—who began to assemble in force by the mid-1960s—insisted with Frederic Jueneman that "Velikovsky's efforts are not the labors of a pseudoscientist, because his work has touched on too many things which preemptively have been proven correct, or with furthered knowledg[e] might be proven one way or another."[17] But Velikovsky was not just a combatant in the pseudoscience wars. He and his doctrines were ground zero.

At this point, you may very well be scratching your head. Depending largely on your age, the name Velikovsky recalls fond memories of college, waves of outrage, or a complete and utter blank. In my informal (and profoundly unscientific) polling of individuals over the last few years, I have almost never found a person under the age of fifty who has heard the name. (The exceptions were astronomers, intense science-fiction fans, or aficionados of scholarly arcana.) And yet, in the 1970s, his writings were mainstays of college bookstores, and the man himself cycled through campuses, the pages of popular journals, and the columns of newspapers. He was, as such things go, a household name, a celebrity from the world of scientific controversies whose books went through over seventy editions in English alone during his lifetime (and were translated into dozens of languages).

In April 1950 the Macmillan Company published *Worlds in Collision*, which rocketed to the top of nonfiction best-seller lists nationwide.[18] In this book, Velikovsky argued that ancient mythological, scriptural, and historical sources from a variety of cultures contained repeated homologous descriptions of major catastrophes: rains of fire, immense earthquakes, tsunamis, dragons fighting in the heavens. These passages had long been interpreted by rationalist readers as metaphors or ecstatic visions. Not so, argued Velikovsky: when compared and synchronized, they pointed toward real and massive global catastrophes. Velikovsky tracked two of these: one that happened around 1500 B.C., during the Exodus of the Children of Israel from Egypt; and another in the eighth century B.C., which changed the length of the year from 360 days to its current 365¼ days, stunning the prophet Isaiah and depicted in Homer's *Iliad* as the battle between Athena and Ares.

This was the first salvo of the pseudoscience wars, an invasion into the heavily fortified domain of American science, at the time enjoying peacetime prosperity and elevated prestige due to its triumphs during World War II. The incursion was not completely without warning—earlier attempts to dissuade Velikovsky or his publishers from releasing the book had broken down, a failure of diplomacy—and the defensive maneuvers were rapid and, in retrospect, surprisingly vigorous. Scientists expressed significant doubts about the reality of such catastrophes in historical times, but the greatest sticking point was his mechanism for their occurrence. Velikovsky claimed that the first (at the time of the Exodus) was caused by a comet that had been ejected from Jupiter and almost collided with Earth, remaining trapped in gravitational and electromagnetic interaction with this planet on two separate incidents separated by fifty-two years, raining petroleum from its cometary tail, igniting the heavens, and tilting Earth's axis. Eventually, the comet stabilized into the planet Venus. Thus, Earth's nearest planetary neighbor was a comet born in historical times, as attested by proper interpretation of the records of the collective memory of humanity. Venus's movements had, however, displaced Mars, which threatened Earth in the second series of catastrophes. Velikovsky's arguments presupposed a reformulation of geology, paleontology, archaeology, and celestial mechanics, not to mention ancient history.

From the point of view of the defenders of science on the front line, Velikovsky had not only set up beachheads in their domains, but he incited a fifth column of humanist intellectuals and the broader public, who eagerly read his book and called for scientists to take his arguments seriously. This they were not about to do, and after a series of literary volleys—including a threatened boycott of Velikovsky's publisher—*Worlds in Collision* was transferred to a more commercial press and the guns began to go silent. The beachhead remained, however, and Velikovsky dug in during the 1950s, attempting to recruit allies among mainstream scientists through a series of renewed diplomatic overtures to broker a longer-lasting peace. For the most part, scientists ignored these efforts, opting instead for a return to normalcy, as if Velikovsky were not continuing to launch his books across the demilitarized zone.

In the mid-1960s, hostilities re-erupted, but this time not because Velikovsky rolled out his tanks for another assault. Rather, insurgents behind the scientists' own lines—undergraduates at their institutions, the "counterculture," and even humanist academics—marched forth under Velikovsky's colors, in many ways appropriating for themselves a cause that was different from the author's own. For two decades, a heated debate persisted in the United States: Was Velikovsky right? Had the discoveries of the space age confirmed or refuted his picture of the solar system's history? Counterinsurgency measures, and even one high-level attempt to negotiate with Velikovsky's forces at the American Association for the Advancement of Science annual meeting in 1974, came to naught, and the quagmire deepened. Unlike in 1950, however, this time there was no clear front line. The pseudoscience wars persisted as low-intensity conflicts, and they burned in a plethora of redoubts of American culture.

And then, sometime in the early 1980s, Velikovsky dropped out of the collective consciousness, and his name is now a distant memory—when it is a memory at all. The war did not so much end as fade away. A main emphasis of the pages that follow is on finding out why he became well known in the first place. How did *Worlds in Collision* assume such a prominent position in the public imagination? Why was Velikovsky the target of so much ire from the scientific community? And what does his story tell us about science in American culture during the height of the Cold War?

This is not a biography of Velikovsky, or an attempt to debunk him or exonerate him, or even a judicious weighing of the arguments in favor of and against the picture of the world that he built up in many writings over the course of his career. An interested reader can turn to many other places for such accounts.[19] Rather than merely reprise the Velikovsky debates—as fascinating as those are—I mean to explore this notion of the "pseudoscience wars." Every chapter after the first (which lays out the context of the controversy over *Worlds in Collision*) juxtaposes Velikovsky's case with that of one or more purported pseudosciences: Freudianism, *Welteislehre*, Lysenkoism, eugenics, parapsychology, creationism, orgone theory, ancient astronauts, and finally contemporary debates about science and public policy. In order to see how different theories became imbricated with his, set the stage for the reaction to his ideas, or in some instances provided an alternative trajectory toward greater legitimacy than he ever achieved, I will at times veer rather far off the Velikovskian path. This book is primarily an exploration of the concept of pseudoscience in postwar American culture, and for that purpose Velikovsky provides an exceptionally sharp analytical lens, one that enables us to scrutinize *science* by looking at that which scientists reject as resembling themselves, but not quite. I take Velikovsky and other struggles over so-called pseudoscience as entry points into what philosophers have called the "demarcation problem."

The term "demarcation problem" was coined by a young Austrian philosopher named Karl Popper in 1928 or 1929, a decade after he had already begun to muse over what differentiated what he considered the most impressive scientific achievement of his day—the confirmation in 1919 of Albert Einstein's general relativity through the measurement of the deflection of starlight around the sun during an eclipse—and a rather more local scientific practice: psychoanalysis. Popper was distinctly impressed with the *audacity* of Einstein's case. This physicist boldly set forth a quantitative prediction of the consequences of his theory, as if daring scientists to prove him wrong. Had the deflection not been measured, so Popper reasoned, Einstein's general relativity would have been proven wrong, and the theory would have died. With Sigmund Freud's and Alfred Adler's psychoanalysis, on the other hand, Popper saw something different. These doctrines did not thrive on prediction, but on *confirmation*: they would examine a case of neurosis, and then explain it in terms of their own theoretical framework (Oedipus complex or inferiority complex, say). The difference between the two examples interested him, and by the late 1920s he believed he had come up with a solution to demarcate science from non-science; "pseudosciences" were doctrines that claimed to be sciences but failed a crucial test.

Popper's demarcation criterion was publicly articulated in a 1953 lecture at Peterhouse, one of the constituent colleges of the University of Cambridge, and published in his 1963 volume *Conjectures and Refutations*. Ever since, its popularity has grown, and it has been widely quoted to me (especially by undergraduates) as a solution to the problem of how one identifies a pseudoscience. According to Popper, "*the criterion of the scientific status of a theory is its falsifiability, or refutability, or testability.*"[20] The notion is appealing in its simplicity. For a variety of technical philosophical reasons, it is not possible to simply confirm that a theory is true; all we can know is that we have another confirming instance of what we suspect to be true. (This was, for example, Popper's problem with psychoanalysis: its claims were amply *confirmed*, but he considered the bar for what counted as confirming the theory to be set unacceptably low.) All that we can say about scientific theories, Popper argued, is that they were *not yet shown to be false*. Thus, science progresses by advancing claims and subjecting them to rigorous efforts at falsification. A statement that claimed to be scientific but was immunized from such rigorous examination by ad hoc hypotheses or vague articulation was, for Popper, unfalsifiable, and would thus be clearly marked as pseudoscientific.

The problem with this elegant proposition is that it utterly fails. First, there is a logical conundrum: How do you determine whether a theory has been in fact falsified by a particular experimental result? Suppose you are using a mass spectrometer to test a specific claim about the composition of a compound and get an anomalous result. Is the claim now proven false, or is your mass spectrometer on the fritz? In practice, we do not actually test single statements, but rather groups of statements and assumptions that travel together, embedded in our instruments and experimental setup.[21] The clarity of falsifiability thus becomes a lot murkier. The standard also proves problematic in that this is not what scientists actually *do* when they conduct experiments or make observations. What Popper dismissed as the "unscientific" generation of ad hoc hypotheses to immunize a theory turns out to be one of the most common practices of scientific work. It would be silly to toss out a theory just because you found a single experimental result at variance; better to assemble more data and reserve judgment.[22]

Second, the falsifiability criterion does not perform the task demanded of it. If all a theory has to do in order to count as a scientific is make bold claims that might be proven false, then many doctrines widely deemed pseudoscientific pass muster. This was true even of older doctrines, like alchemy, which had their heyday before Popper wrote, but became even more problematic after falsifiability achieved broad currency. Now that there was a standard, advocates of fringe doctrines just had to make sure they met it. Creationists, for example, routinely make predictions about what kinds of geological structures one should find; parapsychology is nothing but a series of falsifiable statements; and, as we shall see, Velikovsky staked a great deal on the predictive claims of his cosmic catastrophism. Even more embarrassing for Popper's bold attempt, many sciences, such as the more "historical" sciences of evolutionary biology and geology, explain natural phenomena with tools and theories that do not fit nicely into Popper's schema, despite various Procrustean attempts to save the situation. Popper's falsifiability test provides a poor map of the kingdom of pseudoscience. The criterion is neither necessary for demarcation nor, as it turns out, sufficient.

If Popper—despite his popularity outside the realm of philosophy of science (where falsificationism has been long abandoned)—is no help, could we come up with another bright line to distinguish the scientific from the pseudoscientific? After the 1962 publication of Thomas Kuhn's widely read historical-philosophical manifesto, *The Structure of Scientific Revolutions*, many looked to his central argument of paradigms as a possible site of demarcation. According to Kuhn, science consists of periods of stasis ("normal science"), in which scientists solve puzzles within the framework of a general schema of reasoning, which he called a paradigm. As anomalies—experimental findings that prove difficult to reconcile with the dominant paradigm—pile up, occasionally a rupture occurs ("paradigm shift"), and the old paradigm is replaced by a new one, and normal science then continues apace in this framework.[23]

Kuhn rarely invoked the demarcation problem—when he did it was to criticize Popper's solutions—and for good reason.[24] If paradigms *by definition* decide what is scientific and what is not, then any statement outside the paradigm could conceivably be designated pseudoscientific. The only problem is that the later paradigm would also meet that criterion and thus be ruled out of court, which is a nonsensical result. Likewise, individuals that are widely accused of being pseudoscientific could simply claim that they represent a new paradigm, and thus are not to be judged within the frame of reference of contemporary science.

They not only *could* do this; they in fact did and do, as the case of Velikovsky demonstrates. In the 1960s and 1970s, Kuhn lived in Princeton, New Jersey, and taught at the town's eponymous university. Immanuel Velikovsky also lived there (although with no university affiliation), and he offered Kuhn the opportunity to use his personal files to examine the scientific status or merely the history of his revolutionary claims. "You may have access to all these papers," Velikovsky wrote Kuhn, "whenever you wish."[25] Kuhn, for his part, studiously avoided commenting on the Velikovsky affair.[26] Lynn Rose, a philosopher of science at the State University of New York–Buffalo and an extremely vocal supporter of Velikovsky's theories, damned Kuhn for his silence. "It is questionable whether Kuhn would be able to recognize a scientific revolution even if there were one in his back yard. As a matter of fact, he has *already* overlooked a major scientific revolution right in his own back yard," he quipped. "Kuhn will perhaps be remembered, if at all, as the orthodox and unimaginative student of scientific revolutions who lived for a number of years in Princeton, New Jersey, never even noticing the Velikovsky Revolution that was centered there."[27] And, meanwhile, some Velikovsky acolytes tried their own hand at remedying Kuhn's silence with Kuhnian readings of cosmic catastrophism.[28]

Even if Kuhn was no help, the transformation in the academic field of the history and philosophy of science that his book wrought opened a new avenue to potentially cracking the demarcation problem. Since Popper's strictly semantic and logical formulation would not hold water, maybe aspiring demarcationists should look instead to the community of science as a whole and observe (with a philosophical eye) how it decided what was scientific and what was not. What ensued post-Popper were a series of attempts to create not a single demarcation criterion, but a host of checklists. These consisted of characteristics that seemed to belong to many of the doctrines deemed pseudoscientific (assuming there was an elephant), from which a number of rules or properties, something like family resemblances, were extracted. These could be ticked off when trying to determine whether a candidate theory was pseudoscientific or not.[29] Such criteria included, for example, isolation from the scientific community, vigorous resistance to criticism, exaggerated claims of revolutionary innovation, the invocation of supernatural forces—and, yes, unfalsifiability. No individual characteristic was either necessary or sufficient for demarcation, but if you garnered "enough" of them, you could be suitably tossed into the dustbin of crankishness.[30] The same problems bedevil this approach as Popper's: How do you know when you should tick off a criterion, when the alarm has in fact been tripped? And, more problematically, quite a few of these characteristics are displayed by perfectly legitimate (if somewhat cantankerous) representatives of the scientific community, and plenty of supposed pseudosciences met the test of "scientific naturalism" (Velikovsky, for example). Demarcation eludes us once again.

By the late 1970s, philosopher of science Larry Laudan had had enough: "The fact that 2,400 years of searching for a demarcation criterion has left us empty-handed raises a presumption that the object of the quest is nonexistent."[31] Pseudoscience was a problem, he suggested, but it might not be amenable to a philosophical solution. In a 1983 article—controversial in that it appeared in the context of legal wrangling over the scientific status of creationism and whether it should be taught in the public schools—

Laudan laid out what a proper demarcation criterion ought to do in order to be worthy of the name: "Minimally, then, a philosophical demarcation criterion must be an adequate explanation of our ordinary ways of partitioning science from nonscience and it must exhibit epistemically significant differences between science and nonscience."[32] By Laudan's estimation—and that of many philosopher colleagues—there was never going to be a simple criterion such as Popper had imagined, and checklists were simply collections of Popperian-style criteria. Laudan has come under his own share of criticism, and debates continue decades later on this question, but even his most vocal critics concede that "we should not expect a sharp, bright pinline of demarcation."[33] Meanwhile, Laudan insisted on a transformation of language: "If we would stand up and be counted on the side of reason, we ought to drop terms like 'pseudoscience' and 'unscientific' from our vocabulary; they are just hollow phrases that do only emotive work for us."[34]

As "pseudoscience" gained currency as a term of abuse, one finds a wave of attempts beginning largely in the 1980s among academic observers of science (if not among scientists themselves) to find alternatives to the term. Some of these contenders continue to be popular: unorthodox science, non-establishment science, cryptoscience, parascience, emerging science, protoscience, unconventional science, and anomalistics.[35] (One *does* find people self-identifying under these categories.) At the same time, it became increasingly clear that "pseudoscience" picked out different phenomena from frauds or hoaxes—those displayed levels of conscious insincerity that appeared to be wholly absent in the often-cited cases of alleged pseudoscience.[36] Another common term, "pathological science," is also inappropriate. Coined by Nobel Laureate Irving Langmuir in an often-cited 1953 lecture, this described scientific claims that hovered on the edge of perception, faint effects that drifted perilously close to experimental error and were magnified by wishful thinking.[37] Pseudoscience must also be distinguished from "anti-science," movements such as those of the 1970s arguing that scientific and technical reasoning were leading civilization down the wrong path. As science studies scholar Helga Nowotny observes, this was not pseudoscience's domain, for "in many ways, the pseudo-sciences aspire to become scientific."[38] And despite the popular (in the sense of populist) character and participatory nature of many alleged pseudosciences, the notion also has to be separated from "amateur science," which is ordinary science performed by those other than professionals.[39]

What if we have been looking for demarcation in the wrong places? Philosopher Philip Kitcher, for example, has recently suggested that we should focus on the people who advocate the doctrines, not the "pseudosciences" themselves. "Pseudoscience isn't marked out by any clear criteria that distinguish it as a body of doctrine," he writes. "It is simply what pseudoscientists do in their more-or-less-ingenious pretenses."[40] That is to say, there are certain individuals who simply resist contrary evidence or refuse to revise their views, and those individuals have a *psychological* predisposition to advocate certain doctrines. Some people with this recalcitrant mind-set enter the sciences, and if they are lucky they become partisans of an accepted domain of science. But if they end up in a more dubious domain, such as UFOlogy, their natural predisposition combines with a degree of intense advocacy as they "pretend" to be scientists. The more you probe them, the more they reiterate the same arguments or dodge various objections. In such instances, one would often find that values, rather than epistemic concerns, motivated these individuals. There may very well be such a psychological profile that can be picked out, but it is hardly a demarcation criterion, and Kitcher does not intend it as one. For how would you know *in advance* that someone with these characteristics was advocating in a disingenuous way without investing the time to interrogate them and uncover the pretense? (It is the whole point of demarcation criteria to save the time and effort involved in such investigations.) It may serve in particular instances, but it leaves the general philosophical project in limbo.

Among this plethora of distinction-making, one rough area of consensus emerged among the philosophers, historians, and science watchers: whether or not you believed that a category of pseudoscience existed, there was certainly no bright line to demarcate science from pseudoscience. Martin Gardner, the writer who probably did more than anyone else in the postwar period to turn discussions of alleged pseudoscience into debunking crusades, observed of rigid attempts to demarcate that "clearly no such criteria are precise. Pseudoscience is a fuzzy word that refers to a vague portion of a continuum on which there are no sharp boundaries."[41] But even calling it a continuum makes the problem seem simpler than it in fact is. As noted earlier, pseudoscience is presumed to be beyond the bounds of even the most incompetent execution of proper science. Even those philosophers who continue the quest to demarcate now start from the assumption that "the boundaries separating science, nonscience, and pseudoscience are much fuzzier and more permeable than Popper (or, for that matter, most scientists) would have us believe. There is, in other words, no litmus test."[42] And this should not surprise us, for if there is indeed no elephant of pseudoscience, but rather the characteristics that are used to pick out certain doctrines as pseudoscientific were all created for specific ends, then why should we expect unity?[43]

We are reduced to a variant of Justice Potter Stewart's famous dictum about obscenity: We know pseudoscience when we see it.[44] But do we really? If the historically intertwined narratives that follow indicate anything, it is that people cannot pick out pseudoscience unproblematically. Although every individual seems certain that he or she is doing an admirable job of demarcation when surveying the realms of knowledge, consensus is hard to find. For example, astrology and phrenology were once considered sciences and only later were cast out as pseudoscientific, so they clearly at one time passed the Potter Stewart test. (If you believe these fields were always "pseudosciences," then you have a problem here.) So what do we do? Should we stop this effort altogether? Scholar of rhetoric Charles Alan Taylor, who analyzes efforts at demarcation, encapsulates our dilemma: "To say that the 'demarcation game' should not be played by philosophers' rules need not entail that it should not be played at all. It is a game which can be (indeed is) played in practical, rhetorical terms every day."[45]

But how do we play this game—and by "game" I do not minimize the seriousness of the issue—if we must? We are confronted with what I call the central dilemma of pseudoscience. Imagine a bar that separates out "reasonable" hypotheses and scientific claims from those that are unacceptable or "fringe." (This is a bar set by consensus of the scientific community, not a sharp philosophically rigorous demarcation criterion.) One could set that bar pretty high, allowing only relatively uncontroversial, well-established scientific claims through, but then one would hamstring innovation. Such a high bar might have excluded Albert Einstein's special theory of relativity, Gregor Mendel's notions on heredity, or Alfred Wegener's continental drift. Without new ideas, we are mired in intellectual stasis. On the other hand, one could set the bar relatively low, letting in all sorts of unconventional ideas. The problem with a low bar, of course, is that it is impossible in advance to distinguish many unconventional but promising notions from what some might call "crankish" ones. Calibrating the exact level of this metaphorical bar is the central dilemma, and it is inescapable in any scientific venture. This point is well expressed by science-fiction author and avid debunker of fringe claims L. Sprague de Camp: "Thus orthodoxy acts as a kind of filter for new ideas. Ideally it would provide just enough resistance to them to make their proponents extend themselves to confirm them, and to show up weaknesses of those that were not sound, but not enough resistance to suppress any valuable new discovery."[46] It is not simply enough to allow everything in and just sort out the good ones afterward, for the expenditure of time detracts from other scientific projects, and investigation generates publicity that many scientists would rather not give these doctrines.[47] Every discussion of demarcation at its core hinges around this fundamental tension between innovation and crackpottery.

Despite the ubiquity of the central dilemma, it would be a mistake to walk away thinking scientists rack their brains about whether the bar should be set high or low. Scientists spend remarkably little time and effort attempting to characterize pseudoscience in any philosophically coherent way.[48] Yet we nonetheless must pay attention to this topic, even if scientists fail to do so explicitly, for two reasons. First, as philosopher Paul Thagard has argued, even if a bright line cannot be constructed, it is important for science policy and scientific literacy to have a healthy debate about what is considered to be science and what is not.[49] The second reason is yet more significant: even if scientists do not overtly argue about the definition of pseudoscience on a daily basis, sociologist Thomas Gieryn notes, "demarcation is routinely accomplished in practical, everyday settings." That is, "demarcation is not just an *analytical* problem: because of considerable material opportunities and professional advantages available only to 'scientists,' it is no mere academic matter to decide who is doing science and who is not."[50] Since this is a weighty social and political problem—not least because defining "science" is the crux of important legal disputes, such as those surrounding the teaching of evolution in public schools in the United States—as well as an intellectual one, I argue in these pages that we should subject these categories to *historical* analysis, in order to see how they have been articulated at a particular time and place, and their implications for the present. In this, I concur with leading historian of creationism Ronald Numbers: "As a historian I am much more interested in how persons and parties *used* 'science' and 'pseudoscience' to further their ends than in judging whether they employed these labels appropriately by the standards of the 1990s."[51] The point is no less true two decades later.

But why undertake the study of such matters, when there is so much accepted science that begs for sustained historical investigation? As was once said in the context of psychoanalysis: "A successful pseudo-science is a great intellectual achievement. Its study is as instructive and worth undertaking as that of a genuine one."[52] I agree with this statement, but I do not mean it in the same way. A pseudoscience is indeed an achievement, but it is not achieved solely by those who espouse the doctrines (as the quotation would have it). Rather, pseudosciences are the products of actions and categorizations made by scientists, and particularly important products at that. But if it is conceded that the topic of demarcation must be discussed—and I believe that there are few topics more significant for understanding the place of science in society and culture—why structure the story largely around Immanuel Velikovsky? There are a few specific features of Velikovsky's case that make this story rather different, and potentially more illuminating, than a history of phrenology or mesmerism or parapsychology.[53]

Consider what scientists and philosophers have often portrayed as the usual pattern by which pseudosciences are labeled: some practitioners (mainstream or not) advance controversial claims; these are debated within the scientific community and (sometimes) by the public at large; and then the new doctrine is either accepted or it is demonized, labeled pseudoscientific, and cast out of respectability. Velikovsky's case is more interesting. He was trained as a medical doctor and a psychoanalyst, so he was not completely outside the scientific community, although he was certainly not a trained astronomer or geologist. But from the moment of publication of *Worlds in Collision* in April 1950, Velikovsky was branded a crackpot. There was no careful consideration, no engaged debates about the book's status within the scientific community. Velikovskianism was, so to speak, born pseudoscientific. This makes his instance a particularly intriguing locus of inquiry, since the positions of scientists on the issue of demarcation were quite explicit and heated—more so than in the "standard" variant described above—raising buried assumptions to the surface.[54]

There is an additional attraction to the Velikovsky case, again best illustrated by a quotation interpreted against the grain of the author's intentions. Velikovsky supporter Frederic Jueneman at one point objected to analogies that looked at the controversy as in some way analogous to the Galileo affair (a popular touchstone for many fringe doctrines) or any other historical exemplar. Rather, he "would venture to say that the case of Velikovsky is *unique* in the annals of scientific inquiry. To my knowledge there has never before been in the history of rational thought such a massive documentation of the events relating to a single individual, his adversaries, and the sociological pot-boilers which have been written defending his views."[55] The crucial point for Jueneman was the focus on a single individual, but that is neither that unique in the history of science (or pseudoscience) nor part of my project here. The real issue is a term he let slip casually: documentation. Massive documentation.

Immanuel Velikovsky believed that he had produced a scientific and historical doctrine of world-changing significance, one built on careful attention to scraps of evidence buried in the mythologies of humanity. He organized the writing of his books and his campaign in their defense around similar methods: gathering large amounts of historical data, quoting it extensively, and circulating copies of correspondence to establish his positions, whether about the nature of Venus or the perfidy of Harvard astronomers. Such assiduous attention to detail demanded a phenomenal memory and a rather comprehensive personal archive. The former he was born with; the latter he built over the decades and stored at 78 Hartley Avenue in Princeton, New Jersey, where he lived from 1952 until his death in 1979.

In 2005 Firestone Library at Princeton University announced that it had acquired the papers of Immanuel Velikovsky, cataloged them, and opened them to researchers. The name "Velikovsky" dislodged a memory from the browsing of bookshelves in the public libraries of my childhood, and I decided to spend a few hours looking at some of these documents. A few hours turned into a few weeks, months, years. The Velikovsky Papers are one of the most comprehensive personal archives I have ever seen. They consist of sixty-five linear feet of material: drafts of manuscripts, fan mail, hate mail, correspondence with publishers, with friends, with enemies, and much, much more. The Velikovsky archive was not only what he used to wage his own battles in the pseudoscience wars; to the historian, it is an unprecedented entry into how a demonized theory was built from the ground up, and then torn down by both internal and external assaults. It chronicles the process of demarcation in practice with microscopic detail, and most of the book that follows is drawn from its pages.

There are, however, some obvious challenges of writing history out of the archives of a single man, especially an archive that was constructed to be a weapon in his quest for legitimacy, and it is necessary to be candid about them. Velikovsky assembled this archive, which means that the evidence itself bears the impression of his own idiosyncrasies. Obviously, I can only use what Velikovsky chose to keep, not what he discarded, and one might suspect that Velikovsky deliberately sanitized the archive to present a rosy picture. Considering, however, the massive quantity of negative and downright slanderous material that he retained (some of which he preserved so future historians would excoriate his critics, but other parts present the man himself in a petty or vindictive light), it is hard to believe that what one finds in the archive is only what Velikovsky consciously chose to keep.[56] Based on descriptions of the household during the 1970s, piles of these documents were scattered around the house, and any systematic purge of "compromising material" seems quite improbable. Nonetheless, the Velikovsky archive is not a random or disinterested find (like a vein of ore in a mountainside or the holdings of the Library of Congress), but a specific slice through the documents that passed through his life. Within that slice, I have made my own cuts, focusing on the issue of demarcation above all, and leaving aside detailed inquiry into other fascinating questions like the nature of his fan base, the dynamics of publishing on the fringe, and many other topics that await their historians.

Velikovsky thought of this archive as a repository for future historians of science. He wrote to George Sarton, the first president of the History of Science Society, about the richness of his documentation already in 1950, when it was still in its infancy: "For the historian of science, the files that I possess with the reactions of those who made efforts to suppress the book will be of great interest. And since you are a historian of science, I am writing this letter to you."[57] Over the years, Velikovsky asked his correspondents to send him copies of anti-Velikovsky letters by third parties so he could store them, because the "archives should have all the material that may interest the future historian of science."[58] In the end, he hoped "to make it accessible to all working in the field of history or sociology of science."[59] It is clear, therefore, that Velikovsky wanted it open for research. (At one point, as he wrote to the archivist of Boston University, he had a plan for four duplicate archives: one in Western Europe, probably in Edinburgh, because he had studied there; one in Eastern Europe, probably at Moscow University, for the same reason; one in the Middle East, in Jerusalem; and one in the United States, location unspecified.[60]) To anthropomorphize for a moment: the material wants to be used.

But will the fact that these documents were gathered by Velikovsky in the first instance distort the book that follows? Well, of course—but no more than any archive would, whether assembled by a single person or a government agency or a municipality. The same point holds in more extreme instances. Historians routinely use records of the Inquisition, for example, or Stalinist coerced confessions—if you read them carefully, you can sift historical information while being aware of the limits.[61] There is no view from nowhere, at least not among the flotsam and jetsam of the past. There is even a virtue to the kind of distortion one finds in Velikovsky's case: the structure of the archive itself is a kind of evidence; since Velikovsky used it for his own purposes, what he retained shaped his own arguments, and we can read some of his intentions out of how he filed the material.[62] One could, of course, maintain a bottomless skepticism about Velikovsky's assertions about his life—for example, did he really have a medical degree from Moscow University?—but most of his doings are quite well corroborated by outside sources. When there is cause to be suspicious of certain documents or assertions, I note it, and when corroboration (or refutation) can be had from other persons, I provide it. Complete reliability, however, will elude us here, as it does (to a less noticeable degree) in all history.

In the interest of preserving the tenor of the original disputes, I need to reiterate one point: I do not set out to debunk Immanuel Velikovsky and his theories. Such works already exist in profusion, and you will find citations to them throughout this book. I also do not intend to defend Velikovsky from attack, though I provide citations to the writings of those who do so. It is impossible to write about this debate without raising the arguments for and against his cosmic catastrophism, since whether the events Velikovsky hypothesized really happened was the central question, but in doing so I do not pretend to resolve them.[63] My goal is historical: to chronicle what happened, to explain when possible why, and to reveal the passions excited by calling something "science" across this temporal period. That is where the bar for this history is set, and I trust it is neither too high nor too low.

1 · The Grand Collision of Spring 1950

During the first few months of 1950, it seemed that American scientists had completely lost their heads. Were they apprehensive at the announcement of President Harry S. Truman on January 31 that the United States would undertake a crash program to develop a hydrogen bomb, launching the thermonuclear age? Was it the shock of the arrest on February 2 in Great Britain of German-born physicist Klaus Fuchs for wartime espionage at the Manhattan Project, whose secrets he had shared with the Soviet Union? Or perhaps anxiety about the February 9 speech of the junior senator from Wisconsin, Joseph McCarthy, in Wheeling, West Virginia, in which he declared that he had a list of card-carrying members of the Communist Party in the State Department, and some of those he named were prominent scientists?

Not exactly. The uproar was over a book that had not even been published yet. The initial agitation began in January and concerned the advance press for a trade book due to appear in April from the Macmillan Company, at the time the most distinguished publisher of scientific works (especially textbooks) in the nation. It was entitled *Worlds in Collision*, by a man named Immanuel Velikovsky.[1] This text does not lend itself to being summarized without a significant degree of distortion. Writing in the 1980s, Velikovsky critic Henry Bauer noted that it was impossible to condense the book "in a completely unbiased fashion; one selects what seems most significant, and opinions about what that is will inevitably differ."[2] In a somewhat higher register, Velikovsky for the rest of his life resisted all attempts to represent any of his books' arguments in brief, and especially not that of *Worlds in Collision*. As he noted in his memoir of the events discussed in this chapter (published posthumously in the 1980s): "I cannot compress *Worlds in Collision* any more than it is in its present form in a book—there I have not left a sentence that I deemed superfluous."[3] Indeed, the first fusillades were launched not by the book but by summaries, serializations, and condensations. A few decades ago, a writer on Velikovsky could expect that almost everyone had read or heard of *Worlds in Collision*—for then it seemed that everyone had. Now, as in 1950, we begin with a précis.

"This book is written for the instructed and uninstructed alike," Velikovsky declared on the first page of his preface. "No formula and no hieroglyphic will stand in the way of those who set out to read it. If, occasionally, historical evidence does not square with formulated laws, it should be remembered that a law is but a deduction from experience and experiment, and therefore laws must conform with historical facts, not facts with laws."[4] This was the central claim of *Worlds in Collision*: that if one examined the global store of myths—Chinese, Mayan, South Asian, Norse, Aztec, but principally those legends of the ancient Near East, and especially those presented in the Hebrew Bible—one found repeated, and disturbing, patterns. Over and over again, the chronicles referred to a rain of fire, to battles in the heavens, to extended days or extended nights, catastrophic floods, barrages of stones from above, earthquakes, and so on. "The events were called miracles and were explained as subjective apperceptions or as symbolic descriptions because they could not be otherwise accounted for," but Velikovsky posed the hypothesis, and then claimed he had demonstrated as fact, that something had indeed happened to Earth, a series of global catastrophes that remained metaphorically veiled within the collective literary heritage of humanity.[5] If one finds a correlation among many different peoples, there are several logical explanations: it could just be coincidence, or perhaps a case of diffusion, the story spreading from a single origin point. But Velikovsky opted for a third alternative: "In more than this one instance it is possible to show that peoples, separated even by broad oceans, have described some spectacle in similar terms. These were pageants, projected against the celestial screen, that, a few hours after they were seen in India, appeared over Nineveh, Jerusalem, and Athens, shortly thereafter over Rome and Scandinavia, and a few hours later over the lands of the Mayas and Incas."[6]

What were these pageants, a light word to describe something represented as horrifically destructive in these fragmentary reports (if indeed they were reports)? Velikovsky presented his argument as three nested claims, each more specific than the last: "(1) that there were physical upheavals of a global character in historical times; (2) that these catastrophes were caused by extraterrestrial agents; and (3) that these agents can be identified."[7] After presenting a fascinating montage of extracts of legends from around the world, pointing to the common symptoms of this global onslaught, he reached his first conclusion: sometime around 1500 B.C., at the moment of the Exodus of the Children of Israel from Egypt under Moses's leadership, a massive comet was ejected from the body of Jupiter and hurtled toward Earth. It became trapped in a gravitational but also electromagnetic interaction with the planet, rupturing its crust, tilting its axis, and showering meteors on the population. It then stayed in an unstable interaction with Earth for several decades before finally settling into an orbit around the sun. "Under the weight of many arguments," Velikovsky wrote, "I came to the conclusion—about which I no longer have any doubt—that it was the planet Venus, at the time still a comet, that caused the catastrophe of the days of Exodus."[8]

The ten plagues, the parting of the Red Sea, even manna from heaven, all of these things were true—they just were not miracles. ("Of course, there is no person who can [part the Red Sea], and no staff with which it can be done."[9]) They were natural phenomena, the results of this catastrophe that nearly destroyed Earth. An event so terrifying, so cataclysmic, would of course be remembered in the oral legends of the world's peoples, and in the case of the ancient Hebrews became the central event of their monotheistic religion. Venus was the goddess of rebirth in the ancient world, the newcomer, Lucifer, the angel of destruction. (In Greek myth, Velikovsky was careful to identify the planet Venus with Athena, who burst fully formed from the head of her father, Zeus, instead of with Aphrodite, goddess of love.[10]) For almost four hundred pages, Velikovsky laid out his evidence for this Venus catastrophe, as well as another near collision of Venus with Earth, this time caused by a Mars displaced by the Venus-Earth interaction, chronicled, according to Velikovsky, as the struggle between Athena and Ares in Homer's *Iliad* and as a series of calamities in the book of Isaiah. In its wake, due to a change in the length of the year, global calendrical reforms ensued between 747 and 697 B.C.[11] He begged only one indulgence from the reader: he used an unconventional chronology when he synchronized Egyptian history with the events of the Hebrew Bible, which he promised he would defend in a subsequent work.[12]

The book was, and remains, an enthralling read. It also required, to account for the events described—near collisions of planets, comets the size of Venus, the transformation of a hydrocarbon/petroleum tail of the comet into carbohydrate manna for the Israelites—outright contraventions or at least severe modifications of the conventional understandings of celestial mechanics, physics, chemistry, geology, paleontology, and biology. Astronomers could explain the solar system with astonishing precision using a series of gravitational equations that could not accommodate the shenanigans of planetary bodies, and many also claimed there was an unbroken chain of eclipses going back to the third millennium B.C., which excluded shifts in the poles or the orbital orientation of Earth. This was a book that courted controversy—and controversy it found.[13] Scientists from numerous disciplines wrote both publicly and privately, excoriating the content of Velikovsky's theory and, especially, the publisher who lent to a man they perceived as delusional the imprimatur of Macmillan's respected name. This was one of the greatest publishing scandals of the postwar period, and it triggered the pseudoscience wars. The tone and volume—meaning both the mass of commentary and the loudness with which it was delivered—of these rebukes was as unusual and idiosyncratic as the Venus catastrophe related by Velikovsky. There was simply nothing quite like it in postwar America.

Why so much outrage? An answer to this question must come in parts. I will defer explanations of the motivation of the scientists' behavior—and there are Velikovskian and non-Velikovskian variants—to the next two chapters, but there is one consistent refrain in this anti-Velikovskian discourse: publicity. As becomes clear from a dispassionate reading of all the reviews and letters, the scientists objected to more than the content of Velikovsky's theories. I do not want to be misunderstood here: these astronomers, geologists, and other scientists despised those claims, insisted they were deeply and utterly wrong, and minced no words in saying so. But the main target of their fury was not Velikovsky—he was a secondary target, to be sure—but Macmillan itself. The scientists objected to the press's involvement with this book and took additional umbrage at its vigorous (and effective) publicity campaign, which they unwittingly abetted through their angry outpourings. The debate ranged across the professional responsibility of the press and the authority to interpret science for the general public. It was about science in the postwar public sphere.

THE DAY THE EARTH DIDN'T STAND STILL

The January 1950 issue of *Harper's Magazine* hit the newsstands in late December, on the same cycle as every other month. A staple of American public discourse, this magazine included a spectrum of articles on different subjects, including a discussion of physicist Edward Condon's troubles with the House Un-American Activities Committee. But this was not the piece that struck readers of this particular issue. *Harper's* writer Eric Larrabee had penned an article entitled "The Day the Sun Stood Still," a reference to the battle on the plains of Gibeon when the biblical Joshua ordered the sun to tarry in the sky so that the Israelites could be victorious—according to Velikovsky, a consequence of Venus's violently shifting Earth's axis—and echoed the following year in the popular science-fiction movie *The Day the Earth Stood Still*. Larrabee's article was "an attempt," he wrote, "necessarily condensed and incomplete, to offer a preview of Dr. Velikovsky's findings." In a few pages, Larrabee rather accurately described the book, the density of its footnotes (which struck many readers), and the fact that Velikovsky questioned such fundamentals of contemporary science as the predominance of gravity in the solar system. The basic point: "Dr. Velikovsky presents historical evidence that these ancient records were not incorrect at the time when they were made."[14]

A feature article in *Harper's* is fairly high profile for an unknown author's first English-language book, and it came about in a circuitous fashion. As Larrabee recounted after Velikovsky's death, the initiative for this report came from the editor in chief, Frederick Lewis Allen, who was friends with Velikovsky's editor at Macmillan, James Putnam. Putnam had related Velikovsky's account of the Joshua story while the book was still in press—along with the surprising claim that Velikovsky believed that Central American myths of an especially long night, halfway around the world from the Middle East, were correlated with the Joshua story, thus indicating a common event. Allen enjoyed relating the anecdote at cocktail parties. When Macmillan began circulating materials about *Worlds in Collision* in late 1949, another editor at *Harper's*, Merle Miller, recalled the story, obtained the page proofs of the book, and assigned Larrabee to serialize it. When Larrabee demurred at the difficulty, they opted to present a summary instead.[15] Macmillan had indeed planned an extensively coordinated publicity campaign, but this first volley happened at *Harper's* request.[16] (Velikovsky would later claim that he was barely involved in the advance publicity.[17])

This was only the first of several condensations to be reprinted across the nation. Two (out of an advertised three) serializations appeared in the broadly distributed weekly *Collier's* on February 25 and March 25, accompanied by somewhat lurid illustrations of ancient Egyptians pelted by meteorites. The entire packaging by *Collier's*, even more than *Harper's*, attempted to inject *Worlds in Collision* into a debate about science and religion, with the editorial foreword noting that Velikovsky's theory, "among other things, challenges the Darwinian theory of evolution" by questioning uniformitarian, gradualist assumptions about Earth's history. The text, attributed to Velikovsky, declared that "the great Architect of nature sent a celestial body—a comet almost as large as the earth itself—close to our planet."[18] *Worlds in Collision* itself studiously avoids this kind of religiously laden language, yet historian James Gilbert notes that Velikovsky's "science had little meaning or importance outside its religious context."[19] This was clearly a key set of references for Velikovsky's readers, and *Collier's*, as well as a different adaptation by Fulton Oursler in *Reader's Digest*, promoted Velikovsky as rescuing biblical literalism.[20] To further the point, a text box embedded in the first *Collier's* article entitled "The Greatness of the Bible" by Norman Vincent Peale, the famous pastor of Marble Collegiate Church, declared that "Dr. Velikovsky's work interestingly draws the attention of thoughtful people to the substantial basis of fact upon which the Old Testament was written."[21]

Velikovsky was enraged by the treatment of his work in *Collier's*[22] (although not, at least according to his archives, by the *Reader's Digest* version). The process of serialization was not smooth. Velikovsky, worried about the way previews of his theory might prejudice its reception, wanted complete control of the images and the writing. The process was the opposite of the *Harper's* case in almost every way, leading to recriminations and bad feeling on both sides. Over fifteen years later, one of the two excerpters recalled the drafting sessions with horror:

> He [Velikovsky] was infuriated by *everything*, by our introduction, by our excerpting, by our footnoting. All because we did not present his stuff as unquestioned truth supplanting all previous "erroneous" theories. He was not only a supreme egomaniac, but he was evangelical about it. He was Mahomet, John the Baptist, and St. Paul rolled into one. He had found the long-secret truth and would proclaim it to the world, which *should* bow down and thank him, but which instead doubted, questioned, and tried to silence. He was a pretty good paranoiac, too. . . . [F]inally he leapt from the table, when I was being insistent on one point, and came back with a pistol or revolver, which he placed on the table beside him, saying something like, "Now we'll see how this will be handled."[23]

It is impossible to know what actually happened over that kitchen table on New York City's Upper West Side as both sides labored to a deadline, but *Collier's* certainly did not harm Velikovsky's sales, bringing him a vast readership among sets less self-consciously intellectual than the *Harper's* crowd.

That intellectual set, however, also included some readers less delighted by Velikovsky's proposed innovations. One such reader was Harlow Shapley, director of the Harvard College Observatory and, next to Albert Einstein and J. Robert Oppenheimer, one of the most widely recognized scientists in America. (He had been president of the American Association for the Advancement of Science in 1947 and was a prominent advocate of liberal causes in an America increasingly ensnared in anti-Communist politicking.) In mid-January 1950, Shapley sent a letter to the Macmillan Company, listed in the *Harper's* article as the prospective publisher of the book, in which he stated that he had heard a happy rumor that Macmillan would *not* in the end publish *Worlds in Collision*. He hoped this was true and wanted to confirm with Macmillan. He mentioned that he had discussed the argument presented by Larrabee with several scientists, including the president of Harvard University, chemist and science administrator James Bryant Conant, and all were "not a little astonished that the great Macmillan Company, famous for its scientific publications, would venture into the Black Arts without rather careful refereeing of the manuscript." If they had not yet vetted the book and realized that it was nonsense, he urged them to do so now. In conclusion, he called Velikovsky's theory about the sun standing still "the most errant nonsense of my experience, and I have met my share of crackpots."[24]

In fact, as he surely recalled, he had met this one. On April 13, 1946, Shapley was the speaker at a forum at the Commodore Hotel in New York, discussing world government (one of his many political interests). Velikovsky, wanting to consult with the great astronomer about his new theory of the solar system, approached Shapley and began a conversation. We only know the content of their discussion from letters written in spring 1950 in the aftermath of Shapley's démarche to Macmillan, so what follows must be taken with a grain of salt on both sides. According to Velikovsky, after outlining the basics of his theory, he said to Shapley: "I wish you would agree to read the book manuscript; and if you will be satisfied at its reading that my thesis is supported by sources to an extent that it deserves some laboratory investigation, would it be possible to undertake one or two rather not complicated spectroscopic analyses?" Shapley claimed that he was busy, although if Velikovsky would have someone of stature recommend the text to him, he would take a look at it, and that Velikovsky should write about the tests to Fred Whipple (himself an astronomer of no mean distinction), who was then in charge of experiments at the observatory, and "if possible, we will do it for you." Velikovsky suggested Horace Kallen as the referee (about whom more soon), and in parting Shapley said: "And believe me, if you have proved in your book that in historical times there occurred a change in the constitution of the solar system, there is no thing in my power I would not do for you."[25] Two days later Velikovsky wrote to Shapley asking for spectroscopic tests for the presence of argon and neon in the atmosphere of Mars, and two days after *that* letter, he proposed a search for "gaseous hydrocarbons . . . in the absorption spectrum of Venus."[26] (Velikovsky believed these gases should be present as a result of contact of atmospheres during the near collisions chronicled in his manuscript.) A month later Shapley, through his secretary, declined to do any tests for Velikovsky.[27] And there, apparently, the matter rested.

In the meantime, Velikovsky behaved just as he had promised Shapley he would: he contacted Horace Kallen, philosopher and dean of the New School for Social Research in New York City, and asked him to recommend the manuscript to Shapley. Kallen and Velikovsky maintained a close correspondence—an alliance between one of America's most respected intellectuals and one of its most reviled outcasts—until the former's death in 1974. Both were immigrants (although Kallen came to the United States much earlier and had studied under William James at Harvard), and both were committed Zionists. They had met, in fact, through Judge Morris Rothenberg, a leader in Jewish affairs.[28] Kallen had even enthusiastically read, as early as 1941, several early versions of the manuscript that would become *Worlds in Collision*, first as an analysis of Freud's heroic figures and later as a historical manuscript called "From Exodus to Exile."[29] He happily endorsed the revised cosmological manuscript to Shapley in 1946, though Shapley had already declined to do the experiments for Velikovsky.[30]

At this point, Shapley was still relatively well disposed toward Velikovsky, despite his lack of interest in conducting any empirical investigations on the man's behalf. He regretted his detachment in his response to Kallen: "The sensational claims of Dr. Immanuel Velikovsky fail to interest me as much as they should, notwithstanding his exceedingly pleasing personality and evident sincerity, because his conclusions were pretty obviously based on incompetent data." Recent changes in the solar system were simply not in the cards. "In other words, if Dr. Velikovsky is right, the rest of us are crazy," he concluded. "And seriously, that may be the case. It is, however, improbable."[31] He suggested that Velikovsky contact some other astronomers about the atmospheres of Mars and Venus, and the latter dutifully did so, writing to Otto Struve at Chicago, Rupert Wildt at the University of Virginia, and Walter S. Adams of the Mount Wilson Observatory in California. (Struve brusquely spurned the inquiry, but Wildt and Adams wrote back with polite and detailed refutations.[32]) In May 1946 Kallen's recommendation to Velikovsky, in the wake of the latter's disappointment in Shapley, was to beef up the science:

> Dr. Shapley's reactions are those that were to be expected. You will have to talk to him in his own language and not expect him to talk in yours. His language is that of the mathematician and astronomer, yours that of the philologist and historical critic. I suspect that you will need to complete your final study before you get serious attention from natural scientists.[33]

Judging from Shapley's letter to Macmillan in early 1950, Velikovsky had not changed his language enough. He got some serious attention, just not the sort he was looking for.

Velikovsky's editor, James Putnam, had the job of addressing Shapley's missive. (George Brett, the head of the press, also responded.) Putnam did not defend the book's content but merely argued for the place of such a book in Macmillan's catalog. "As I am sure you realize, we are publishing this book not as a scientific publication, but as the presentation of a theory which, it has seemed to us, should be brought to the attention of scholars in the various fields of science with which it deals," Putnam wrote. "I cannot believe that our publication of this book, which is presented by us as a theory, will affect your feeling toward our publications in the scientific field."[34] Shapley's response sounds quite sinister with hindsight: "It will be interesting a year from now to hear from you as to whether or not the reputation of the Macmillan Company is damaged by the publication of 'Worlds in Collision.'"[35] From this point until the end of Macmillan's relationship with Velikovsky, as far as Velikovsky's, Shapley's, and Macmillan's archives can show us, Shapley left the press alone.

He did not, however, surrender his opposition to Velikovsky's theories. Larrabee's January article was reprinted widely in other periodicals, including the *Daily Compass*, edited by Ted Thackrey, a friend of Shapley's. In a jokey, informal letter, Shapley wrote to Thackrey suggesting that the latter had been hoodwinked into reproducing a piece of claptrap. "In my rather long experience in the field of science, this is the most successful fraud that has been perpetrated on leading American publications," he wrote, adding that "I am not quite sure that Macmillan is going through with the publication, because that firm has perhaps the highest reputation in the world for the handling of scientific books."[36] Thackrey retorted hotly: "It seems to me that you are making both a personal and professional mistake—a gravely serious and dangerous one—by the totally unscientific and viciously emotional character of your attack upon Dr. Velikovsky and his work."[37] Shapley answered that, far from attacking *Worlds in Collision*, he had intervened to stop the Council of the American Astronomical Society from sending a protest to Macmillan, on the grounds that this would only generate more publicity.[38] Then he slightly modified his earlier position, shifting from Velikovsky's theories to the press: "Our trouble about the Macmillan Company and Harper[']s, if you call it trouble, was that such publications seem to throw doubt on the care with which they referee other manuscripts on which we want to depend."[39] In fact, in the entirety of the debate over Velikovsky, Shapley made only one public (in the sense of published) comment on *Worlds in Collision*, in the *Science News-Letter* (which he also happened to run) of February 25, 1950. Amid the remarks of several scientists on the advance press, Shapley called the Velikovsky story "rubbish and nonsense."[40]

We are left, then, with a microcosmic presentation of what would follow: Velikovsky's theories about the interpretation of ancient myth to uncover the history of the solar system make a huge public splash; a scientist objects, to a limited extent publicly, but mostly through correspondence; and the issue shifts from the veracity of *Worlds in Collision* to the publisher. Over the first six months of 1950—the duration of this first "Velikovsky affair"—one indeed observes a shift from Velikovsky to Macmillan.

THE MAKING OF MACMILLAN'S
WORLDS IN COLLISION

How, one might wonder, did a press of Macmillan's stature come to be involved with a book like *Worlds in Collision*? George Brett and the rest of Macmillan's senior staff came to ask themselves this very question in earnest in spring 1950, but the answer is not very complicated. Macmillan acquired the book much the same way books were acquired by publishers in that day, and in this one.

By 1946 Velikovsky had completed a manuscript that closely resembled the spectrum of evidence and the general argument of *Worlds in Collision*, although some significant stylistic and structural changes took place before the Macmillan version appeared in April 1950. Eight publishers had already rejected the book, and a total of nine had rejected its companion historical monograph, *Ages in Chaos*.[41] Horace Kallen had attempted to interest Alfred A. Knopf (the man and the publishing house) in a version of the manuscript as early as 1941, declaring that reading it was "the most exciting intellectual experience I have had in several years."[42] Knopf looked at it but sent it back to Velikovsky, writing that "it just isn't suited to the list of a general publisher."[43] And again, in June 1946, Velikovsky approached the Appleton-Century publishing house, largely (he stated in his cover letter) because they were Charles Darwin's first publisher in the United States and so were historically unafraid of controversy and associated with innovations in science. He extended the argument further: "If I have proven in it what I intend to prove, it will be regarded as one of the great achievements of science, comparable to 'De Revolutionibus' of Copernicus."[44] Appleton-Century also passed.

A common complaint in these early rejection letters was that the copious footnotes, somewhat Hebraized English, and choppy presentation made the book unsuitable for a trade publication. But trade is precisely what Velikovsky sought, and somehow he managed to contact Clifton Fadiman, mainstay of the New York publishing world and a judge for the Book-of-the-Month Club. Fadiman's advice was crucial in turning the book into the kind of text that attracted a broader readership. After sending out queries to determine whether Velikovsky was legitimate, Fadiman read through the lengthy manuscript and offered detailed and serious suggestions—adding a preface, signposting the complex argument, and tightening the style.[45] He also advised Velikovsky on the process of submission. Fadiman's final suggestion, interestingly, was to get support from leaders in science: "However, even before doing this, I feel it necessary to get some testimonials from at least three or four respected scientists in such fields as astronomy, geology, and geophysics. Even if these authorities merely give you a statement to the effect that your hypothesis is by no means a wild one, certain benefits would be assured."[46] This is exactly what Velikovsky had earlier tried to do with Shapley. After a year of correspondence, Fadiman was optimistic for the book, even hyperbolically so: "In other words, if your thesis can withstand both the analysis of experts and, more important, the test of time, your book may well turn out to be as epochal as *The Origin of Species* or the *Principia* of Newton."[47]

Macmillan had long been a prospective target for Velikovsky's publishing projects, and he sent them this manuscript in due course. Crucial in Macmillan's eventual acceptance was an article by science editor John J. O'Neill in August 1946 in the *New York Herald Tribune*, which reprised Velikovsky's major claims in a few paragraphs at the end of a broader survey of catastrophic theories.[48] This article seems not to have attracted much attention, but it intrigued Gordon A. Atwater—curator of the Hayden Planetarium at the American Museum of Natural History in New York and organizer of a series of public shows on astronomy—enough to consider Velikovsky's theory as potentially appropriate for such a show, which could accompany publication of the book.[49] When Macmillan learned that free advertising on such a scale might be possible, they agreed to look at the manuscript.

It was often alleged, in the firestorm surrounding *Worlds in Collision*, that the book was not refereed properly—noted physicist and popular science writer George Gamow wrote to Velikovsky's editor and demanded to know who the referees might have been, because they could not have been competent—and that this represented a evasion of professional responsibility on the part of George Brett and his publishing house.[50] But the manuscript *was* refereed, in two entirely separate rounds.[51] The first stage, in 1947, consisted of three referees, John J. O'Neill of the *Herald Tribune*, Gordon Atwater himself, and a third who remained anonymous and whose report is lost to history. O'Neill's initial report noted that Velikovsky's data on historical catastrophes, culled from ancient legends, were striking, but that his explanation (that these were caused by a near approach of Venus) was hopelessly inadequate. But this mattered little:

> There will remain, however a solid foundation of historical records of cataclysmic events which will withstand all criticism and which will necessitate the opening of a new era of research in this field, and probably bring about an explanation with a more solid scientific explanation than that offered by Dr. Velikovsky. . . . A correct scientific explanation of the cataclysmic events, however, is not essential to acceptance of the data establishing the reality of the events themselves. The book is well worthy of publication despite criticism of his explanations.[52]

In an addendum, O'Neill mentioned that he was "very favorably impressed by the originality of the author's conception—holocausts and terrestrial convulsions caused by comets striking the earth, although I held his thesis untenable," and appended in both reports some factual errors.[53] Atwater concurred: "I believe this book is a positive must for publication. . . . The corrections which I have made in the script deal with astronomical facts of a minor nature. . . . I believe the author has done an outstanding job. In fact, he has gone beyond what might normally be expected of a single individual."[54] With two positive reports, Velikovsky received a regular contract to replace his advance contract in May 1948, and he left with his wife to visit their eldest daughter in Israel, returning to the United States in early February 1949.[55]

The second bout of referees (Velikovsky would later call them "censors") came in February 1950 and was a direct response by Macmillan's George Brett to Shapley's suggestion that Velikovsky's book might be crackpot literature.[56] All were New York scientists in different fields, selected because Macmillan editor Henry B. McCurdy knew them personally (a suggestion to add Albert Einstein to the list came to naught).[57] Edward Thorndike, head of the Department of Physics at Queens College, called the book "quasi-scholarly" and noted that "the physics is not good." As a "scholarly book" he would "rank it low," but as "a book to sell to the general reader, I would rank it higher. Certainly the idea is one to capture the interest."[58] C. W. van der Merwe, professor of physics at New York University, responded along similar lines: "'Worlds in Collision' is not a text on science and I am sure its author never intended it to be. It is not scientific fiction, it seems to me that it is more nearly fictional science and I am not saying this in any disparaging sense. . . . The running catch-as-catch-can fight between the Earth, Venus and Mars sounds unscientific." This did not mean, however, that Macmillan should reject the book, for "on its own merits it cannot fail to appeal to the reader, whether he be a man of science or not," and he recommended publication, hoping there would be "good demand" for it.[59] The only univocally positive report was Clarence S. Sherman's, associate professor of chemistry at Cooper Union, who rejected the idea that this was the work of a "crackpot with a message," and concluded that "as a scientist, particularly a chemist, I can find no great flaws in his deductions."[60]

There are several interesting features here. First, Macmillan, as befitted a scholarly press, had a normal refereeing process, although in the first round they selected their referees to fit their vision of the book: this was a trade publication designed to appeal to the public, and so professionals in science popularization should be the judges. Intriguingly, even the referees of the second round, which consisted entirely of professional scientists, made this very same distinction, in two cases rejecting the science entirely but *still* endorsing the book because trade books were not expected to be accurate. Later, these assumptions were called into question, but the fundamental point is that refereeing was a relatively recent phenomenon in science publication—it was certainly not widespread before the 1930s in the United States—and these scientists saw their task as advising the *press* about the qualities of the book, not "protecting the public" from erroneous views.[61] And so *Worlds in Collision* cleared peer review and was released to the American public.

THE BOOK REVIEW GAUNTLET

Now it faced the book reviewers. Scientists' reviews of *Worlds in Collision* were unanimously negative. In fact, the negative reviews preceded the book's publication by months and started with attacks on Larrabee's article in *Harper's*. The most widely cited and discussed critique was penned by Harvard astronomer Cecilia Payne-Gaposchkin, who worked with Harlow Shapley. Apparently, Shapley had been asked to review the article, but, as Payne-Gaposchkin recalled, he "(very sensibly) tossed a hot potato into my lap," but not before writing Macmillan to ask for page proofs of *Worlds in Collision* so that she could review the entire book.[62] No page proofs were forthcoming, so Payne-Gaposchkin made do with Larrabee. The review, which had circulated in manuscript among astronomers for some weeks, appeared in the *Reporter* on March 14, 1950. (The foreword made it clear that this was not a review of the book itself.) Aside from astronomical objections, one of Payne-Gaposchkin's main complaints had to do with religion: "The most insidious part of the argument is the appeal to Biblical sources. There always have been, and always will be, well-meaning people who defend the literal interpretation of Scripture."[63] This point, as noted earlier, was not so much articulated by Velikovsky as by the advance press, but that press had the effect of structuring the book's reception.

Payne-Gaposchkin actually reviewed Velikovsky's work on two further occasions. She addressed the full book in *Popular Astronomy* that summer. The editors of the magazine decided to devote an unusual amount of space to *Worlds in Collision* because the book "has been brought to the attention of a large reading public by having been mentioned favorably in several popular magazines."[64] Payne-Gaposchkin's review was, if anything, even more damning than before, but now her attention focused not so much on the astronomy as on Velikovsky's sources: "They are selected with bias, rearranged *ad libitum*, and often misinterpreted. It is notorious that anything can be found in the Bible; the same is true of classical literature. . . . Comparative mythology cannot be made to yield facts acceptable to physical science." After cataloging what she classed as errors of philology and errors of astronomy, she had just about had enough: "We have here an extraordinary achievement in a very difficult type of marksmanship—four (or even five) hits in a couple of thousand years, and all (by a lucky chance) at crucial points in the history of Israel. It is not only impossible. It is ridiculous."[65]

Her final intervention in the Velikovsky matter came in 1952, when she narrowly missed confronting the man himself. (As far I have been able to determine, the two never did meet.) During its annual meeting of April 24–26, 1952, the American Philosophical Society in Philadelphia—America's oldest scholarly organization—held a symposium devoted to unorthodoxy in science.[66] In addition to pieces on dowsing and extrasensory perception, Payne-Gaposchkin was to present a paper on Velikovsky. But she could not attend in person, so Donald Menzel—also of the Harvard Observatory—was supposed to present in her stead. He, in turn, was preparing for a talk at the upcoming American Physical Society meeting, so the paper was finally read by that society's executive secretary, Bell Labs scientist Karl K. Darrow, one of America's most respected physicists.[67] Payne-Gaposchkin's paper reiterated many of her earlier claims, although this time at a somewhat more measured pace, since, as far as she could tell, the Velikovsky affair was over. "A critic," she notes in the published version, "is faced not only with the comparatively easy task of showing that the results are untenable, but also with the Herculean labor of laying a finger on the flaws in an argument that ranges over the greater part of ancient literature." The problem rested not with astronomy, but in Velikovsky's evidence, the very point that (unbeknownst to her) John J. O'Neill had singled out as the strength of the book: "[Velikovsky] has not only chosen his sources; he has even chosen what they shall mean."[68]

Velikovsky begged to differ. Upon learning of the symposium, he took the train to Philadelphia from Princeton, New Jersey (whither he had recently moved), and crashed the meeting. His very presence sparked a frisson of excitement in the learned audience, and he asked the chair whether he could rebut the allegations. He was permitted and—having misplaced his notes—spoke *ex tempore*. As he described it in a letter to his daughters: "And I enjoyed very much the speech because never a thought was missing.

I built it mainly on the idea that they the scientists have segregated fact from theory—they observe that the sun is round but acc[ording] to their laws it cannot be round, because as result of rotation it should become flat to some extent." The talk was politely received, but he endured some mockery afterward by dismissive scientists in the cloakroom, including some who gloated that they had not read his book but would condemn it anyway. Velikovsky considered them more with pity than anger: "To me such behavior is more like that of Galileo's contemporaneous scientists who would not believe that he had discovered with the help of a then newly used telescope four satellites of Jupiter—and *refused* to look with the telescope."[69] The American Philosophical Society declined to publish his rebuttal to Payne-Gaposchkin, which he then sent to the History of Science Society's journal, *Isis*, where it was also rejected.[70] But this is getting ahead of the story.

At the start of April 1950, *Worlds in Collision*, its path trumpeted by laudatory summaries and controversy stoked by Payne-Gaposchkin's damning pre-review, hit bookstores. It was met with a host of reviews in the major publications—mostly, but not all, by scientists—which condemned the work as errant nonsense.[71] Edward Condon, director of the National Bureau of Standards, dismissed Velikovsky briefly, but would not accede to the suggestion of some that *Worlds in Collision* was an elaborate hoax: "If it is a hoax, it is an extraordinarily good one, and if it is later revealed to be a hoax, it will be interesting to learn which of the persons so far involved are the hoaxers and which are the hoaxed."[72] Paul Herget of the Cincinnati Observatory was similarly dismissive and could not "bring himself to justify the presentation of this collection of material to the unsuspecting public for acceptance at face value."[73] Otto Struve of the University of Chicago dubbed the work pseudoscience: "It is not a book of science and it cannot be dealt with in scientific terms. . . . Mr. Velikovsky's book is not unusual. The pseudo-scientific fringe of the academic world is well populated and many books of this kind are printed each year, by one means or another." Only the stature of Macmillan and the enthusiasm of the advance press made this case different: "Its publishers are a firm which has established a reputation for its excellent textbooks in many branches of physical science, and the book has had an unprecedented buildup by three popular magazines which have printed sensational condensations in advance of its release by the press thereby spreading its ideas among their millions of readers."[74] This point was underscored by geochemist Harrison Brown in the *Saturday Review of Literature* on April 22:

This book will be, for years to come, a shining example of book- and magazine-publishing irresponsibility. I do not object to publication of this book—or for that matter any book. But the reader may rightly be offended, as was this reviewer, by the irresponsible publicity, including magazine articles which preceded publication as part of the build-up. The publisher, who in this case is usually most meticulous in the publication of scientific treatises, should have sought the advice of reputable scientists before launching its sensational fireworks.[75]

More popular reviewers were not necessarily kinder. Waldemar Kaempffert, the science editor of the *New York Times*, wrote a featured review (adorned with an illustration of an occult mage in the cosmos) damning Velikovsky's science. Among several other criticisms, he invoked Roche's limit—2.44 times the radius of a planet, within which any approaching body would be torn apart by tidal forces—to explain why a near approach of the Venus comet was physically impossible. "Were it not that it took years to compile and collate hundreds of citations and footnotes," he concluded, "a critical reader might well wonder if this quasi-erudite outpouring is not an elaborate hoax designed to fool scientists and historians."[76]

Particularly revealing were those reviews that saw the publication of *Worlds in Collision*, for good or ill, as some sort of diagnostic marker of the contemporary age. Alfred Kazin in the *New Yorker* considered the book ridiculous, but it "plays right into the small talk about universal destruction that is all around us now, and it emphasizes the growing tendency in this country to believe that the physicists' irresponsible scare warnings must be sound, for they are the greatest experts of all, and how shall we ever create a world order except by first threatening everyone with world destruction?"[77] (This was hard upon the heels of Truman's initiation of a crash program for the thermonuclear weapon.) Likewise, the *New Republic* featured a peculiar page-long piece by Harold Ickes, former secretary of the interior:

If Venus, the planet, is as feminine as she sounds, and is intent upon acquiring a mate, then all of our scientists, including Americans and Russians, should devote their exclusive attention to making a match between Venus and Mars. That would leave those of us on Earth free to entertain such minor worries as the hydrogen bomb. . . . In the circumstances it seems more than a little childish for Russia and the United States to continue a race in armaments that would melt into molten metal before the ardency of an amorous Venus.[78]

Perplexingly, later Velikovskians would take this commentary as an endorsement, yet it is hard not to concur with Shapley that Ickes was writing a tongue-in-cheek satire on the enthusiasm for *Worlds in Collision*.[79]

And meanwhile the negative reviews kept coming, and the books kept flying off the shelves to the tune of a thousand copies a day. That was a large part of the problem: every negative review seemed only to enhance the controversy and increase excitement, and *Worlds in Collision* rocketed to the top of the best-seller lists and stayed there. The sting of the drubbing in the reviews was quite well compensated, for Macmillan at least, by the volume of sales. But it was too soon to celebrate the publishing coup of the season, for in parallel to these reviews, letters began to arrive at the editorial offices, and these gave Macmillan something to really worry about.

THE UNMAKING OF MACMILLAN'S
WORLDS IN COLLISION

Harold S. Latham, a senior editor at Macmillan in 1950, would recall in his memoirs penned fifteen years later "the hundreds, perhaps thousands of letters the publisher received."[80] Based on an exploration of the Macmillan archives, that number seems high, even accounting for the fact that many would have been lost or destroyed in the shuffle of time. Velikovsky himself, in a press conference in July 1950, denied such massive numbers: "There was no flood of letters to Macmillan," he declared.[81] In any event, there were certainly dozens, although the set received before June 1950—which, as we shall see, is the important context for Latham—was around thirty. Up until June 1950, almost every letter Macmillan received about *Worlds in Collision* came from a member of the scientific community and was a harshly worded statement of protest.

Dean McLaughlin, a professor of astronomy at the University of Michigan, was perhaps the most vehement of any of these writers and reserved his harshest words for the press itself: "In the case of the Velikovsky book, the author is no more to be blamed than the insane are to be blamed for their acts or their delusions. He is a deluded man. . . . It is the Macmillan Company that is guilty,—either through intent or through negligence,—of an irresponsible act towards society. Their responsibility as publishers was perfectly clear." Asserting that there must have been no peer review—for how could this drivel have survived?—he found himself "seriously wondering which of two hypotheses is nearest to the truth. Has the Macmillan Company been 'sold a bill of goods' by a quack? Or did they, *knowing* the author's claims to be absurd, recognize material for a best-seller? On the one hypothesis they are dupes and gulls; on the other they are deliberate partners in a fraud . . . a deliberate conspiracy to misinform the public." McLaughlin proposed a threefold plan for the redemption of Macmillan's name: withdrawal of the book, canceling any future Velikovsky books (the companion *Ages in Chaos* had been advertised), and publishing "conspicuously a statement that they recognize the misleading character of the book, even though it was prepared in good faith by the author; that he has been shown to be mistaken in his conclusions."[82] A mantra of attack against *Macmillan*, and only incidentally against Velikovsky, runs through the other protest letters that have survived.

For example, Roy Marshall of the Morehead Planetarium at the University of North Carolina at Chapel Hill argued that "the informed will not be misled by the lunatic reasoning in the book; the uninformed will now, thanks to Macmillan, be completely misinformed. It is a disgraceful thing, that makes me sorry I have ever purchased a book with the Macmillan imprint."[83] In another letter, he declared that he felt "almost sick at contemplating the possibility that a responsible publisher will put this dreadful stuff between covers."[84] Frank K. Edmondson, director of the Goethe Link Observatory of Indiana University, went further, dubbing Velikovsky's work "annotated clap-trap," but aiming his greatest censure for Macmillan's advertising campaign. Specifically, he considered the listing of Velikovsky's work in the science catalog (and not simply in the trade catalog), either "a gratuitous insult to the thousands of men and women who have contributed to the real advance of human knowledge, or it is evidence that the Macmillan Company is a screwball outfit."[85] Objections came from outside professional astronomy as well. The president of Moravian College in Bethlehem, Pennsylvania, was shocked that such a book could be published without securing the bona fides of an accredited biblical scholar, joining a professor of religion and philosophy from Greensboro College, North Carolina (who was incensed at this "incredibly stupid fantasy on astronomy" being "foist[ed] on the public"), and a freelance mathematical consultant ("To my mind, this is a new low in the ethics of the publishing business!").[86] All of them, whether from the physical scientific side or the biblical hermeneutic side, lamented the rejection of expertise in these important matters, a standard that was supposed to be upheld by the press, not subverted by it.

The scientific public was Macmillan's primary market, so having them angry was by no means a smart business strategy, but it was far from fatal.

Paul Herget of the University of Cincinnati (whose review of *Worlds in Collision* was quoted earlier) raised a more serious issue in his own letter to the press. Objecting to the "publicity campaign of sensationalism" and the use of trusted public names like Clifton Fadiman's as endorsements, he alleged that Macmillan's behavior "has all the appearances of a well laid campaign to make suckers of the public at a rate of $4.50 [$42.32 in 2012 dollars] each. This is unquestionably unethical conduct on the part of any reputable publisher." So far, his censure was like McLaughlin's and the others'. But there was something further: "There is a strong sentiment among many individuals to 'black-list' you in the selection of books for class use, and also in the choice of a publisher for new books."[87]

A boycott? Now *this* was bad. Boyd T. Harris, science editor of the College Department of Macmillan—and thus responsible for the textbooks that earned most of the company's profits—handwrote across the top of that letter: "This man is no crackpot."[88] Harris immediately wrote back, asking for Herget to referee the final manuscript of an astronomy textbook by Wasley Krogdahl of Northwestern University and confessing that he "deeply regret[ted] that Macmillan published the Velikovsky book. It was a straight trade department publication over which the College Department had no control." So why boycott? "The unfortunate thing about the current boycott campaign which you refer to is that its effect will be to harm a department of this company and individuals in that department who had nothing to do with the situation."[89] Herget countered: "This would not be the first time that the rain has fallen upon the just as well as the unjust. I am, however, thoroughly in favor of having your company boycotted by all prospective and future authors of scientific books, if there is no ameliorating action taken by your company about this matter. This is a position which I believe I would be willing to maintain indefinitely."[90] Independently, Krogdahl agreed, writing to Harris in mid-May that the chairs of two midwestern universities told him that they were going to stop buying Macmillan books. "Such a stand might be criticized as wholly unreasonable," he continued, "were it not for the manner in which the book is being advertised by Macmillan. For one thing, it is taken as a professional affront that the book has been placed in the Science section of Macmillan's spring catalogue."[91] Worries for his own book were clearly foremost in his mind (as they had been for his chair, who had warned about the consequences of the Velikovsky publication for Krogdahl's book as far back as February).[92] The fullest intellectual case for a boycott was made by Columbia University professor (and later 1955 Nobel Laureate in Physics) Polykarp Kusch:

It appears to me that the recent publication of Velikovsky "Worlds in Collision" by Macmillan has done more to impair rational scientific thought in America than any other similar enterprise within my knowledge. No probable amount of sound scientific publication by a single publisher will remove the air of astrology, mysticism, dogmatic assertion and emotionalism which "Worlds in Collision" has injected into popular scientific thinking.

I would no more contemplate using a text published by the publishers of "Worlds in Collision" than I would contemplate using a text published by any publisher of science fiction under the guise of science. I believe that a fundamental question of cultural integrity is involved.[93]

Rhetoric aside, this was not much of a campaign: the letters were disorganized, uncoordinated, and threatened different things—some not to buy books, some not to referee manuscripts, others not to write them. But, unlike the reviews, these threatened textbook sales, comprising by some accounts 70 percent of Macmillan's revenue, and that meant they had to be taken seriously. Macmillan could not afford to call a bluff.

But what to do about Velikovsky? After all, they had a contract and remained the conflicted publisher of a national nonfiction bestseller. George P. Brett Jr., the chairman of the American division of Macmillan Publishing, had to do something, but what? Macmillan's first policy was to laugh it off. In mid-March Boyd Harris wrote to the traveling salesmen, who were starting to meet some resistance—even hostility—about Velikovsky, with tactics for calming down enraged scientists. First, they should cry censorship: "The right to publish unpopular theories has been won by the scientists at some cost over the years. Any attempt to denounce us for publishing Velikovsky smacks of censorship and no thinking scientist should want to pursue the subject further if he were to think about the matter in this light." If that failed, switch to humor: "We hope that it is all a bit of fun and perhaps you can implant the spirit of fun in the men whose first reaction is to confront you with accusations. If the book is fiction in good part why should anyone worry about it?" Finally, echoing a sentiment that Ickes and Kazin would offer later, put the book into some perspective: "After all we haven't dropped a hydrogen bomb on the world."[94] A form letter was drawn up to answer hostile correspondents, insisting that "Dr. Velikovsky's work is being issued in our general trade department, since it is designed for the general public and presents the author's ideas about what may have happened in the past and what may happen in the future."[95] Nothing, in short, to worry about when issued from "a department store of publishing," and "whether right or wrong we feel that the publication of the book is justified on several counts, one of them being that it will serve to awaken the lay mind."[96]

This approach failed quickly. By May several scientists and departments—it is impossible to tell how many, and it was quite possibly only about five, but enough to be ominous—had turned away Macmillan representatives and sent back books. So Brett turned to his lawyers. F. Sims McGrath informed him that it would be possible under paragraph 6 of the Velikovsky contract to assign the book to another publisher if the author suffered no financial damage, and if Velikovsky was to complain, they could always say that Macmillan had been misled into thinking this was a scientific book.[97] (The angry letters in the file, which were soon shown to Velikovsky, could establish that.) Brett rapidly arranged a deal with the Doubleday publishing house to take over the Velikovsky contract—a coup for them, since they would be acquiring a best seller at no cost and had no textbook department to leave them vulnerable to boycotting. Brett now had to sell the idea to Velikovsky.

That was harder. On May 25, 1950, at three in the afternoon, Velikovsky was brought into Brett's office to discuss the contract. We have two versions of what happened in that room. First, there is Velikovsky's, written sometime in the mid-1950s and published posthumously. According to him, Brett was distraught, pleading with Velikovsky to take mercy on the firm and free them from the boycott. *Worlds in Collision*, on the other hand, was a cash cow. Fifty-four days after publication of the book, 54,000 copies had been sold (including prepublication). Brett showed Velikovsky the hostile letters, explained the Doubleday deal, and tried to get him to sign immediately so he could head off to Europe on a business trip. Velikovsky said he would think about it and get back to him. By June 7, he had signed the new contract. In his own account, Velikovsky was philosophical, even magnanimous, in his condescension to Brett. What other options remained open to the poor publisher? "He would have ruined his textbook department—and for what? For a book that, if right, would make many books in his textbook department obsolete."[98]

Shortly after the meeting on May 25, however, Brett wrote down his own account of the discussion, which differs from Velikovsky's in almost every particular. Brett had his solid legal advice, backed by the protest letters, and Doubleday gave him an out without incurring legal liability. Brett had no patience for Velikovsky, who (in his eyes) dragged out an unpleasant meeting and behaved childishly: "The man went through the gamut of everything that a man could do without becoming violent. He threatened to sue us for half a billion dollars. When he saw threats were no good, he pleaded."[99] According to Brett, Velikovsky demurred for two weeks mostly out of pique (there is, of course, no way for us to know the real reason), but on June 7 the deal was done, and he telegraphed all their outlets and sales personnel: "CONTRACT VELIKOVSKY WORLDS IN COLLISION CANCELLED THIS AFTERNOON. MAKE NO FURTHER SHIPMENTS OF BOOK. REPORT EXACT SALES SINCE PUBLICATION AND STOCK ON HAND TO MY OFFICE BY AIRMAIL AS SOON AS POSSIBLE."[100] A new flood of letters from private citizens from all walks of life arrived at the offices—by a rough estimate, more than had been sent earlier—castigating Macmillan for succumbing to the astronomers' pressure.[101] There was, at this point, nothing Macmillan could do. At least the scientists were finally happy. "You and the Company," Krogdahl wrote to Latham, "are to be congratulated for your demonstration of extraordinary business ethics."[102]

There were—rather later—some rumblings that the Velikovsky problem had not gone away. John Pfeiffer asked in *Science* in July 1951: "Has the shift of publishers from Macmillan to Doubleday (which has no soft underbelly in the form of a textbook department) improved the situation in any fundamental sense?"[103] The straightforward answer would have to be: Yes, it did. There were some letters to Doubleday protesting the publication of *Worlds in Collision*—including one from Fred Whipple of the Harvard Observatory concerning his own contract with the Blakiston Company, a subsidiary of Doubleday—but these were quickly defused by counter-accusations of censorship, that "it is important that opinions are expressed and not suppressed."[104] As Ken McCormick, the editor in chief of Doubleday, put it in response to one such letter: "We have not forced the book on anyone nor do we offer it as a textbook. In no way do we present it as a great contribution to science."[105] And the rumblings died down. If the purpose of the boycott was to suppress Velikovsky's theories, it had very much failed. If, on the other hand, the point was for the scientists to assert control over Macmillan, then it had unquestionably succeeded.

In the meantime, the grand collision seemed to be over. Velikovsky, incensed at how he had been treated by both the reviewers and Macmillan, wrote a lengthy history of the affair and attempted to get it published in March 1951—that is, about a year after the controversy started. The *New York Herald Tribune* rejected it, however, because "the controversy over your book ended, for all practical purposes, last year."[106] Although the Velikovsky affair would flare up again on several occasions in the decades to come, it is important to recall that no one had an inkling of the future, and for many this was a flash in the pan, something to recall dimly in the years to come as a regrettable episode of histrionics on both sides. As psychologist Edwin G. Boring predicted (rather poorly) at the American Philosophical Society meeting that Velikovsky had attended in 1952: "He is probably a nova and will soon fade to the dim status of an historical instance of the instability of an intense implausible conviction."[107] That might have been the case had a different narrative of 1950 not taken over the discourse—a story that the scientists had suppressed a heretic, someone who uncomfortably challenged astronomers' assumptions and offered a real alternative to their science.

A QUESTION OF SUPPRESSION?

On July 5, 1950, just under a month after the transfer of rights over *Worlds in Collision* to Doubleday, *Newsweek* published a short article entitled "Professors as Suppressors," which argued that Velikovsky had been silenced by the scientific establishment and strongly insinuated that a Harvard cabal (and hence, implicitly, Harlow Shapley) was behind the campaign to suppress a new theory of the solar system.[108] By the end of the year, the *Saturday Evening Post* had identified the supposed "suppression" of Velikovsky as one of the signal events of this year's "silly season": "One of the most astonishing episodes of the summer idiot's delight was the effort of American scientists to suppress a book, Worlds in Collision, by Dr. Immanuel Velikovsky."[109] There was, of course, evidence for this interpretation, and the weapon of the boycott was interpreted by many, including Fulton Oursler (the *Reader's Digest* author) as "book-burning by intellectuals."[110] (Velikovsky, it should be said, was not at first enamored of this narrative: "To me it is most important to show that I was right. Who cares, besides the defenders of civil liberties, if a wrong idea is suppressed?"[111]) During the following decades, this interpretation of 1950 became dominant, perhaps best expressed in 1989, ten years after Velikovsky's death, by one of his critics: "The campaign against Velikovsky was well-orchestrated, but it failed in its main purpose, which was to have the book suppressed."[112] Here are two key assumptions: that the campaign was organized and that it failed. But the objections to Macmillan were scattered; in fact, in hindsight, one could easily say that the press overreacted to light pressure. The second point, about failure, requires a proper understanding of the boycott's purpose.

It is tempting to look back on the uproar of spring 1950 and see it as so much hot air. That would be a mistake, for two individuals who publicly backed Velikovsky lost their jobs, and their stories became central markers in the development of Velikovskian auto-mythology. The first was the case of Gordon Atwater, the planetarium director and one of the original referees of *Worlds in Collision*. Besides endorsing Velikovsky's book before publication and giving a (very noncommittal) blurb to the back cover, he also planned a planetarium show that would coincide with the book's release and summarized Velikovsky's views for the high-circulation *This Week*.[113] While that article was in press, Wayne M. Faunce, vice director of the museum, approached Atwater on March 9 and ordered him to stop talking about Velikovsky. On March 10, the administration revised Atwater's planetarium show ("Our Battle-Scarred Earth") to remove all mention of Velikovsky. Atwater was then told that he had to be out of the building by the first of April, two days before the publication date of *Worlds in Collision*. When Atwater protested, museum officials offered him full pay until October, provided he resigned and did not compel them to fire him. He took the deal and packed up his office, to his great regret in later years. Not only was he unable to get a job in the scientific world again, but he lamented that the failure to do a Velikovsky show hurt the public's understanding of a crucial theory that he believed was an accurate recounting of the recent past of the solar system.[114]

The second martyr was James Putnam, Velikovsky's editor for *Worlds in Collision*. Frank Edmondson of Indiana University, in his angry letter to the press, fingered Putnam in particular: "Your own Mr. James Putnam deserves the heaviest blame and the strongest censure, for he had the opportunity to secure some competent opinions before the manuscript was accepted."[115] (And so he did, as the referee reports attest.) Indeed, after Brett's meeting with Velikovsky on May 25, the author insisted that he had warned Putnam that the book was dynamite and he should have it vetted, and so Macmillan had prior warning. This was good enough for Brett, who consulted with Harold Latham and J. Randall Williams (the director of sales for the Trade Division), and terminated Putnam after twenty-five years at Macmillan, with a year's pay in lieu of notice.[116] By August 1950, however, Putnam had a plum new job at the Trade Department of the World Publishing Company and later moved to the post of general secretary at the PEN World Association of Writers.

Both of these stories contain more than meets the eye. Consider Putnam first. Convinced of the full suppression narrative by the 1960s, Velikovsky was very interested in elucidating the exact details of his own victimization, and he wrote Putnam for specifics of the firing and the transfer of the book, and "whether it was obvious that one was connected with the other? In other words, whether we can say in our 'recollections' that you have been a victim of the opposition that my book encountered among certain groups of scientists?"[117] There was no answer in the file, but Randall Williams later noted that although Putnam was a "scape-goat," Brett had "good grounds on which to feel dissatisfaction with Jim Putnam's work in general, quite unrelated to the Velikovsky embarrassment."[118] Latham told Velikovsky the same: "[Putnam] had been on very insecure ground for a long time and only my intervention had prevented earlier dismissal. There were a number of reasons for the official attitude which have no relation to any responsibility he may have had for the sponsorship of your book. I should prefer not to go into these and I can only ask that you believe me when I say that they were in no way connected with you or WORLDS IN COLLISION."[119] Velikovsky disagreed, but one may presume Latham was in the know. Although *Worlds in Collision* was undoubtedly a pretext for letting Putnam go, it was only the final straw in a series of personnel difficulties. And, besides, Putnam's career was not derailed for any great length of time.

But Atwater never worked in astronomy again. Many astronomers thought that he should never have worked in astronomy in the first place. Atwater was a navy man and he taught navigational courses to draftees during World War II at Harvard, and then later in New York, a position that resulted in the planetarium job. Roy Marshall, who was also engaged in public education in astronomy, declared that Atwater was "not an astronomer or, for that matter, a trained scientist of any kind. He was a Navy navigator, not an astronomer. His opinion of the work is worth approximately as much as that of a college student in astronomy."[120] Likewise, Krogdahl thought Atwater was "a gross incompetent in his present position. His appointment is presumed to be explicable only as an unfortunate consequence of New York City politics."[121] So, was the dismissal justified? Donald Menzel thought so: "He was not much of a scientist, as his verbose endorsement of Velikovsky indicated. The management dismissed him for the simple reason that he was disseminating, in the name of science, nothing but nonsense. He was dismissed for the same reason that an educational institution would dismiss a professor in a medical school if he started teaching and advocating a return to voodooism or witch doctors."[122] Atwater, who had been injured in a boating accident in late

1949, found it hard to get work, and his later career does indeed evoke sympathy.[123] In this case, unlike that of Putnam, the tentative endorsement of Velikovsky did exemplify why scientists considered Atwater to be unscientific, even though the criticisms had been of long standing. Velikovsky was to a great extent a pretext, not a cause for martyrdom.

Nonetheless, the general understanding of the Atwater and Putnam cases, broadcast in articles like *Newsweek*'s, evinced a strong sense of a doctrine persecuted by establishment science, a latter-day Galileo affair, and for there to be a scandal of suppression, there had to be a villain. As the most visible American astronomer of the day, who had an unquestioned connection with Macmillan's abandonment of *Worlds in Collision* and was also the employer of many of those (Cecilia Payne-Gaposchkin, Donald Menzel, Fred Whipple) who criticized the book, the casting job was relatively easy: Harlow Shapley was thrust into the role, completely ignoring his earlier tepid encouragement of Velikovsky. As the latter neatly summarized the case to Harvard orientalist Robert H. Pfeiffer: "It is no secret that the emotional outburst against 'Worlds in Collision' started at the Harvard College Observatory. It is even possible that a number of reviews were written at the suggestion of Shapley. A review by [J. B. S.] Haldane [a British geneticist], otherwise a clear thinker, gives me the impression of being written in the pattern already known as originating from the Harvard Observatory."[124] Shapley denied vigorously that he was behind any such campaign, and he sent his correspondence with Velikovsky and Kallen to his estranged friend Ted Thackrey to prove it, but the very denials later came to be interpreted by some as so categorical that the very absence of evidence of a Shapley-organized campaign was taken to be proof of that campaign's insidiousness.[125] In one of his final writings, dated 1973, Horace Kallen wholeheartedly endorsed the Shapley-as-suppressor theory, ascribing to him a "lifelong vendetta" against Velikovsky.[126]

The only problem, again, is evidence. There is simply no hoard of letters indicating a centrally organized campaign against Velikovsky's book in either Shapley's papers at Harvard, Velikovsky's at Princeton, or Macmillan's in New York. Asked in the 1980s whether they were encouraged to protest by Shapley, both Krogdahl and Edmondson flatly denied that they had been approached by *anyone* to write their letters.[127] Donald Menzel, a direct employee of Shapley's, did not even participate in the boycott.[128] And even Velikovsky supporters in 1950, although they were suspicious, did not adhere to this particular version (although some would later). James Putnam, although he wanted to believe in a conspiracy against himself and Velikovsky, wrote to a friend that "it would be very difficult to prove what Professor Shapley's role was in the whole matter."[129] Eric Larrabee, who had started off the whole hullabaloo with his *Harper's* article, told Harold Lavine, the political associate editor at *Newsweek* responsible for the original "suppressors" article, that he himself only had spotty information on the campaign against Velikovsky, but he "didn't think there was any evidence that there was an organized campaign or that Shapley led it."[130] This was Shapley's position as well. "The boycott to which you refer," he wrote in 1951 to Frederick Lewis Allen, the editor of *Harper's*, was "in itself largely imaginary and made up by the publicity specialists," and claims of a campaign organized by himself were "persistent New York lies."[131] Yet it was too late. The suppression narrative had already congealed by the end of 1950, and it structured the debates over Velikovsky for the decades to come.

INTENTIONS

This understanding of the 1950 response to Velikovsky—that, while it included a vociferous reaction against the contents and methods of the book, it was primarily focused on Macmillan's behavior in publishing and advertising it—makes more sense than the very common later view of a censorship campaign to repress a heretic. After all, the method deployed in letters to the publisher was a boycott campaign (if we can consider half a dozen letters a "campaign"), which is an *economic* tool designed to punish a commercial enterprise. It was not aimed at the author; he was the target of the reviews. Scientists—the chief consumers for Macmillan's products, and also in many cases the suppliers of content for their books—threatened a boycott, however haphazard and uncoordinated, to assert control over the press. This is why Macmillan authors like Krogdahl greeted the transfer to Doubleday so warmly: as long as *his* press was not tainted with Velikovsky, let the man publish wherever he could.

Ironically, the person who explicated this position best was Harlow Shapley himself. He wrote to Ted Thackrey in June 1950, declaring that he was perfectly content with an outcome that left Velikovsky in bookstores:

Certainly you and he and his publishers should be quite satisfied with his leadership of the best sellers for week after week, and I ought to be satisfied that I have not yet met an astronomer, or in fact a scientist or scholar of any sort, who takes "Worlds in Collision" seriously. Some referred to the clever promotion; some referred to the rather charming literary style; and some, while fully exonerating Dr. V. (who should do as he pleases in this free country), are unrestrained in their condemnation of the once reputable publisher.[132]

The very next day, Macmillan abandoned *Worlds in Collision*, but that seemed not to have been Shapley's particular interest (although it obviously was for many of the letter writers discussed earlier). According to Yale geologist Chester Longwell, the scientists' "chief concern is to focus attention on the publisher, rather than on the book or the author."[133] That concern is worth taking seriously.

Central to it, and animating the chapters that follow, is an abiding anxiety about science's relation to the "public," a nebulous and undefined but vital term for all the players. For Velikovsky and his supporters, the public had a right to hear novel theories about the nature of the universe, and attempts to hinder the flow of information were criminal, almost totalitarian. For the establishment astronomers—or at least for some of them—the public was easily misled and a certain form of noblesse oblige demanded that responsible individuals screen material that could hinder popular enlightenment. The mechanism of communication for either side was, of course, publication, and it is no accident that the controversies around Velikovsky began with a discussion of the role of publicity, a publisher, and popularity—all terms that reveal in their etymology the bone of contention. The *form* of the scientists' objection, the boycott, was not only a direct mechanism to demonstrate to Macmillan that market forces worked both ways, but also a form of mobilizing (if only rhetorically) a particular public: the scientific one.

There remain, of course, fundamental questions about this opening volley in the pseudoscience wars. What, precisely, was it about Velikovsky's vision of the universe that so enraged the astronomers (and their non-astronomical scientist colleagues, such as physicists, who joined in with them)? Why were they so certain that he was wrong? And, perhaps more directly, why did the scientists react so vehemently to this publication, in language and behavior that asymptotically approached hysteria, when the typical response to "pseudoscience" to date had been to ignore it altogether—why, that is, respond to Velikovsky's border incursion with full-scale warfare? The following two chapters approach the earlier questions through this last one, offering first Velikovsky's account of the hostile scientific reaction, an account that draws on the totality of his synthetic vision. Then we will turn from the Velikovskian version to explore the broader context facing the American scientific community, and why it felt particularly under threat when *Worlds in Collision* was published. Both avenues go a long way to rectifying the dialogue of the deaf that stormed through American popular culture in the early months of 1950.

2 · A Monolithic Oneness

It takes at least two sides to have a war. The astronomers were sure they knew the battle lines: there were scientists on one side and pseudoscientists on the other. That, obviously, was not how Immanuel Velikovsky saw it. His version of the debates of 1950 only had scientists in it: himself versus obscurantist dogmatists too blind to see the merits of his new theory. But everyone seemed to concur that the battleground was science itself.

That supposition deserves further scrutiny. What kind of work was Velikovsky's *Worlds in Collision*? A book about recent catastrophic changes in the solar system? An inquiry into the origins of astral religion? A compendium of catastrophic folklore from around the globe? Or an establishment of fixed points (a global catastrophe caused by a near approach of Venus) that would enable scholars to accurately determine the chronology of the ancient world? All these interpretations, along with the claim to be a revolution in science, were maintained by Velikovsky at various points in his career, but the historical reconstruction and the repair of chronology motivated the entire project. As he wrote to Horace Kallen as early as 1946, using locutions he would repeat many times in the years to come: "In my opinion a historical fact cannot be denied because of a physical theory, and if such a fact is established, the physical law must suit the fact, not the fact the law."[1] As hard as it might be to see with hindsight, the Venus scenario emerged out of an inquiry into *history*, not science, and Velikovsky often returned to historical questions, especially the historical significance of the Jewish people and the reliability of the Hebrew Bible as a source.

It is impossible to overstate the centrality of Jewish history, both recent and ancient, to Velikovsky's outlook. The significance is partly biographical. One can hardly ignore his birth into an intellectual Jewish family in imperial Russia at the dawn of Tsar Nicholas II's reign, the early development of a lifelong commitment to the Zionist project, and the persistent tropes of Jewish history that emerge in his writings. Velikovsky's career before he moved to the United States in 1939 produced ripples that shaped his future development, when he would become, as he would later put it, "the prisoner of an idea," the idea of cosmic catastrophism.[2] While *Worlds in Collision* correlated myths and legends from around the world, the central episode with Venus was also the central event in Jewish religious history—the Exodus from Egypt—and Velikovsky's repeated practice was to use the story of the Bible as bedrock and correlate all ancient material, especially Egyptian, to that standard. He saw his approach to history as a path to reconciliation. As he stated in the final book published in his lifetime, *Ramses II and His Time*: "The centuries both preceding and following the decades described in this volume constitute together, in the reconstruction of ancient history, a monolithic oneness."[3] History would provide the glue that held humanity together.

The debates around Velikovsky cannot be cast as simply history (or the humanities or the social sciences) on one side and "science" on the other.[4] While Velikovsky did approach the writing of *Worlds in Collision* from the vantage point of history, he treated historical questions from the perspective of psychoanalysis. Whatever the current attitude toward Sigmund Freud's psychological theories, there is no question that during the first half of the twentieth century they were considered scientific by many physicians, including one trained in Moscow named Immanuel Velikovsky. A psychoanalytically informed view of history was therefore at least partly grounded in science.

In his interpretation of pan-human cosmic psychic trauma, rooted in Freudian as well as Jungian debates, Velikovsky envisioned a "collective amnesia" that served differing functions throughout the six books he published in his lifetime, as well as his posthumous and unpublished works. This collective amnesia, rooted in a model of the mind indebted to Freud (while at the same time forming a criticism of it), comprises the second reason why we cannot dismiss the "scientific" features of the affair. For amnesia, at first an explanation for the peculiarities of his source base, became Velikovsky's explanation for the scientific community's violent reaction to his book. He proposed a putatively scientific account of the scientific reaction, buttressing the coherence of his system.

Velikovskianism became a front in the pseudoscience wars—and not, say, a battle with folklorists or ancient historians. Velikovsky published a book that was interpreted in Macmillan's promotional campaign as confronting science, but also unearthing the origins of religion. The astronomers and others who reacted to his arguments, however, transformed what was fixing to be a controversy about religion into one about scientific expertise. The scientists, through their condemnation, reoriented the conversation away from history and toward science. They fired the first shots of the pseudoscience wars.

FROM THE PALE TO PSYCHOANALYSIS

Velikovsky appeared in 1950 like a comet from nowhere, bearing a message so foreign to contemporary scholars that it was almost as though they could not understand it. But others did, and for decades Velikovsky would reach greater and greater popularity among young people looking for answers about the solar system, religion, history, and the mind. The oddness of this appeal to youths is accentuated when one realizes that he was already fifty-five years old when *Worlds in Collision*—his first English-language book and his first publication of any kind to exceed seventy pages in length—appeared. Immanuel Velikovsky was a child of the nineteenth century, from a time and a place that could not be further removed from the United States in the early years of the Cold War.

He was the youngest of three boys, born to Simon Yehiel Velikovsky and Beila Velikovskaia (née Grodenskaia) in the city of Vitebsk on May 29, 1895, according to the Old Style Julian calendar then employed in the Russian empire.[5] (According to the New Style Gregorian calendar, adopted after the Russian Revolution of 1917, the date would be June 10.) Located in present-day Belarus, Vitebsk was then part of an imperial Russia that stretched from the borders of Germany to the Sea of Japan, and the town formed a central economic node of the Pale of Settlement, where a majority of Russia's sizable Jewish population had been confined since the late eighteenth century. (Vitebsk was also the hometown of two titans of modern Jewish culture: author S. An-sky, born Shloyme-Zanvl Rappoport [1863–1920], and painter Marc Chagall [1887–1985].[6]) This was a time of transition for the Russian empire, as the institution of autocracy faced the challenges of modernization, and also a time of transformation for Europe's Jews with the advent of Zionism and secularism. Looking back at his year of birth, Velikovsky characterized the moment with his typical literary flair: "One is under the influence of the spirit of the time. The dream of Herzl, the intuitions of Freud, and the rays of Roentgen in 1895 were the earthly constellations which marked the direction in which I was to wander—ideas, like men, need time to grow and to find their place in the world."[7]

The situation into which Velikovsky reached young adulthood would have been literally unimaginable for his grandparents. Their offspring, the young Simon and Beila, met in Starodub in northern Ukraine, where Simon had been on business. The Velikovskys prospered, and in either 1900 or 1901, Simon moved to Moscow, and soon Beila and brother Daniel followed, with Alexander and Immanuel coming later. In the wake of emancipation reforms of previous tsars, Immanuel was born into a period of heightened opportunity compared to those of previous generations of Russian Jews, although of course he still faced significant constraints.[8]

Those constraints were most palpable in education, the area that young Immanuel cared about most. Faced with tsarist-era quotas on the number of Jews allowed in higher medical education, in 1914 he traveled to Edinburgh to take courses in the natural sciences in preparation for a medical degree, which he eventually received from Moscow University. After the Russian Revolution of 1917, he and his parents experienced tremendous dislocations along with many of their compatriots (regardless of confessional background), and they spent much of the period up to 1920 in the city of Kharkov (today in Ukraine and called Kharkiv). It was a period of intense activity for him, centered around the project of settling the Jews of Europe in Palestine. Indeed, before Velikovsky became a name synonymous with cosmic catastrophism, the polestar of his activities was Zionism. It was a family passion; Velikovsky's father had been a leading member of the Jewish community in Vitebsk and had met Theodor Herzl, the architect of modern Zionism, while serving as a delegate to the Second Zionist Congress in Basel.[9]

In the fall of 1917, as Russia crumbled into revolution and civil war, Velikovsky composed a thirty-two-page pamphlet entitled *The Third Exodus* (*Tretii iskhod*) under the pseudonym Immanuel Ramio. (The Exodus theme appears as a trope throughout Velikovsky's writings.) An impassioned defense of the Zionist project, Velikovsky's distinctive argument was to push against the secular, socialist strands within the movement in favor of a more religiously motivated, pious variant. The title page notes that the book was published "1848 years since the destruction of the temple" and was dedicated to "Jewry." Aside from these external markers, the text is rich with biblical allusions even as its Russian language marked it as a missive to assimilated Russian Jews. The general theme concerned virtue and retribution. Venereal disease among Jews, for example, was attributable to the diaspora, and would be remedied by sober emigration, while the Great War was a punishment for Europe.[10] Velikovsky yearned to settle in the Holy Land himself; he had already traveled there for five weeks in 1912, aged only seventeen, and felt the pull to return.

After the end of the Russian civil war, in 1921, Velikovsky obtained permission for himself and his parents to emigrate, and the three arrived in Berlin.[11] (His brothers remained in Soviet Russia, and Immanuel never returned.) Aspects of his Russian origin never left him, and he would later write to a Soviet astronomer that it had been "over 40 years since foreign languages replaced my native Russian as my daily speech, though probably I still often think in Russian."[12] He certainly continued to write in Russian in a Zionist vein, although after a more belletristic fashion. In 1920, while on a trip to the Caucasus, he composed a haunting sixty-eight-page booklet, *Thirty Days and Nights of Diego Pires on the Bridge of Sant'Angelo*, which chronicled the torments of a Portuguese Marrano who had converted to Judaism and, under his more famous moniker Solomon Molcho, proclaimed the advent of a new Messiah, which led to his burning at the stake for heresy on December 13, 1532.[13] Broken into short vignettes, Velikovsky (under the pseudonym Emanuil Ram) presented each day through stream-of-consciousness monologues in poetic Russian. According to Velikovsky, Russian émigré writer Ivan Bunin (the first Russian awarded the Nobel Prize for Literature, in 1933) encouraged him to publish this work, which he did in 1935. The story of Molcho, a heretic punished for his beliefs, stayed with Velikovsky, and he left in his archive a curious typescript entitled "Three Fires," which tells the life stories of three martyrs spanning the sixteenth century: Molcho (1500–1532); Michael Servetus (1511–1553), a Spanish theologian and pioneer in anatomy; and Giordano Bruno (1548–1600), often cited as a hero of science, who died at the hands of the Inquisition at least in part because of his heliocentric views.[14] Their stories haunted Velikovsky.

That was the end, however, of Velikovsky's Russian-language world, and in 1921 he emerged into the rich culture of Weimar Zionism in Berlin.[15] Velikovsky wanted to make a name for himself as a scholar while still promoting the Zionist cause, and he proposed that his father—who had left Russia with significant financial resources—support the publication of several volumes of scholarship by prominent Jewish intellectuals, printed simultaneously in a language of scholarship and translated into Hebrew, to "demonstrate the role played in the scientific world by Jews, who were then known only as citizens of their adopted countries."[16] Simon Velikovsky endorsed the idea, and Immanuel approached Heinrich Loewe (1869–1951)—a prominent German-Jewish journalist, publicist, and bibliographer—to assist in the project. Loewe's seniority was surely vital in recruiting many of the contributors to what was eventually named the *Scripta universitatis atque bibliothecae hierosolymitanarum*.[17] As the Latin name indicates, these writings were intended to be a seed for the university planned for Jerusalem. Loewe and Velikovsky soon realized that the best work was clustered in two categories, "Orientalia et Judaica" and "Mathematica et Physica." For the latter of these, they recruited their most famous collaborator, Berlin professor Albert Einstein, as editor. (Sigmund Freud demurred, claiming that his readers would not know to look for his works in any publication other than his journal, *Imago*.[18])

The *Scripta*, which appeared in 1923, were a success in that they brought Velikovsky to the attention of a circle of prominent Zionists. (Publication had been delayed because of a trip Velikovsky had taken to Palestine in February 1922. He remained there for five months visiting his parents, who had already emigrated, and the *Scripta* stalled in his absence.) Velikovsky claimed that Chaim Weizmann, later the first president of the State of Israel, asked him to be "the father of the university"—the request was quoted in Hebrew—when Velikovsky was a ripe twenty-eight years of age, a story that led to his being labeled in later press accounts as one of the "founders" of the Hebrew University.[19] It has proven impossible to find any testimony to this effect from Weizmann, although Velikovsky did write him a letter in 1935 (twelve years after the statement was supposedly made) declining to run the university.[20] Velikovsky never had any official connections with the Hebrew University (although a copy of a portion of his archive is now held there).

Zionism provides one of the few continuities across the dramatic transition from Russian physician to cosmological heretic. When he moved to New York in 1939, Velikovsky quickly joined the community of American Zionists. With noted Columbia University anthropologist Franz Boas, he arranged a meeting to propose an academy of sciences for Jerusalem, a project that disintegrated upon Boas's death in 1942.[21] While composing *Worlds in Collision*, Velikovsky avidly followed the news of the collapse of the British Mandate, and he wrote over forty anonymous articles (bylined "the Observer") for the *New York Post* passionately defending the creation of a Jewish state. (A full collection is preserved in his archive in Princeton.[22]) His spirited defense of Zionism continued until his death, and he engaged in a lively meeting of minds with *New York Times* columnist William Safire, who proclaimed himself a fan of Velikovsky's scientific as well as geopolitical claims.[23]

It is difficult to extrapolate from these public and political pronouncements to a firm judgment about Velikovsky's private beliefs about Judaism as a religion. He remained until his death quite cagey about his precise level of spiritual belief. As he wrote in 1962 to his friend Horace Kallen, describing his fan mail: "People in prisons demanded to know whether I am a believer or not. I am but not in the accepted meaning of the term."[24] He did not go to synagogue regularly, but he maintained a kosher household, and his daughter remembers him as devoted to the Bible. "My father believed in God," she wrote, "and in time of indecision opened the old testament for answers."[25]

Returning to our young man in Berlin and his successful foray into the publishing world, we should note that the *Scripta* also brought him success of a more personal nature. While working on the project in Berlin, he met a young violinist studying under Adolf Busch named Elisheva Kramer, originally from Hamburg. She began helping Velikovsky on the *Scripta*, and as soon as it was completed, they married and moved to Palestine, settling in Mount Carmel outside the northern port city of Haifa. In December 1930 the Velikovskys (including his daughters, Shulamit and Ruth, born in 1925 and 1926, respectively) moved to Tel Aviv and remained there until the summer of 1939, as Immanuel plied his medical trade and managed his father's real estate investments. There he might have remained, if not for the influence of another Jewish physician—Sigmund Freud. In a classic (and unwitting) Freudian juxtaposition, he later wrote: "In 1928, after the death of my mother, I turned my interest to psychoanalysis."[26] Psychoanalysis would occupy him for the next two decades and become the second major thread running throughout his long life.

In 1930–31, Velikovsky traveled to Zurich and then to Geneva to study neurology at the Monakow Brain Anatomy Institute with psychiatrist Eugène Minkowski, although he soon split with his mentor because of a disagreement over whether there was a connection between physical and psychic phenomena.[27] (Velikovsky was convinced of such a link, a topic he had begun investigating by reading the classic nineteenth-century spiritualist investigators William Crookes and Oliver Lodge, as well as the monumental history of the movement by Arthur Conan Doyle.[28]) He continued to read deeply in psychoanalytic theory, and he traveled to Vienna in spring 1933. After a brief period of study with distinguished Freudian Wilhelm

Stekel, the latter declared Velikovsky fully trained, and Velikovsky was free to practice as a psychoanalyst.

He produced a series of psychological and psychoanalytic publications. His first major piece displayed the residues of his dispute with Minkowski and concerned the phenomenon of telepathy. Prefaced with a foreword by Swiss psychoanalyst Eugen Bleuler agreeing with the piece's argument, Velikovsky's article contended that since nervous energy was a form of energy, and all energy was conserved, telepathy could be caused by the transfer of psychic energy from one mind to another. According to this argument, thoughts have some materiality, and hallucinations might actually be the reception of someone else's thoughts. The piece was littered with citations to earlier scientists who had found spiritualistic phenomena produced at séances convincing.[29] On June 24, 1931, Velikovsky received the highest compliment: Freud wrote him that his views on this matter generally concurred with Velikovsky's.[30]

The remainder of his work concentrated in two specific areas. First, he published a series of studies, based on observations in Tel Aviv, on the role of the Hebrew language and Hebrew texts in the psychoanalytic tradition. A 1933 article argued that rabbinic sages had practiced dream interpretation in their oneiromancy.[31] More directly based on patient experience, in 1934 he published an essay on wordplay in dreams that contended that immigrants to Palestine made puns in their dreams based on Hebrew, a language acquired later in life, demonstrating that it was possible for such languages to become expressive of the subconscious.[32] His second area of interest was the application of psychoanalytic findings to medical therapeutics, especially a multiply reprinted piece on "psychic anaphylaxis," deploying an analogy with allergic shock.[33] As late as 1977, two years before his death, Velikovsky collected all his psychological writings into a book manuscript for eventual publication.[34]

On December 16, 1937, the heaviest personal blow of his life to date struck the forty-two-year-old psychoanalyst: Simon Velikovsky, his beloved father, died. This led to some soul-searching:

> Looking back on the almost sixteen years spent in Israel, I could note but little achievement. I treated many psychoanalytic patients, and usually succeeded. I published a few philological works of my father (*Sfotenu*), and two issues of *Scripta Academica* after his death. I wrote several psychological papers, as well as a treatise on philosophy and biology called *Introgenesis*, which was accepted for publication by Presses Universitaires of France but

was left incomplete because of the war. . . . But at the age of 43 I had already lost the faith of achieving something great as a scholar.[35]

And then Sigmund Freud came once more to the rescue, setting Velikovsky on a path that would result in *Worlds in Collision*.

INVENTING *WORLDS IN COLLISION*

Freud's final work, *Der Mann Moses und die monotheistische Religion*—which appeared in German in 1937 and was translated into English two years later (and a year after the author's death) as *Moses and Monotheism*—is surely one of the most controversial works by the father of psychoanalysis. For our purposes, the significance of the book lies in its first two chapters, which argue that the figure known as Moses was actually an Egyptian priest from renegade Pharaoh Akhnaton's monotheistic sun religion who then recruited the Hebrew slave population for his cause. (I maintain the non-standard transliteration "Akhnaton" because this is what Velikovsky used.) Denying the ethnic kinship of the single most important figure of the Jewish religion would have been bad enough, but Freud went on to argue that after the Exodus, Moses's puritanical strictures provoked so much resentment among the Hebrews that they murdered him and then covered up their crime, blending the Egyptian Aten (the sun disk as god) with Jehovah (a Midianite volcano deity). The murder of the father figure, that all-purpose engine of Freudian dynamics, thus lay at the heart of the world's oldest surviving monotheism.[36] Much as Freud had developed an entire psychology around interpreting slips of the tongue, dream images, and casual jokes to reveal an individual's deeper psyche, he found in Hebrew lore and the Bible stray references that he interpreted—undoing the "dream-work"—to uncover the horror of the murder of Moses, the Egyptian patriarch. The book has provoked outrage and inspired furious debates ever since.

Immanuel Velikovsky was plenty mad. He had purchased a copy of *Moses and Monotheism* in spring 1939 in a Tel Aviv bookstore, and he was both drawn in and repelled by an argument that, on the one hand, proclaimed the power of psychoanalysis and, on the other, denigrated the origins and destiny of the Jewish people. He pondered the book intensely and arranged to take a year of leave in New York City so he could use the tremendous library resources of this new continent and compose a response. He and his family boarded the steam liner *Mauritania* and arrived in Manhattan on July 26, 1939. Within six weeks, Hitler's Germany had invaded Poland, and Europe—and soon the world—was engulfed in war. Velikovsky, his wife, and his two teenage daughters settled into an apartment on Riverside Drive, and Velikovsky shuttled back and forth to the New York Public Library. They expected to return to Palestine within a year.

Velikovsky was composing several different books in the wake of his father's death. The first of these, apparently begun in the summer of 1937 but never completed, was entitled "The Masks of Homosexuality" and argued that subconscious homosexuality lay behind many neurotic behaviors.[37] Elements of this project survived in Velikovsky's posthumous publication *Mankind in Amnesia*, and we can reconstruct some features of the argument from that work, when he claims:

> My view, derived at that time from psychoanalytical thinking, saw in repressed homosexuality of entire nations the source of hatred and of lust for doing bodily harm on a mass scale, of the massacres and the triumphs of a race motivated by male homosexuality, against and over an effeminate nation. . . . I still believe that suppressed homosexuality has much to do with aggression.[38]

The only publication from this work-in-progress was an analysis of Leo Tolstoy's novella *The Kreutzer Sonata*, which appeared in Freud's journal *Imago* and contended that the protagonist Pozdnyshev, who murdered his wife in a fit of jealous rage, was actually suffering from repressed homosexuality.[39] This marks a minor current in his writings, which are peppered with pejorative comments about homosexuality.[40] For example, and rather scandalously, Velikovsky's last secretary, Jan Sammer, recalled that Velikovsky planned to write a book entitled "Son of Man," which "was to have been a psychoanalytic study of the historical Jesus, in whom Velikovsky saw numerous indications of repressed homosexuality, along with other assorted psychological disorders."[41] In general (and somewhat atypically for a psychoanalyst), Velikovsky was "prudish" (in his associate Alfred De Grazia's terms) about sex in general, but especially about homosexuality.[42]

One major argument of the proposed book, as Velikovsky recalled, was to ascribe Arab anti-Semitism to latent homosexuality fostered by an overly male-bonded society. The Zionist theme embedded in Velikovsky's intellectual project was nowhere more explicit than in a book he planned in 1940 called "The Hatred," a psychoanalytic study of the origins of anti-

Semitism. He sent a proposal of this text to Macmillan (of all publishers), describing this book as "a new approach to the social psychology and the psychology of the unconsciousness. It reveals the most important part which the not finished struggle between the masculine and feminine ingredients of an embryo continues to play in the life and activities of a single personality and of the collectives."[43] We know no more, although he claimed that the proposal had been accepted by Presses Universitaires de France in Paris (Alcan & Co.).

These book projects were quickly sidelined or abandoned by the major campaign: an all-out assault on *Moses and Monotheism*. The idea for this book, entitled "Freud and His Heroes," foregrounded the role of Judaism in Freud's thought and uncovered a deep vein of self-hatred that tormented the Viennese psychoanalyst.[44] The nature of the project and its conceptual sweep are described most succinctly in a 1940 book proposal: "There are revelations in it relating to three different fields—psycho-analysis, the cultural history of Greece, and the history of the XVIII Dynasty in Egypt. All these different sections, though apparently so diverse, are bound together by the unity of the psychological method of investigation; in all probability this is the first time it has been possible to make important historical discoveries by way of psychological interpretation."[45] He was in negotiations in March 1940 to publish the book with the Allicon Publishing Corporation and then return to Palestine, but miscommunications sidelined the manuscript—which no less a reader than Franz Boas considered "striking," although he hesitated to make statements about its accuracy—and Velikovsky stayed in New York.[46]

The book was divided into three parts. The first hazarded an analysis of Freud himself. Focusing on the small number of dreams from Freud's 1899 masterwork, *The Interpretation of Dreams*, which the author had admitted were his own, Velikovsky submitted them to an interpretation that contradicted Freud's self-analysis and emphasized a constant anxiety about converting to Christianity. As Velikovsky stated in an article derived from this section entitled "The Dreams Freud Dreamed," published in 1941: "The most important determination of almost all the dreams mentioned by Freud is his inner struggle for unhampered advancement: In order to get ahead he would have to conclude a Faust-pact; he would have to sell his soul to the Church."[47]

The second and third parts of "Freud and His Heroes" investigated two of Freud's figures: Oedipus and Moses. Velikovsky was convinced that the myth of the king of Thebes—who killed his father and married his mother, eventually blinding himself and dying in exile—was based on a historical memory stemming from Egypt: that of the monotheistic pharaoh Akhnaton himself. These perceived parallels eventually became the core of Velikovsky's short 1960 monograph, *Oedipus and Akhnaton*. The book makes for fascinating reading, yet Velikovsky admitted (in a footnote) that he massaged some of the quotations from various stele and presented them as though they were continuous, and he engaged in somewhat unorthodox methods such as a scene where he psychoanalyzed Akhnaton as if the pharaoh were (literally) on his couch.[48] Despite the considerable charm of *Oedipus and Akhnaton*, it was not received well by classicists.[49]

The text bore marks of its origin as part of a monograph on Freud's *Moses and Monotheism*. After devoting almost two hundred pages to laying out parallels between Egyptian history and Greek myth, he suddenly switched to an attack on Freud himself—and especially his "degradation of Moses. He degraded him by denying him originality; simultaneously he degraded the Jewish people by denying them a leader of their own race, for he made Moses an Egyptian; and finally he degraded the Jewish God, making of Yahweh a local deity, an evil spirit of Mount Sinai." Freud's sin was inexcusable: "On the eve of his departure from a long life he had to blast the Hebrew God, demote his prophet, and glorify an Egyptian apostate as the founder of a great religion."[50] Over two decades after reading the book, Velikovsky was still struggling with Freud's Moses.

Which brings us to part three of "Freud and His Heroes." When Velikovsky set off for his sabbatical, he had not yet determined his approach to the all-important conclusion, the refutation of the Egyptian Moses, and he set himself to the task during the eight months he traveled almost every day from Riverside Drive to the New York Public Library on Fifth Avenue. In April 1940 he came across a reference to the Dead Sea as being (geologically) recent. It immediately occurred to him that it might have been created in a catastrophe. The rest, as they say, was history. As he wrote to French archaeologist Claude Schaeffer in 1958: "This was the most fruitful idea of my life, when in the spring of 1940 I realized that the Exodus took place amidst a natural catastrophe."[51] He began following the trail of documents, and in a book on the Hyksos, a foreign people who invaded Egypt and devastated it before leaving centuries later, he found a reference to a papyrus composed by an Egyptian sage named Ipuwer that recounted a series of catastrophes. The translation of this Papyrus Ipuwer, which was to prove so important to Velikovsky, was not in the public library's stacks, so he took his first trip to the library at Columbia University (closer to his home, especially after he moved to 525 Riverside Drive, and then in 1946 to 526 West 113th Street) and looked it up there.[52]

"Freud and His Heroes" was abandoned, or, rather, completely transformed into a different book. "All these finds were made by me in a matter of days in June 1940. At that time I thought to call the book 'From Exodus to Exile' since the reconstruction at that time reached the fall of Jerusalem and the Babylonian Exile," Velikovsky recalled later. "I thought that I would finish the book in a matter of a few months."[53] In October 1940 he began to read ever more broadly in world mythology, collecting references to common events he found in Ipuwer and in the Bible, beginning with accounts of sun-stopping. Within about two weeks, he came to suspect Venus was involved. Instead of a few months to complete, it took just under ten years. It was called *Worlds in Collision*.

This book that stunned the publishing world and catapulted Velikovsky to fame began as an inquiry into Freud's last work. Later critics would often mock Velikovsky's credentials as a psychoanalyst and claim that he surely did not have the requisite skills in ancient history or astrophysics to substantiate his Venus scenario. But psychoanalysis was precisely the relevant expertise. *Worlds in Collision* was fundamentally, at its core, a book engaged in the Freudian project, even if those sources were, like the Zionist inspiration, so deeply buried as to be almost unrecognizable. Velikovsky's method in "From Exodus to Exile"—soon separated into *Worlds in Collision* (the scientific claims) and *Ages in Chaos, Volume I* (the historical claims)—was thoroughly psychoanalytic. Velikovsky approached the world's literary heritage by interpreting their traces as masking a hidden trauma. He was no literalist: "To uncover their vestiges and their distorted equivalents in the physical [in later editions: "psychical"] life of peoples is a task not unlike that of overcoming amnesia in a single person."[54] With an admirable degree of candor, Velikovsky noted in his unpublished memoirs that he now saw "how my years of sessions with patients prepared me for my future work by allowing me to see similarities in things that do not at first glance appear related."[55] By collecting an earthquake reference here, an account of a dragon in the heavens there, Velikovsky assembled a dream journal for humanity.

Later presentations by both Velikovsky and his followers emphasize this moment in 1940 when he began to see the Venus story in its full outlines as a single inspiration. But Velikovsky was nothing if not a voracious reader, and in the course of his extensive spelunking into the two great New York libraries, he came across more than a few hints that his framework was not entirely original. He at first denied all knowledge of these predecessors; later in life, however, he would concede that he did indeed have precursors, whose work he had come across in the 1940s. Three stood out: William Whiston, Hanns Hörbiger, and Ignatius Donnelly.[56]

William Whiston was by far the oldest. In 1696 Whiston published *A New Theory of the Earth*, a natural philosophical treatise in the mathematical mode exemplified by his contemporary, Isaac Newton, whom he succeeded as the Lucasian Professor at Cambridge in 1702. Whiston's book argued, in dense calculations of orbital trajectories, that the Noachian Deluge was caused by a tremendous comet that collided with Earth. Whiston topped off the account with a prediction that the world would likewise end with a divinely inspired comet collision, again demonstrated with both higher mathematics and biblical exegesis.[57]

The clear differences between Whiston's project and Velikovsky's made him the least problematic to acknowledge, and Velikovsky did so explicitly in several of his unpublished manuscripts, as well as his posthumous account of the *Worlds in Collision* controversy.[58] Whiston was not interested in correlating ancient myths, for the testimony of the Bible was more than sufficient for his own book, and as a result the striking encyclopedism of Velikovsky is absent. But the chief difference is the focus of biblical attention: the Flood versus the Exodus. (This contrast also differentiated Velikovsky from the later movement of scientific creationism, with which he would tangle in the 1960s.) Whiston was a safe precursor, so distant that the whiffs of the fringe that clung to him in his own time did not contaminate Velikovsky's project. Whiston was useful in the footnotes to *Worlds in Collision* in another way: his translation of the ancient Jewish historian Josephus was the one Velikovsky used.

The case of Hanns Hörbiger was likewise simple. Velikovsky did not care for Hörbiger's cosmology, made a point of almost never citing it except to attack it, and disavowed any knowledge of the work while he was producing *Worlds in Collision*. This distancing move has the virtue of probably being true, but it also made a great deal of sense for an author with Velikovsky's preoccupations. Hörbiger was a furnace engineer in late nineteenth-century Austria-Hungary who had a number of successful inventions to his credit, but the connection to Velikovsky (such as it was) stems from the *Welteislehre* (World Ice theory), which he published in 1912. In very brief outline, the theory posited that the cosmos was filled with ice, and that satellites of Earth, such as our present moon, were pulled down by frictional drag and eventually crashed (and will continue to crash) into

Earth, leaving catastrophic traces on Earth's surface. Hörbiger supported his claims with invocations of Norse mythology, and under Hitler's Third Reich (which seized power two years after Hörbiger's death in 1931), *Welteislehre* enjoyed a vogue of popularity and some political support, imbuing the doctrine with a strain of virulent anti-Semitism.[59] This was no doctrine for Velikovsky. He was openly critical in his 1955 geological text, *Earth in Upheaval*, calling the theory "bizarre."[60]

Ignatius Donnelly was different. Born in 1831 to an Irish immigrant to the United States who married a second-generation Irish American, he grew up in Philadelphia and received an excellent education, being admitted to the bar in 1852. Tainted by scandal, he moved west and entered politics, serving as lieutenant governor of Minnesota during the first half of the Civil War, as a Republican congressman from 1863 to 1868, and then a stint as a state senator. Donnelly also took on a lively second career as a literary author, penning books on the lost continent of Atlantis, the authorship of Shakespeare (he credited Francis Bacon), and a striking 1883 volume called *Ragnarok*.[61]

Almost everything about that book recalls Velikovsky. A comet terrorizes Earth and eventually hits it, and we know this through careful reading of similarities in ancient myths. "The legends seem to represent the diverging memories," Donnelly wrote, "which separating races carried down to posterity of the same awful and impressive events: they remembered them in fragments and sections, and described them as the four blind men in the Hindoo story described the elephant;—to one it was a tail, to another a trunk, to another a leg, to another a body;—it needs to put all their stories together to make a consistent whole."[62] As a book, it has pronounced differences from Velikovsky's. Donnelly used plenty of illustrations and placed much less stress on biblical evidence than on other ancient myths (especially Scandinavian). When he did discuss the biblical evidence for catastrophes, Donnelly ascribed his comet, as did Whiston, to Genesis rather than Exodus. But the similarities are no less remarkable. Donnelly even ascribed the legendary fires to ignited cometary hydrocarbons, prefiguring one of Velikovsky's more prominent claims.[63]

Donnelly's *Ragnarok* is cited only once in *Worlds in Collision*, at the end of a lengthy footnote that stresses the differences: "[Donnelly] placed the event in an indefinite period, but at a time when man already populated the earth. Donnelly did not show any awareness that Whiston was his predecessor. His assumption that there is till [a stiff clay] only in one half of the earth is arbitrary and wrong."[64] The rest of Velikovsky's footnotes, however, indicate that he followed Donnelly's citations as a bibliographic guide to world legends.[65] Even the sole reference to Donnelly was inserted late in the composition of the text, in response to science journalist John J. O'Neill's referee report for Macmillan. He had come across *Ragnarok* after having reviewed the Velikovsky manuscript. "What amazes me," O'Neill commented, "is that no mention is made by Dr. Velikovsky of Donnelly's work. Velikovsky's research has been so thorough that it seems very unlikely that he should have failed to encounter it in his reading, especially since Donnelly published two other volumes along the same line."[66] In fact, we know from Velikovsky's associate Alfred De Grazia that Velikovsky had found *Ragnarok* in 1940 at the New York Public Library, and that he "was depressed by the discovery, according to his own words."[67]

And so Velikovsky spent the 1940s in New York, revising his manuscript into the form that Eric Larrabee would summarize in *Harper's* in January 1950 and that Macmillan would publish that April. He supported his family on savings, rents from an office building inherited from his parents in Tel Aviv, and occasional patients referred to him by émigré Austrian psychoanalyst Paul Federn—although the last was illegal, as Velikovsky was not certified to practice in the United States.[68] (The family of four amazingly managed to survive on roughly $6,000 a year.) Publication would end his financial woes and his psychoanalytic career, as the royalties from the phenomenally successful book became his primary source of income. In 1952—in order to be closer to his older daughter and son-in-law, physics graduate student Abraham Kogan, and their new baby—Velikovsky and Elisheva moved to 78 Hartley Avenue in Princeton, New Jersey. He lived in this house for the rest of his life, at first quite modestly until the 1970s, when he received a $100,000 advance for selling the rights to his books to the paperback giant Dell, which sparked a burst of remodeling and the purchase of a summer place at 300 Catalina Avenue, in Seaside Heights, New Jersey.[69] All that was in the future. Velikovsky's attention in 1950 was, as it had always been, on history.

HISTORY OF THE WORLD, PART I

In *Worlds in Collision*, Velikovsky conceded that his interpretation of ancient texts used "a synchronical scale of Egyptian and Hebrew histories which is not orthodox."[70] That brief clause in the preface might pass the casual reader by, especially one eager to get to the sensational claims about cosmic catastrophes. But those words mark the fact that *Worlds in*

Collision was only one-half of an expansive project, carved off from its historical companion. To understand the essential unity of Velikovsky's picture of the past, the cosmic picture must be reunited with the historical component.

While *Worlds in Collision* was in press at Macmillan, Velikovsky labored on the companion tome, *Ages in Chaos, Volume I*, which was released by Doubleday in 1952, with promises of a sequel to follow soon after. Velikovsky continued to tinker with that second manuscript, eventually splitting it into four parts, only two of which were published, in 1977 and 1978. (The final two remain in manuscript.) *Ages in Chaos* proposed to cover "altogether twelve hundred years of the history of the ancient East." He knew his readers were likely approaching this text—the volume that made sense of his datings of legends in *Worlds in Collision*—looking for the Venus catastrophe, and he informed them that instead of dominating the narrative, "the occurrence of a widespread natural catastrophe serves here only as the point of departure for constructing a revised chronology of the times and lands under consideration." What he offered instead belonged to another highly appealing literary genre:

> Because I had to discover and collate them, this book is written like a detective story. It is well known that in detective work unexpected associations are often built on minute details: a fingerprint on a bar of metal, a hair on a window sill, a burnt-out match in the bushes. Some details of an archaeological, chronological, or paleographic nature may seem minor matters, but they are the fingerprints of an investigation in which the history of many nations in many generations is vitally involved.[71]

Those details mattered: when examined extremely closely, one found strange parallelisms, repeated events to be explained through tremendous historical surgery.

The argument concerned a missing six hundred years, the discrepancy between the chronology offered by the king lists of the Egyptian dynasties and that presented by the royal hierarchies in the Hebrew Bible. This mismatch was no news to ancient historians; they resolved it by claiming that the Hebrew chronology was defective, too short by six centuries, and relied on the authenticated king lists instead. Velikovsky, on the other hand, argued that there were six "ghost centuries" in the *Egyptian* chronology, and that radical amputation was necessary to bring the story in line with the correct biblical narrative.[72] To highlight the scale of his revision, he deployed a contemporary analogy: "One must try to conceive of the chaos which would result if a survey of Europe and America were written in which the history of the British Isles were some six hundred years out of line, so that in Europe and America the year would be 1941 while in Britain it would be 1341."[73] So when Winston Churchill traveled to Canada to meet Franklin D. Roosevelt in 1941, according to European and American sources, we would find no trace of this visit in British chronicles. Rather, we would find an elliptical tale in a medieval manuscript about the travel of a war leader across the oceans. What seemed two separate incidents were actually the *same* historical event. In this book, too, then, we find the hallmarks of Velikovskian reasoning: a hidden commonality behind diverse texts from different cultures.

The method of argumentation was in many ways the inverse of that displayed in *Worlds in Collision*. In that book, he found resonances in the mythologies and histories of the world and used those to argue for a series of great catastrophes. The histories served as the fixed point to characterize and date the disaster. In *Ages in Chaos*, the logic was reversed, and the crux was the Ipuwer manuscript, that reference that propelled Velikovsky from the New York Public Library uptown to Columbia University in 1940. This collection of dire lamentations was in Velikovsky's reading "the Egyptian version of a great catastrophe." In fact, it was the Exodus catastrophe. "The evidence, when found," he continued, "brought forth more analogies and showed greater resemblance to the scriptural narrative than I had expected. Apparently we have before us the testimony of an Egyptian witness of the plagues."[74] The only difficulty was that Ipuwer was dated by Egyptologists (on the basis of style, context, and textual references) six hundred years *earlier* than the most likely date for the Exodus. In the words of its translator: "But on the whole the language of the papyrus (and, we may add, the paleography) makes us wish to push back the date of the composition as far as possible," but, he added, "it is doubtless wisest to leave the question open for the present."[75] For Velikovsky, the Venus comet itself now became the fixed point that revised the dating of Ipuwer. Once Ipuwer was placed at 1500 B.C. along with Moses, we then possessed corroborating textual evidence of events that were once considered mythical—and possibly the solution to several puzzles of Egyptian and biblical history. For example, with this new dating, Velikovsky confidently asserted that the mysterious Queen of Sheba in Kings I was the famous female pharaoh Hatshepsut, and that the Hyksos who terrorized Egypt were not the Israelites (as had been claimed by ancient historians such as Manetho, an assertion

Velikovsky considered one of the historical roots of anti-Semitism), but the Amalekites, *after* the Exodus.[76]

Velikovsky took the story to the advent of Akhnaton (d. 1336 or 1334 B.C.)—in this narrative happening several hundred years after Moses, and so another nail in the coffin of Freud's *Moses and Monotheism*—and concluded with a promise that the second volume would end with Alexander the Great (356–323 B.C.), at which point the Egyptian and Hebrew chronologies coincided. Nearly every claim went against the entire chronology of ancient history established over the previous hundred years since the decipherment of hieroglyphs. This made clearing peer review somewhat of a hurdle, even more than for the astronomical arguments in *Worlds in Collision*. Velikovsky sent the manuscript to Oxford University Press in 1945, which turned it down because of one strongly negative referee report.[77]

The same thing had happened earlier at Harvard University Press, although in the process Velikovsky found an unlikely backer: Robert H. Pfeiffer of Harvard's Semitic Museum, whom Velikovsky considered the "good" Harvard professor to counterbalance the calumnies of Harlow Shapley. Pfeiffer refereed *Ages in Chaos* for Harvard as far back as 1942 (the press evaluated it again in 1945, after revisions) with strong negative criticisms: Velikovsky ignored current scholarship, gave inexact dates, and other such matters.[78] After this, Velikovsky visited Pfeiffer, and their conversation convinced the professor of the writer's sincerity and led to a warm correspondence. Pfeiffer then wrote to Velikovsky: "I regard your work—provocative as it is—of fundamental importance, whether its conclusions are accepted by competent scholars or whether it forces them to a far-reaching and searching reconsideration of the accepted ancient chronology."[79] Three years later Pfeiffer was the second referee for Oxford, and he submitted a strong review but still qualified along the lines of his letters to Velikovsky: "On the other hand, I am at present unable to accept these conclusions, possibly because the standard views have been so inextricably and so long at the basis of my thinking and of my research. My present opinion is that the chances that Dr. Velikovsky is right are about 10 per cent, but I admit I am prejudiced and I am eager to see his book published: it should prove to be not only sensational, but also stimulating to historians."[80] (Pfeiffer also reviewed the book for Macmillan, writing an almost identical report.[81])

With Doubleday releasing the first volume, Velikovsky needed to complete the second. As the chronology extended into more recent periods, the problem of synchronization became substantially more challenging.

It was hard enough to correlate the Hebrew and Egyptian chronologies; it was altogether vastly more difficult to incorporate the vast textual and archaeological evidence from Greece. Velikovsky was convinced that he was on the right track, and in 1958 he wrote to French archaeologist Claude Schaeffer—who had argued in his 1948 *Stratigraphie comparée* for a series of disasters befalling the ancient Near East—that he was not willing to budge with respect to criticism: "Today, six and a half years after the publication of the first volume of 'Ages' I have nothing to change there; and no critic, you included, could show me even on one single instance that the correlations of vol. 1 are not convincing."[82] Throughout the 1950s, volume 2 sat on Velikovsky's desk in page proofs, and he continually emended it. But there were distractions. First there was the desire to pursue his legitimation among scientists (chronicled in chapter 4), and then there was the composition of his geological tome, *Earth in Upheaval*.[83] Yet the biggest problem, as is evident from his correspondence, was that the presentation in the manuscript as it then stood could not sustain the criticisms of even sympathetic readers like Schaeffer.[84]

Velikovsky never admitted that his picture might be faulty, and he struggled to complete the synchronized history that he had articulated, to his satisfaction, as far back as 1945 in a series of 284 *Theses for the Reconstruction of Ancient History* that he published in a new *Scripta* series.[85] It took a full twenty-five years after the publication of *Ages in Chaos* for the sequels to appear, first as *Peoples of the Sea* in 1977, which concludes the story with Alexander the Great (and changes the "gap" between the chronologies from six ghost centuries to eight), and then *Ramses II and His Time* in 1978. In *Peoples of the Sea*, Velikovsky did not ascribe the delay to any difficulties in the project itself, but rather to the pesky Soviet satellite launched in October 1957: "The first Sputnik and the years that followed with Mariner and Apollo flights deflected my interest toward astronomical problems."[86] To a Swedish fan, he offered a different explanation in 1967: "Intentionally, I postponed the publication of the sequel in order to give time to scholars in their field to study my work and investigate its consequences."[87] The two remaining volumes—"The Assyrian Conquest" (which was to occupy the space between *Ages* and *Ramses*) and "New Light on the Dark Ages of Greece" (which argued that the problem of the Dark Ages separating the Minoan and Attic civilizations in Greece could be removed as ghost centuries)—remained in manuscript.[88]

The decades after *Ages in Chaos* were occupied by many projects, not simply completing his ancient history or defending the account in

Worlds in Collision (although those were his primary goals). One of the most striking—and the subject of many inquiries over the years—was his interpretation of the book of Genesis, a relatively slender manuscript entitled "In the Beginning." Velikovsky later claimed these pages were originally included in *Worlds in Collision* but excised from the manuscript under the advice of readers who suggested he focus on only one or two catastrophes.[89] The manuscript, based on an interleaving of Genesis with various Jewish legends, sets forth a series of bold claims: there was a time when humans lived on a moonless Earth; the Deluge was caused about ten thousand years ago when the planet Saturn exploded and the "hydrogen of the planet combined with the oxygen of the terrestrial atmosphere in electrical discharges and turned into water"; the Tower of Babel recounted an electrical discharge from a Mercury fly-by, wiping out the memories of the survivors; and the destruction of Sodom and Gomorrah was produced by an electric bolt shot out from Jupiter, cauterizing the region.[90]

As important as the historical works were to Velikovsky's vision, he always interweaved them with more narrowly scientific arguments, and some of these would have major consequences in shaping his later reception. One early publication stands out, a twenty-two-page pamphlet published in 1946 in his *Scripta* series entitled *Cosmos without Gravitation*, Velikovsky's first venture into the physical sciences. Characteristically, the thesis is daring: "The fundamental theory of this paper is: Gravitation is an electromagnetic phenomenon. There is no primary motion inherent in planets and satellites. Electric attraction, repulsion, and electromagnetic circumduction [revolution around the sun] govern their movements."[91] There follows a list of phenomena that Velikovsky claimed were incompatible with the conventional understanding of gravitation: the first notes that air is a mixture, yet the heavier gases do not sink; the second that ozone is heavier than oxygen, yet is found high in the atmosphere; the third is that water droplets in clouds are heavier than air; followed by twenty-two others. It is an unusual text—it has no historical argument, it was not published for a broader audience, and Velikovsky rarely mentioned it in later life. His followers would later claim that that it was "generally known" in the 1970s that Velikovsky withdrew from it.[92]

Yet this pamphlet had a great deal of historical significance in Velikovsky's career. It was a clear attempt to situate himself among the scientists, for however brief a moment, before returning his primary affiliation to historians. Velikovsky treated it, in the late 1940s, as a calling card to scientists. He sent copies to philosopher Bertrand Russell and astronomer

Arthur Stanley Eddington in 1946 (the latter had already been dead for two years), as well as a copy to Harlow Shapley in 1947, even after the Harvard astronomer had rebuffed his request to conduct inquiries on the atmospheres of planets. (Shapley gave it to Donald Menzel, who composed a page of technical but non-mathematical objections, which he sent to Velikovsky.[93]) Shapley rediscovered the pamphlet while corresponding with Macmillan in 1950, and Rupert Wildt of Yale Observatory specifically cited it in a lengthy review of *Worlds in Collision* in the *American Journal of Science* to demonstrate that Velikovsky did not know basic physics.[94] So, however abortive this foray into gravitation, there is little doubt that it colored the early scientific reception of *Worlds in Collision*.[95] Perhaps more significantly, in later decades Velikovsky emphasized that his major contribution to physics was to posit an electromagnetic cosmos. This notion finds its earliest incarnation in *Cosmos without Gravitation*:

> I arrived at this concept early in 1941 as the result of my research in the history of cosmic upheavals as they affected the earth and other members of the solar system. A number of facts proved to me that the sun, the earth and other planets, the satellites, and the comets, are charged bodies, that the planets and their satellites have changed their orbits repeatedly and radically, and that gravitational attraction or the weight of objects has changed during human history. I thus recognized the fact that not gravitation, but electric attraction and repulsion and electromagnetic circumduction govern the solar system.[96]

Velikovsky made other scientific claims in the 1940s—that King Solomon had been aware of the properties of radium, that the oppressive 1949 summer in New York City should be alleviated through weather modification, and a proposed experiment to measure the velocity of light (which he argued, *contra* Einstein, was not constant)—but none with such staying power.[97]

THE PERSISTENCE OF MEMORY

"Probably in the entire history of science," Velikovsky wrote to Claude Schaeffer in 1963, "there was not a case of a similar violent reaction on the part of the scientific world toward a published work."[98] An extraordinary effect demanded an extraordinary cause, and Velikovsky explained the reaction to his theories in a way that fell within the unified science-history of catastrophism he had outlined in *Worlds in Collision*. To comprehend the vehemence of the scientists' reaction to his work, Velikovsky repurposed a notion of "collective amnesia" deployed in his 1950 blockbuster.

The original concept was meant to explain some peculiar properties of the legacy that Velikovsky plumbed for his accounts of celestial disasters. The obvious question occurs to the reader of *Worlds in Collision*: if in fact Venus almost wiped out humanity, why is the only evidence of it a few scattered references dressed up in poetic language, so dispersed that it took a decade of labor for Velikovsky to cull them all together? Collective amnesia was the answer: "The memory of the cataclysms was erased, not because of lack of written traditions, but because of some characteristic process that later caused entire nations, together with their literate men, to read into these traditions allegories or metaphors where actually cosmic disturbances were clearly described."[99] The move is double: first many of the survivors repressed explicit discussion of the trauma; and then later readers of these accounts were beguiled into interpreting them non-literally, induced by a subconscious compulsion to deny the violence that had befallen humanity. Here Velikovsky drew on the method of *Moses and Monotheism*, where Freud argued that "the archaic heritage of mankind in-cludes not only dispositions, but also ideational contents, memory traces of the experiences of former generations."[100] For Velikovsky, many of the accounts he drew from legends, such as those of Ragnarök or other apoca-lypses, were simply memories of the past displaced and projected onto the future in order to dull the pain.[101] With proper psychoanalytic interpreta-tion, the evidence can be uncovered from behind the veil of amnesia.

This account was not without its critics. After all, Velikovsky was not the first catastrophist, and if there had been cosmic catastrophists before (such as Whiston and Donnelly), why were they somehow immune from forgetting?[102] Velikovsky's answer was that certain individuals, such as poets (like Dante and Shakespeare) and visionaries (presumably including himself), were endowed with an ability to pierce the veil of the trauma.[103] This argument did not satisfy many, such as Bob Forrest, a British math-ematics instructor, who spent countless hours tracking down every single one of Velikovsky's sources (and claimed many of them did not say what had been asserted on their behalf): "As things stand at the moment, col-lective amnesia simply puts Velikovsky in a 'heads-I-win, tails-you-lose' situation as regards his cosmic drama: if a piece of evidence fits, it can be claimed as a hit; if a piece of evidence doesn't fit, it can be claimed as a collective amnesiac 'disguise,' and thus denied the status of a miss! Such a vetting procedure must surely be regarded with the greatest suspicion."[104]

When the furor from the scientists sprang up in all directions, collective amnesia came to hand as a means to understand it. If one believed that the human race inherited a racial memory that deliberately suppressed what had happened, Velikovsky contended, then the uproar was surely to be expected. In his posthumous *Mankind in Amnesia*, Velikovsky offered perhaps his clearest exposition of this interpretation:

> In my analytical practice I would never have perplexed a patient with sudden revelation of the hidden motifs underlying an affliction, without a preceding lengthy preparation in which I would carefully guide the patient to his or her own insight. Only after such preliminary work had been done could a startling revelation be risked and even then, in some cases, the effect might be almost shattering—but by that time the avenues of retreat into ignorance would have already been blocked; by that time also the patient would have understood the good intentions of the analyst and a link of transference would have been forged. But in offering an anamnesis, or the story of the development of the repression, to a collective suffering from amnesia, I have not followed the same procedure—and I could not. Should I have told first a curtailed story of great upheavals of the past—a watered-down version—or administered it in small doses, a teaspoon after breakfast? Should I have presented the story as only possibly but not necessarily true? Should I have offered it as science fiction? Should I have printed it seriatim or dismembered it among obscure magazines?
>
> I did as I did, realizing that a strong reaction would be generated in everyone who would come into contact with the disclosure, whether directly or through hearsay. In some, the reaction would take the form of vociferous denial, protest, accusation and the organization of opposition. In others—overwhelmed by a revelation—there would be an equally strong reaction of acceptance, acclamation and a rush of missionary zeal to convert others. The demarcation line that divided the camps ran with hardly any deviation between those who did not read the message published as *Worlds in Collision*, in 1950, and those who did.[105]

This method of explaining the reaction has displaced all others among later generations of Velikovskians, including the view that the reaction was anti-Semitic.[106] Collective amnesia evolved into orthodoxy, one of the few aspects of Velikovsky's synthesis that was not originally in place in the 1940s. Velikovsky had not anticipated that the reaction would come quite as quickly and quite as violently as it did, but, according to the amnesia theory, he should have. As one of his most ardent supporters put it: "The Velikovsky theory *implies* the Velikovsky Affair. If the Velikovsky theory is true, then there had to have been a Velikovsky Affair. Thus the occurrence of the latter is one confirmation of the former."[107] Even among these acolytes, the *scientized* version of collective amnesia as a theory of mind had displaced its original function as historical methodology. History was once again sidelined.

THE ECLIPSE OF HISTORY

Velikovsky's trajectory was not, then, primarily about the development of an alternative science of the solar system. It was, rather, motivated by a quest to rewrite the history of the ancient Near East so as to reconcile discordances that had some bearing on the history of the Jews. He was, as he emphatically told WHRB, Harvard's student radio, in 1972, "a historian," not a scientist.[108] The deeply historical nature of Velikovsky's project, surprisingly consistent across his long career, raises the question: Why has the Velikovsky controversy always been understood under the rubric of science, as part of the pseudoscience wars? Where are the historians in these debates?

Velikovsky opened *Ages in Chaos* by calling it his "second front": "After having disrupted the complacent peace of mind of a powerful group of astronomers and other textbook writers, I offer here major battle to the historians."[109] What if you threw a war and nobody came? It is true that an occasional review by a humanist, usually a professor of ancient languages or biblical studies, would skewer Velikovsky, as did William F. Irwin, for being "pre-Herodotan, swallowing gullibly every story that comes to him, with no exercise of that disciplined skepticism which the Greek historian was the first to invoke as a conscious method," but for the most part professional historians were cagey about tangling with Velikovsky.[110] For example, Kenneth A. Kitchen of the University of Liverpool, writing in the late 1970s to one of Velikovsky's admirers, pointed to the dilemma of engaging with Velikovsky's reconstruction of ancient history:

> My problem in replying to you is that Velikovsky always tosses with a two-headed coin. If ordinary orientalists like myself simply leave him aside &

get on with real work, he complains of their disdain (& the public are left unprotected). If, conversely, orientalists like myself (who happen to be burdened with several thousands more facts than Velikovsky even dreams of) actually dare to stand up & expose him, then of course he snidely implies that we are some sort of closed caucus with interests at stake. It's always "heads you lose, tails I win". . . . Would you, I reflectively wonder, publish a book that insisted on the identity of Harold Wilson and Harold of Hastings, of Napoleon, Bismarck & Charlemagne, and on the role of your firm as secret HQ of the IRA, all as absolutely genuine historical fact, with "proofs" (e.g., all aunts in France are large, because *tante* is the feminine for *tant*)? Because that is, comparatively, the level of historical fraud that V. represents.[111]

So what was one to do? As early as 1950, Carl Kraeling, the director of the Oriental Institute at the University of Chicago, outlined an approach essentially the polar opposite of the astronomers': "There is nothing we as historians can do about Dr. Velikovsky's work other than smile and go about our business."[112] In the end, this was remarkably successful: "The silence of the Middle Eastern scholars is more effective than the scandalous attempt at suppression by the academic astronomers."[113]

The grand exception to the frosty silence that greeted Velikovsky from the historical profession, and which resembled the hostility of the scientific community, was from a group of scholars that, in the early 1950s, defined themselves as straddling the border between those two groups: historians of science. If the question is why the Velikovsky affair was construed as a fight over the boundaries of science and pseudoscience, and not, say, between history and fiction, then the historians of science are an excellent test case, because they could have moved the debate in either direction. In the event, they resolutely defined this as a scientific dispute, with Velikovsky on the wrong side. They did not criticize Velikovsky's historical chronology; rather, they pointed to ancient scientific instruments and the records of ancient eclipses. They treated ancient evidence as sources of *scientific data* that would refute the cosmological claims in *Worlds in Collision*. In order to see how historians of science contributed to the definition of the debate as a scientific one, as well as to provide a detailed case study of how Velikovsky reacted to criticism, let us focus on the case of Otto Neugebauer.

Neugebauer was the most distinguished historian of ancient exact sciences in the postwar world. Born in 1899 and a veteran of the Austrian army in World War I, Neugebauer settled as a mathematics student at Göttingen

University but soon moved into historical topics with a dissertation on ancient Egyptian unit fractions. In 1927 he was appointed to the faculty at Göttingen, but emigrated to Copenhagen in 1934 after Hitler's civil service laws purged Jews from German universities (devastating the Göttingen department), and then eventually became a professor at Brown University, dividing his time with the Institute for Advanced Study in Princeton.[114] With his expertise in ancient sciences and his copious knowledge of the relevant languages, he was the logical person to turn to when Velikovsky emerged. In fact, astronomer Rupert Wildt found the absence of footnotes to Neugebauer's work in *Worlds in Collision* a damning flaw in Velikovsky's historical scholarship.[115]

George Sarton—the Belgian-born Harvard professor who edited *Isis*, the journal of the History of Science Society—asked Neugebauer to weigh in by reviewing *Worlds in Collision*. The result was, to say the least, unflattering. Neugebauer deemed the book "on a level far below science-fiction" and claimed that "it shares all the characteristics of a widespread type of crackpot publication. It attains, however, an exceptionally high degree of distortion of scientific literature. It is this latter aspect which may justify the waste of space of a scientific journal."[116] Neugebauer began with a detailed empirical example. He noted that on page 349 of *Worlds in Collision*, Velikovsky cited the Jesuit Franz Kugler's 1900 book *Die babylonische Mondrechnung* (The Babylonian Moon Calculation), quoting Velikovsky's translation, with added emphasis as "the distances traveled by the moon on the Chaldean ecliptic *from one new moon to the next* are, according to Tablet No. 272, on the average 33° 14' too great."[117] Then he quoted the German original, providing a "proper English rendering" in a footnote: "In order to demonstrate this we must anticipate our discussion of the relation of the Chaldean ecliptic of No. 272 and of the movable ecliptic and mention that the longitudes of the new moons with reference to the first are in the mean 3° 14' greater than with reference to the second."[118] This, to Neugebauer, was unacceptable: "No word of from 'one new moon to the next' but a totally different statement concerning the counting of longitudes in two different coordinate systems. Is the author's knowledge of German so bad that he had to stick in a whole new sentence in a 'quotation'?" Neugebauer then zeroed in on Velikovsky's major claims, about the erratic recordings of Venus in Babylonian texts as evidence that Venus was in fact a comet:

> Much less is known about other planets with the sole exception of Venus, which is the only planet which according to the Babylonian theory moves

with such regularity that its anomaly is disregarded. In other words Babylonian astronomers, as well as their Greek contemporaries and successors, were fully conscious of the fact that Venus is the most regular member of our planetary system—a flagrant textual contradiction of Dr Velikovsky's basic theory, even if we had no celestial dynamics at our disposal.[119]

Neugebauer concluded his two-and-a-half column review with a note that Macmillan had since abandoned the book. There was nothing more to say.[120]

Velikovsky begged to differ. Take a second look at Neugebauer's quotation of Velikovsky and then of Kugler. Besides the discrepancy in the translation, there was also a difference in the number of degrees: 33 or 3. An incensed Velikovsky wrote to Sarton that Worlds in Collision used the correct number (3)—Neugebauer had falsely misquoted his text in order to make it look erroneous![121] In an extended correspondence, he wanted to know from Sarton whether he could respond (yes), how much space he could have (less than Neugebauer's original review), and whether he could have a list of where Neugebauer sent offprints of his review, so he could send his rejoinder to the same people (Sarton had no idea who these individuals were). Sarton, as an editor, was willing to give Velikovsky a hearing, but he made it clear that he was on Neugebauer's side: "I have no intention of entering the discussion between you and Neugebauer, but I think he is right—in spite of misprints."[122]

Velikovsky never wrote the response to Isis, but he continued to harp on the dispute with Neugebauer. In 1951, in an exchange in Harper's with Princeton astronomer James Q. Stewart, Velikovsky mentioned the review but quite mischaracterized Neugebauer's point about Kugler: "Dr. Neugebauer has published a review in Isis. On one point only is he right: I should not have quoted from Kugler's Babylonian Moon table without questioning the age of the tablet, since Kugler ascribed it to a late century and did not consider it a copy. I shall omit the quotation in future editions."[123] This was not the argument in the review, and the Kugler quotation remained as it was in all editions. Velikovsky and his supporters would bring up the misprint again and again, focusing on that one digit and ignoring Neugebauer's point.[124]

Neugebauer felt hounded by Velikovskian supporters. When Alfred De Grazia, Velikovsky's friend and advocate, visited the historian in Providence in 1968, Neugebauer was guarded and defensive. De Grazia noted that Neugebauer "considers that he has been unjustly treated in the years following that incident and is bitter about you personally and about the activities of your supporters in publicizing the incident and its total context."[125] As far back as January 1965, Neugebauer had sent to De Grazia a dossier compiled of extracts from the *Isis* review, and he noted concerning the famous misprinted "3": "The reader will remark that I did *not* object against the number cited but against the fact that Dr. Velikovsky added a whole sentence to a 'quotation,' thus completely distorting the original meaning. I think I have expressed this fact clearly enough by my Italics and my comment."[126] Neugebauer's frustration had reached its limit. As he told another correspondent in 1967—after making a connection to Hanns Hörbiger's theories: "What I really dislike in the Velikovsky business (and a business it is indeed) is the high degree of scientific dishonesty which operates with purposely distorted facts."[127]

Neugebauer directed the initial review away from classic historical issues—like the proper translation from foreign languages, the methods of weighing ancient texts—and into the context of disputes about Velikovsky's *scientific* claims. He had no comments about the redating of ancient chronologies, but he did argue that the Venus observations of the ancient world refuted Velikovsky's physics. The pattern of Velikovsky's reaction was typical: a critical review was written, and Velikovsky insisted that he be allowed access to the journal to respond. (That he did not take advantage of it in this instance is incidental; he did in many others.) More subtly, Velikovsky shifted the issue from a debate over his translation of Kugler to the issue of a misprint, thus presenting a picture to the public of himself being wronged by academics. The irony is that Velikovsky absolutely understood Neugebauer's point, and he admitted to the inaccurate quotation in a December 1950 letter to Pfeiffer and again in 1951 to a professor of physics at Dillard University in New Orleans. He even conceded, in his posthumous memoir about the 1950 controversies, that "it is true that in paraphrasing Kugler, I should not have used quotation marks."[128] Yet despite these admissions, he could write to Gordon Atwater in 1950, and repeat many times after, that "nobody showed that [*Worlds in Collision*] is false or that any single quotation is invented or falsely quoted."[129]

It becomes relatively easy to lose oneself in the minute distinctions between a single digit and translations of German histories of ancient astronomy, and even easier to become absorbed with the vicissitudes of the life and intellectual circles of Immanuel Velikovsky, as he moved from tsarist Vitebsk to Weimar-era Berlin to the slopes of Mount Carmel and the stacks of New York libraries. For all the discussion of seventeenth-

and nineteenth-century predecessors, and the rival chronologies in ancient Egypt, there is a risk of losing the focus on the historical context
in which Velikovsky wrote *Worlds in Collision* and in which his critics and
interlocutors reacted. As Velikovsky would write, somewhat grandiosely,
in his posthumous *Mankind in Amnesia*, "the publication of *Worlds in Collision* [was] a warning against atomic warfare. Disaster may come, not from
another planetary collision, but from the handiwork of man himself, a
victim of amnesia, in possession of thermonuclear weapons."[130] (Well,
perhaps not the thermonuclear weapons—those only became a subject of
public discussion in November 1949, in the wake of the Soviet detonation
of their first atomic bomb late that summer, and by that point Velikovsky's
manuscript was in production.) The more relevant context is that of the
nascent Cold War, not in the geopolitical sense of superpower confrontation and brinkmanship, but in the reverberations of those struggles in
the culture of the United States. That was the world into which *Worlds in
Collision* was born.

3 · The Battle over Lysenkoism

Immanuel Velikovsky argued that the reason the scientists who mobilized against him reacted (or overreacted) with such ferocity was that they were amnesiac, a condition they shared with everyone else. Unaware (at least consciously) of the terrible Venus tragedy that had almost annihilated humanity in antiquity, they had sublimated the trauma and were bound to react with vigorous denial when a wound so deep, so painful, was suddenly exposed by *Worlds in Collision*. If one chooses not to subscribe to the correctness of this explanation—that the Velikovsky scenario necessitates a hostile Velikovsky affair—one is still left with the puzzle of just *why* Harlow Shapley, Cecilia Payne-Gaposchkin, and other scientists lashed out against Macmillan with such rage for publishing this book. It was not about forgetting; it was about remembering.

One of the most crucial aspects of the uproar surrounding *Worlds in Collision* was the timing and the location: the year 1950 in the United States. The American scientific community was at that moment in tremendous flux, having emerged from World War II with greater visibility, greater funding, greater prestige, and greater power than it had ever had before— and consequently significant anxiety. Its position might appear to have been so solid that there was no threat from someone like Velikovsky, and yet the reaction speaks otherwise. We return to the role of pseudoscience as a historical indicator: since it is only applied as a term of abuse against individuals and doctrines one perceives as threatening, it signifies an escalation. What made some American scientists believe that they were threatened by someone they perceived as a delusional crackpot?

The answer lies not just in the United States, but in what American scientists remembered about recent events in the Soviet Union, and what lessons they drew from those memories applied to their present case. The pseudoscience wars broke out because Velikovsky wandered into a minefield that had laid fallow for a year and a half, triggering a backlash one might well think was out of proportion with his transgression. The root of the issue was a man named Trofim Denisovich Lysenko, born in Ukraine in 1898 (three years after and several hundred miles to the south of Immanuel Velikovsky). A trained agronomist and ardent supporter of the Russian Revolution of 1917, Lysenko would seem to have nothing to do with the psychiatrist-turned-cosmologist we have been following, but American scientists in 1950 thought differently. To understand the Velikovsky affair, therefore, we must turn to the Lysenko affair.[1]

Hailing from a proletarian background, Lysenko was ambitious and claimed to be able to massively improve the agricultural productivity of the Soviet Union. His personal biography and optimistic aspirations served him well as he entered the 1930s, when Joseph Stalin began the collectivization of Soviet agriculture: a bloody process that ravaged the countryside with terror and famine and left millions dead or deported by the time it was done. But Lysenko was not involved in those features of collectivization; he was interested in seeds. Starting in 1927, Lysenko began to attract attention for a technique he developed (but did not invent) that he dubbed "vernalization" (*iarovizatsiia*, in Russian). Essentially, this meant treating seeds of plants with cold water or otherwise manipulating them so that they would germinate more quickly, or be better able to endure planting in cold climates. He used very small sample sizes and no controls, but if correct, his practices would be transformative. He was encouraged further by the powerful botanist Nikolai Vavilov, the president of the Lenin All-Union Academy of the Agricultural Sciences (VASKhNIL), collector of seed stocks from around the world, and one of the most prominent of the vibrant community of Soviet geneticists.[2]

The situation began to sour. Teaming up with a sophisticated philosopher, I. I. Prezent, Lysenko moved beyond the practices of vernalization and argued that his collected views on plants comprised a revolutionary new Marxist theory of heredity that he dubbed "Michurinism," after Russian plant breeder I. V. Michurin, often analogized both at the time and since to Luther Burbank.[3] In short, Lysenko argued that "in experiments and in practice, it is possible to alter directionally the heredity of various processes in plant and animal organisms, and to build and to fix a new heredity according to plan"; for example, that one could deliberately alter spring varieties of wheat into winter wheat so they could grow in harsher conditions while retaining high yields.[4] Crucially, Lysenko insisted that changes induced in the new seed stock would be *heritable* to the next generation, in direct opposition to Mendelian genetics—at this point almost universally subscribed to by scientists in the Soviet Union and abroad—which posited that heredity was carried in "genes" that were generally stable across generations. According to Lysenko, genes were "idealist" notions that belied the fact that the entire organism contributed to heredity, not just a tiny invisible part of it—a view reminiscent of the doctrine of the inheritance of acquired characteristics associated with the early nineteenth-century biologist Jean-Baptiste Lamarck—and therefore all of Darwinian theory and genetics needed to be reformulated to account for the "shattering" of heredity produced by vernalization. This would be the end of what he considered "bourgeois pseudoscience."[5]

At first, the Soviet genetics community ignored or tolerated Lysenko, whether because they thought he was harmless or because having a proletarian practical agronomist among their numbers provided good ideological cover in the dangerous atmosphere of the 1930s. But soon Lysenko turned on the geneticists and eventually took over the Academy of the Agricultural Sciences. Its president and Lysenko's former patron, Vavilov, was arrested in 1940 (allegedly for sabotaging Soviet agriculture) and died of starvation in a Saratov prison in 1943.[6] Debates over science and pseudoscience had fatal consequences.

These events in Soviet genetics played out on a world stage, and Western scientists (in particular geneticists) lamented them, especially Vavilov's disappearance.[7] (His fate was not learned until after World War II.) Historians have since repeatedly chronicled Lysenko's depredations in the Soviet Union, often focusing on the microdynamics of his rise to power, the influence of Marxist philosophy on biology, the impact of the eclipse of genetics on agriculture, the imprisonment or firing of geneticists, and Lysenko's eventual fall from power. One common theme in these accounts is perhaps best expressed by noted historian of Soviet science Loren Graham: "Lysenko's views on genetics were a chapter in the history of pseudoscience rather than the history of science."[8] Indeed, Lysenkoism is often held up as one of the twentieth century's most egregious forms of "pseudoscience," a term used at least as frequently, if not more, in Russian-language accounts both during the Soviet Union and (especially) after.[9]

When individuals both at the time and since call Lysenkoism a pseudoscience, they usually mean one or several of the following four claims:

(1) that the biological claims of Lysenko were erroneous, and scientific claims that are sufficiently wrong become pseudoscientific; (2) that the methods Lysenko used did not belong to the canons of science (such as his rejection of statistics or experimental controls); (3) that the elaboration and justification of scientific findings in terms of philosophy or ideology (especially dialectical materialism) was a non-scientific practice; and (4) that external (e.g., state, media, church) intervention on behalf of a particular scientific doctrine automatically contaminates it and renders it pseudoscientific. There is substance behind each of these claims, yet there is even more to be gained from disaggregating them. In particular, scientists' strategies of engagement with perceived crackpots differ enormously depending upon which of the points is at issue. This chapter will explore how American scientists' tactics in responding to Lysenko's rise—and their efforts to assist Soviet geneticists in saving themselves—depended on which of these factors they thought was most offensive in Lysenkoism. And these tactics will bring us back to Velikovsky.

After August 1948, when the Soviet controversy over genetics took a dark turn, the reactions of American geneticists—especially those who leaned left in an increasingly red-baiting milieu—hardened into a rather narrow set of responses, which then became a significant template for how American astronomers (again largely from the liberal end of the political spectrum) believed they ought to react to all "crackpots," an understanding that fed into the reaction to Velikovsky. To respond to the perceived threat of his cosmological theories, scientists mobilized many of the arguments developed in the aftermath of Lysenko's démarche of August 1948; even more surprisingly, so did *Velikovsky*. The Lysenko affair was not an episode in what I have been calling the "pseudoscience wars" in postwar America; those began properly with the publication of *Worlds in Collision*. It did, however, prepare the battlefield, and it left establishment forces armed and on hair-trigger alert. In the particular historical circumstances that made certain American scientists (rightly or wrongly) see themselves as uniquely vulnerable in 1950, Lysenkoism laid the powder and attached a short fuse. Velikovsky unwittingly ignited this volatile situation and bore the brunt of the blast.

THE EMPEROR HAS NO CLOTHES

But that all happened after 1948. Before, of course, no one, either in the Soviet Union or outside, was certain how the controversy over Michurinism (as Lysenko's ideas about heredity were know in the Soviet Union) would be resolved, as the many inscrutable unknowns—especially the Communist Party's position on genetics—remained murky. If this uncertainty obtained in the Soviet Union, it was even more pervasive in the United States, where observers of the irregular growth of Lysenko's power lacked knowledge of the detailed politicking behind it. This fluctuating level of ignorance (and ignorance about the exact level of that ignorance) conditioned the tactics by which establishment scientists believed they should counter the claims of perceived "crackpots." Consider the development over time of Theodosius Dobzhansky's response to Lysenko.

Dobzhansky's perspective is interesting for several reasons. First, he was trained under the titans of Soviet genetics of the 1920s before emigrating to the United States in 1927, where he became one of the architects of the incorporation of laboratory-based neo-Mendelian genetics into Darwinian natural selection—what came to be known as the "modern synthesis"—as exemplified by his monumental *Genetics and the Origin of Species* in 1937.[10] He was, without question, one of the most influential and respected evolutionary biologists of the twentieth century. He was also uniquely invested in Soviet developments, being in personal contact with many Soviet geneticists and having extensive familiarity with Russian-language publications, a rarity among U.S.-based biologists. Second, because of this personal knowledge and access, Dobzhansky had firm views about Lysenko much earlier than did other biologists, a status only approached among Westerners by American geneticist Hermann J. Muller, who had illadvisedly transplanted himself to Stalin's Soviet Union from 1933 to 1937. Third, we have access, through the papers of Dobzhansky's close friend and fellow Columbia University geneticist L. C. Dunn, to Dobzhansky's private reactions to Lysenko from 1936 through to and following the 1948 caesura of Stalin's endorsement of Michurinism.[11]

There is an important difference between private and public in Dobzhansky's case. He wrote a lot about Lysenko, both before and after 1948, but his publications were consciously crafted to generate a particular reaction (mostly among scientists, but also among politicians) to Lysenko. As Lysenko's position and the fate of Soviet geneticists became more clearly known outside the Soviet Union, Dobzhansky's private views, atypically harsh for this affable scientist, did not alter—but the public tactics did, eventually escalating into a broad discussion of the dangers of pseudoscience.

Dobzhansky first mentioned Lysenko in his correspondence with Dunn as far back as December 21, 1936, in the midst of the controversy over the proposed 1937 Moscow Congress of Genetics. (After it was in fact canceled, it relocated to Edinburgh.) In retrospect, he appeared remarkably prescient about the future of Michurinism. "What worries me most is not the congress, but rather the fate of genetics as a whole in Russia," he wrote. "If Lysenko (who is an old moron and a madman at the same time) will have his way, genetics will be declared a counter-revolutionary doctrine, which means that it will cease to exist." His major worry—shared by Dunn—was that opportunists in the Soviet Union would continue to use valid objections to the abuse of genetics in National Socialist Germany as a wedge to discredit the mainstream science. Dobzhansky suggested that someone write a treatise, endorsed if possible by "many geneticists," to show that racist interpretations of genetics were mistaken. "Such a thing may be very effective in showing those in power who are not idiots that they have taken council [sic] of an idiot."[12] This was the genesis of the book that Dunn and Dobzhansky would coauthor in 1946 and reissue in 1952: *Heredity, Race and Society*.[13] Two points are particularly noteworthy in this letter: first, the tone of Dobzhansky's language is significantly more biting than anything he would *publish* on the question until the late 1950s; and, second, Dobzhansky's approach was equally motivated by his objection to Nazi racial policies (to which we will return in the next chapter). A few weeks later, Dobzhansky suggested that Westerners lay low and not comment, hoping that "it may finally regulate itself, as far as Russian genetics is concerned. It seems to me that for the time being it is best not to try to do anything more from this side, and just wait for developments."[14]

After the cancellation of the Moscow Congress, the Dobzhansky-Dunn correspondence fell silent on Lysenko for a few years, but the interest in stopping the agronomist resurged sharply as World War II began to wane. The crucial point to understand was not how strong Lysenko was becoming, but how *weak* he appeared to American geneticists.[15] This seemed a propitious moment to mobilize in a meaningful way, to provide ammunition for Soviet geneticists and their internal supporters in their own campaign against Michurinism. Dobzhansky, echoing views he had heard from Soviet geneticists such as Anton Zhebrak (who had communicated with Soviet émigré biologist I. Michael Lerner at Berkeley while, of all things, serving as the Belorussian representative to the San Francisco Conference on the establishment of the United Nations), thought the Soviet government might listen to competent Western scientists who opposed Lysenko, especially if those scientists were known to be friendly to the

Soviet Union. Open criticism of the Stalinist state might backfire and invite repercussions against geneticists, the very thing the Western critics (and obviously their Soviet counterparts) wanted to avoid.[16] Dunn, as the leading actor in the American-Soviet Science Society dedicated to Soviet-American friendship, was an ideal candidate. As Dobzhansky put it in a letter dated July 4, 1945: "Lysenko's position is less secure than it has been, and Russian geneticists are hoping to get from under. They personally and *confidentially* ask for support of those American colleagues who are known to be friendly to Russia, yourself *in particular.*"[17]

Dunn responded in two ways. The first was to minimize criticism of Michurinism from Western media that might be perceived as inspired by anti-Communism, keeping the focus on the "perfectly orthodox genetics" that was being done in the Soviet Union, even in Lysenko's own institute. As Dunn explained in a letter of May 22, 1946, to *New York Times* science editor Waldemar Kaempffert (whom we met earlier as one of Velikovsky's harsher critics): "The fact that conflicting views and reports continue to come out of the Soviet Union seems to me the best indication that both sides are free to express themselves. I think it is no service either to science or to the Soviet Union to continue the discussion of Vavilov's fate until we have the actual facts."[18] As for the claims made by Lysenko, "one should no more view the whole of Russian science through the lens of Lysenko, than one should view American science through fundamentalist writings on evolution."[19] Everything needed to be kept in perspective; the Soviet situation was not so alien after all.

The second tactic was more covert. Building on the assumption that the essential question in the Lysenko business was *scientific*, Dunn and Dobzhansky thought the best way to support Soviet geneticists without succumbing to the rising wave of anti-Communist hysteria was to simply expose Western geneticists to the content and methods of Lysenko's arguments by translating the man himself into English. Then they would carefully control the reviews so the emphasis would be on intellectual, not political, questions. Opportunity came their way when Dunn's organization was alerted by the McGraw-Hill publishing house to the receipt of a Russian version of Lysenko's 1943 pamphlet *Heredity and Its Variability*, a summary of his views for a semi-popular audience. Dunn then attempted to persuade McGraw-Hill, and—when that failed—the Columbia-affiliated King's Crown Press, to publish a cheap version of the pamphlet, translated by Dobzhansky himself.[20]

The correspondence about the translation provides a fascinating glimpse into Dobzhansky's tactical vision. Even though the goal of this process was to produce a dispassionate scientific discussion that would discredit Lysenko and yet keep political animosity out, Dobzhansky's letters are littered with personal distaste, even disgust. On July 24, 1945, he wrote Dunn that "I shall return to the son-of-a-bitch in a few days. And I think your idea of having a bang-up review before the translation is published is very good."[21] His attitude had not softened a week later when he sent Dunn "Mr. Lysenko's excrement, 87 pages long. . . . Translating it has been one of the most unpleasant tasks I had in my whole life, and surely I would never undertake a thing like that for money—it can be done only for a 'cause.'"[22] After the manuscript was checked by Carl Epling of UCLA for faithfulness in translation, and by Dunn for clarity in English usage, Dobzhansky felt no better: "Your opinion about Lysenko's treatise I share, of course, and you may believe me that this job gave me no pleasure at all. But, if this will contribute even a little bit toward unmasking this imposter, as it very likely may, I shall regard the time well spent."[23] The pain of the translation process stayed with Dobzhansky. In an oral history from 1962, he still winced: "I may say, translating this thing was one of the most difficult and unpleasant tasks I ever had. . . . His writings are undoubtedly actually *his* writings. And it is foolish of him, because Lysenko is basically illiterate, not only in the field of biology, but in the field of rational language as well. His language, his Russian language, is outrageous. The syntactical incoherence is frequently something phenomenal."[24] Nevertheless, Lysenko finally appeared in English, preceded by a reserved one-page preface by Dobzhansky as translator, which stated clearly that it "is being published in order to give an opportunity to readers not familiar with the Russian language to form their own judgment on the merits of the controversy. . . . The translator wishes to emphasize that his undertaking the work of translation does not imply agreement with the contents of the book, and that he reserves the right to criticize it as he sees fit."[25]

Criticize it he did. Dobzhansky and Dunn hoped to channel the reviews of this translation into a volley of intellectual criticism that would persuade the Soviet leadership to dump Lysenko.[26] Striking in the behind-the-scenes management is how clearly Dunn and Dobzhansky emphasized issues of proper scientific method and the resulting incorrectness of Lysenko's views. Science could regulate itself without politics. This is a remarkably clear articulation of the first and second approaches to Lysenkoism *qua* pseudoscience outlined earlier: it was not, properly speaking, science, for the reason that it failed to adhere to the standards and procedures of scientific method and thus achieved false results. At issue was neither dialectical materialism nor state support. Dunn wrote to Lewis Stadler at the *American Naturalist*, before the book had even appeared, to attempt to secure the right kind of review:

> It is very important that the book be reviewed by persons who understand the implications of his work and who will not use it merely for an attack on Soviet science in general. The book was translated in order that Americans could judge at first hand what Lysenko's ideas really are. It is clearly anti-scientific and should be so criticized regardless of the author's nationality.[27]

But for Stadler, reviews of pseudoscience were unnecessary: "I cannot find anything to say about the Lysenko job that seems to me worth ~~saying~~ printing. . . . It is a useful example of unscientific method, and I think I shall want a few copies to give to graduate students. But it doesn't seem to me that there is any place for an extended review of this sort of thing in a scientific journal."[28] Indeed, there was a hint of the unsavory in Dunn's behind-the-scenes orchestration of what was supposed to be a "free" scientific discussion.

Undaunted, Dunn reviewed the book for *Science*, and in the accompanying correspondence made it clear that "because of my position and known sympathy with the development of science in the U.S.S.R., such criticism, as contained in the review, cannot be attributed to animosity or prejudice but merely to a desire to judge Russian scientific work by the same standards by which other scientific work is judged."[29] Many of the reviews Dunn and Dobzhansky arranged for this book pursued this line.[30] Consider Dobzhansky's review of his *own* translation in the *Journal of Heredity*: "It is the considered conviction of this reviewer that the above reluctance [to criticize Lysenko out of concern for U.S.-Soviet relations] is misapplied, and particularly so when a scientific theory has strayed so far from the truth as that of Lysenko so obviously has. . . . In any case one can assert with complete confidence that [the] genetic theories of Lysenko are invalid regardless of the final disposition of his experimental claims."[31]

The same tactics were used with respect to the other major English-language treatment of Lysenko's scientific and philosophical arguments, published in the United Kingdom also in 1946: P. S. Hudson and R. H. Richens's *New Genetics in the Soviet Union*.[32] Hudson and Richens clearly took great care to read and summarize the Russian literature, including a discussion of Lysenko's philosophical critiques of genetics and defenses of

Michurinism, in a non-partisan manner. In Britain, such an approach was strongly indicated because of a sizable group of distinguished Communist biologists—a strongly contrasting feature from the United States, where Dunn had to move through back channels to get Lysenko translated, published, and reviewed at all. Nonetheless, the approach suited his tactics as well as Dobzhansky's, and the latter reviewed the work favorably as a presentation of Lysenko's views, criticized those views on scientific grounds, and made sure—as a gesture to strengthen Soviet geneticists' positions—to note that Lysenko's sway in the Soviet Union "neither is nor was country-wide."[33] Likewise, although Dobzhansky was clearly hurt and angered in his obituary for Nikolai Vavilov, even there he was careful to note that "it is assuredly not true that all genetic research has been suppressed in USSR, as some writers in American journals hastily asserted."[34]

Throughout this entire period—from 1945 to 1948—Dobzhansky and especially Dunn labored to keep the discussion about Lysenko focused on scientific matters in hopes that postwar Stalinism would be more ideologically pliable than it had been in the 1930s, and that this would aid Soviet geneticists in their own efforts. Dunn expanded further on his tactics in another letter to Kaempffert of the *New York Times*:

> Some American scientists thought it better not to dignify Lysenko's rather vague and mystical ideas by serious treatment and criticism. The other point of view rejects this as not conforming to the usual method of science which insists that what is criticized must be thoroughly understood first. Since I belong to this latter group, I believe that objective discussion of the scientific and practical bases of Lysenko's theories will eventually be a worthy service to Soviet science.[35]

While Dunn subscribed both to the goals of strengthening Soviet anti-Lysenkoists and Soviet science in general, this clearly worked in parallel with but quite separately from a discourse on scientific method.

AFTER AUGUST

This situation would change abruptly in 1948. Dobzhansky's Soviet informants like Zhebrak were reporting the truth on the postwar developments in Lysenko's career: the man seemed vulnerable. His brother had defected to the Nazis during the war, which was bound to blot Lysenko's own record, and Western criticism of his scientific findings had begun to tarnish his reputation among certain Party elites. Although it remained true that Nikolai Vavilov, the scientist most capable of mobilizing a forceful opposition to Lysenko, was dead (while his physicist brother Sergei was appointed president of the USSR Academy of Sciences in 1945), other geneticists, such as Zhebrak and Nikolai Dubinin, were able to air their objections directly. As of 1947, the situation looked bad for Lysenko. And then Iurii Zhdanov, a chemist and son of Stalin's ideological second-in-command Andrei Zhdanov, gave a speech criticizing Lysenko. The agronomist complained to Stalin, and it appears that the latter decided that the controversy had dragged on long enough.

A conference on "The Situation in Biological Science" was convened for July 31–August 7, 1948, to discuss the issue of Michurinism versus genetics and resolve it. This was not in itself an unusual move in postwar Stalinist academic politics. In this period, congresses were called in psychology, political economy, and linguistics; physics narrowly averted having one of its own. The goal was to establish a Party line in each science; the most notorious was the conference on genetics.[36] At the conclusion of several days of discussion at VASKhNIL, during which (rare) criticisms of Lysenko were also presented, the man himself, in his position as president of the Academy, responded. In a lengthy speech that we now know was line-edited personally by Stalin, Lysenko rehearsed many of the central tenets of his theory of heredity, such as that "changes in the heredity of an organism or in the heredity of any part of its body are the result of changes in the living body itself," and that "once we know the means of regulating development we can change the heredity of organisms in a definite direction."[37] But this was nothing new; the striking event happened toward the very end of the speech, when Lysenko fatefully intoned: "The question is asked in one of the notes handed to me what is the attitude of the Central Committee of the Communist Party to my report. I answer: The Central Committee of the Party examined my report and approved it."[38] The debate was now over. The state had intervened, and genetics was banned in the Soviet Union.

This speech conveying the official condemnation of genetics at the 1948 VASKhNIL session flipped the separation of the scientific question from the political one for foreign observers. Interestingly, while the views of Dobzhansky and Dunn diverged on how much Lysenko's elevation to official doctrine tarnished the image of the Soviet Union (for Dobzhansky utterly, for Dunn mildly), they still agreed on tactics. For both of them, August 1948 meant that the debate over Lysenko had ceased to be a *scientific* issue and had become completely *political*. Whereas earlier the tactic of focusing on Lysenko as a scientific dispute avoided discrediting Soviet geneticists by declaring Lysenko a bad scientist surrounded by good ones, the new approach excluded Lysenko and his confederates from the scientific community entirely. They were political hacks, not scholars. This is well expressed in a letter written in January 1949 from Dunn to Richard Morford, the executive director of the National Council of American-Soviet Friendship:

> This being so, discussion on the scientific level is not likely to lead to further understanding but only to widen the gap between Lysenko and geneticists generally, since, Lysenko's views having become officially adopted, he is no longer free to change them.
>
> The better practice to follow in attempting to improve American-Soviet understanding would be to attempt to
>
> 1. provide an interpretation of Lysenko's victory over his opponents on other than scientific grounds....[39]

Or, as put more pithily by the editor of *Advances in Genetics*, Milislav Demerec, the following month: "This has been put forward as a dogma approved by the Communist Party—and therefore beyond question wherever this Party is in control—rather than as a hypothesis to be tested and discussed by scientific means."[40]

After 1948, Dobzhansky brought his public tactics directly into line with his private views on Lysenkoism, as in a blistering *Bulletin of Atomic Scientists* essay in 1949: "We are not dealing with two conflicting views, the relative merits of which could profitably be discussed.... It is difficult to discuss solemnly the view that earth is flat."[41] His private tone to Dunn changed as well. After reading a letter of Dunn's in response to Conway Zirkle, professor of botany at the University of Pennsylvania and outspoken critic of Lysenko, Dobzhansky begged to differ about the nature of Michurinism—not in kind, but in degree.[42] Dobzhansky felt the modification of Dunn's strategies after 1948 was not emphatic enough:

> As I see it, the defect of the article is that it does not say in so many words that Lysenko is just ignorant and an ignorant charlatan at that. The uninformed reader may get the wrong impression that what happens is a discussion among scientists, with a [R]ussian school on one side and a "capitalist" school on the other. I think it is our duty to say without mincing words

that it is not a legitimate scientific discussion of any kind but a conflict of science and obscurantism, knowledge and incompetence.[43]

The point was made more emotionally in a lengthy passage in an earlier letter from December 12, 1948, when he insisted that the only position scientists could take in the Lysenko affair was to shift their critique to Soviet politics:

It is evident that "the mainsprings of action in this particular case" are frauds backed by political chikanery [sic]—an[d] any attempt to represent it as anything else does not serve the truth . . . : a contemptible cheat has not only obtained backing for his prescientific and at best 19th century ideas, but has also succe[e]ded in murdering some and bouncing other scientists who were doing first class work and who dared to oppose his charlatanism. This is the core of the situation, and all else is materials for dissertations of future historians. . . . And as to "the scientific issues involved," the situation is likewise unmistakably clear. . . . And if saying that a political regime under which such things may happen is a crime against humanity, I am in favor of "attacks upon Soviet policy as a whole"—the time when one had to refrain from saying the whole truth because of hopes of saving thereby the lives of Dubinin, Schmalhausen, and others has passed. These people are now martyrs of this political regime, and it is just as well to say so aloud. Yes, I know, saying this will be a help to some people much nearer than Moscow whom I do not like to help. But, Dunn, trying to convince oneself and others that the snow in New York falls black and turns white in a few days is a futile as well as in the last analysis harmful procedures [sic]. Let us say the truth, and let the chips fall where they may.[44]

Lysenkoism was no longer bad science or ignorance; it had become *pseudoscience*. New information about the crimes of Lysenkoists was not the cause of the change. For example, the issue of martyrdom was not new: Dobzhansky had known of the death of Nikolai Vavilov and of Soviet meddling in genetics *before* he translated *Heredity and Its Variability*. (And, in this case, he was mistaken, as neither Dubinin nor Schmalhausen was killed, although each was relieved of his job.) Dobzhansky's private opinions had not changed, but he now asserted the right to say them in public.

In a sense, Dobzhansky was not so much changing the discourse over Lysenkoism as throwing his own voice and the voices of his many allies behind a particular interpretation of the Soviet events that already existed. Analyses of Lysenko's Michurinist theories of heredity that blamed everything on totalitarianism or the constraints of Marxist ideology were common—their ubiquity was what had prompted Dunn to organize the translation project to provide support for the internal critique of Lysenko within the Soviet Union. Now the science-oriented articles disappeared, and all that was left was a much-amplified political variant.[45]

For example, Hermann J. Muller—who had received the 1946 Nobel Prize in Physiology or Medicine for his research on inducing genetic mutations through radiation—had never signed on to Dunn's soft-pedaling campaign, finding the approach and arguments far too weak. Muller had a certain claim to authority in these matters, having lived and worked as a geneticist in Stalin's Soviet Union in the mid-1930s, before fleeing after his pro-eugenic views aroused the displeasure of members of the regime. And so Muller publicly and often repeated that "the story of genetics under these related systems of state control is only the most obvious example of the long-range incompatibility of cultural progress with political absolutism and with authoritarianism in general," and that "despite the pretenses of Communist officials and their followers, this matter is not a controversy between scientists or a dispute over the relative merits of two scientific theories. It is a brutal attack on human knowledge."[46] After the August session, Muller considered himself vindicated, and even Dunn wrote to him that now "we do agree about the fundamental facts of the utter wrongness of Lysenko's views and of the politicians who have pronounced him right."[47] In 1949 in the *Bulletin of the Atomic Scientists*, in milder language than Muller was prone to, Dunn seconded these criticisms of the state's endorsement of Lysenko's theories.[48]

The same pattern unfolded in Britain, which had conducted more of a two-sided debate over the correctness of Lysenko due to the presence of a vocal community of Communist biologists. The most prominent biologist to criticize Lysenko before 1948 was cytologist Cyril Darlington, but after 1948 he was joined by a host of his colleagues.[49] One of the most visible commentators on the Lysenko affair after 1948 was Julian Huxley, widely known as an animal behavior researcher, one of the architects (along with Dobzhansky and several others) of the "modern synthesis" in evolutionary theory and, from 1946 to 1948, the first director general of UNESCO. Huxley had earlier traveled to the Soviet Union and expressed an optimistic view of their science policy; but after the VASKhNIL meeting, he turned very negative. His thoughts, however, especially as published in

1949 in *Heredity East and West* (the British edition is entitled *Soviet Genetics and World Science*), centered entirely around the distinction between the scientific and the political readings of Michurinism. It was important to realize, Huxley wrote, "that Lysenko and his followers are not scientific in any proper sense of the word—they do not adhere to recognized scientific method, or employ normal scientific precautions, or publish their results in a way which renders their scientific evaluation possible. They move in a different world of ideas from that of professional scientists, and do not carry on discussion in a scientific way." That meant there was no point in refuting their claims about genetics (although Huxley attempted that as well). The crux was something else: "I hope I have made clear that the scientific aspects of the controversy are subsidiary to the major issue of the freedom and unity of science. Even if Lysenko were right in his claims to have made new and startling genetical discoveries, this could not justify the official condemnation of Mendelism as scientifically false, nor the suppression of all Mendelian research."[50]

For Huxley, the matter was mostly intellectual, as he had no personal stake for or against the British Communist movement. For J. B. S. Haldane, one of the leading geneticists in Britain, the contrast could not be starker. After personal contacts with influential Communists in the 1930s, Haldane formally joined the Communist Party of Great Britain in 1942, and from 1944 began to serve as a member of its executive committee. While in this position, he continued to write a large quantity of popular scientific journalism, often (but not exclusively) in left-wing periodicals. The growing Western discussion of Lysenko both before and after the war put Haldane in a bit of a bind. In 1938 he wrote that "so long as [the Soviet debates] do not lead to the suppression of research such controversies are a sign of healthy scientific thought," and he was not worried (at least publicly) about the outcome of the cancellation of the Moscow Congress. In 1940 he publicly declared himself agnostic on Lysenko's claims until he was able to access more of the scientific material.[51] After August 1948 he at first temporized and then tried to extricate himself from either adhering to the Party line (which meant endorsing Lysenko) or defending his science. In 1950 he left the Communist Party, the only major British Communist to part ways over the Lysenko affair.[52] Another prominent scientific member of the Party, J. D. Bernal, continued to pen virulent and unapologetic defenses of Lysenko.[53] He was essentially alone except for the Party faithful. One of these, biologist J. L. Fyfe, surprisingly took the opposite tack from Haldane: in early 1948 he had argued in left-wing publications that

Lysenko was incorrect on the science, but in 1950 he wrote an apologetic pamphlet entitled *Lysenko Is Right*, which argued that "the gene theory is anti-scientific."[54]

The Americans were aware of the disputes in Britain, but their attention was focused elsewhere: on the Soviet Union and on the escalation of anti-Communism at home (about which more in a moment). Stalin, Lysenko's original backer, died in 1953, and although Lysenko was increasingly criticized within the Soviet Union, he found a new patron in Nikita Khrushchev and maintained his hold on power (albeit not as monolithically as before).[55] Dobzhansky, who had been blacklisted in the Soviet Union during the deepest days of Michurinist control—papers of his could not be cited nor his name mentioned in publications—managed to achieve a different kind of renown among Lysenko's supporters abroad. In a letter dated November 1955 from Rio de Janeiro, where he was collecting biological samples, he ironically noted that in Brazil his major claim to fame among agronomists was as the man who introduced Lysenko to the Western world: "The fact of the matter is that Lysenko did get to be known all over the world. Oh, that great Ukrainian scientist!" In 1960 the same situation confronted him in Indonesia. His interlocutors refused to believe that the man who translated *Heredity and Its Variability* could possibly be anti-Michurinist.[56]

But Dobzhansky's position never wavered. He maintained his liberal politics, but he never again compromised on criticism of Lysenkoism in the name of higher tactics. For example, in a 1952 article, he juxtaposed Lysenkoism with creationism, and it is not clear which belief system he was intending to disparage more by the comparison:

> Honest dissent and unorthodox ideas often promote scientific knowledge. Even though more often wrong than right, unorthodox ideas are apt to stimulate some clear thinking among the orthodox. And from time to time, a doubter makes a basic discovery. But the lysenkoism is quite sterile of ideas and of suggestions for new experiments. It urges a retreat to archaic views, long abandoned with sufficient reason. In this, the lysenkoism is comparable only to the anti-evolutionism in the USA. New arguments and new facts mean just as little to the lysenkoists as they do to the anti-evolutionists.[57]

He even carried Dunn along with him. As the two wrote in their 1952 revision of their 1946 survey of genetics—designed to rebut the perception that genetics was inherently fascist, a view that had abetted Lysenko's rise to power—the matter had now become black-and-white: "But there can be no justification whatever for the use of intimidation, imprisonment, and exile, to establish the ascendancy of a scientific doctrine, whether a right or a wrong one. The science of genetics has in recent years been destroyed in the Soviet Union by just these methods."[58]

AMERICAN LYSENKO?

Lysenko's rise to total control of Soviet genetics had two immediate effects on the American scientific community. First, it heightened concerns over the potential dangers of "pseudoscience," raising the bar for what would count as acceptable or reasonable speculation. Second, it formulated a set of tactics for how scientists should mobilize against any threats that triggered alarm bells. Those alarms would sound sooner than expected, further deepening the function as exemplar that the Lysenko affair had already begun to assume. Which brings us back to the question that inspired this detour into the debates over Soviet genetics: Why did the astronomers react so vociferously to Velikovsky in 1950? Why did they not, as was typical practice for well over a century, just ignore this particular incarnation of crackpottery? Or why not simply engage with the scientific claims he was making and refute them?

The roots of the violent reaction lie in Lysenkoism—or, to be precise, a particular interpretation of the lessons of Lysenkoism as drawn by a certain segment of the liberal, elite, Northeast scientific establishment. Many American scientists took the 1948 VASKhNIL conference as a wake-up call. Until 1948 many leaders of the scientific community had not yet fully come to grips with the transformation in the political and cultural significance of science wrought by World War II. The atomic bomb, radar, the electronic computer, and penicillin all marked a new postwar science that was bigger, more expensive, and more exposed to political scrutiny and interference. Geneticist Richard Goldschmidt, in an article ostensibly about Lysenko and published in the special issue of the *Bulletin of the Atomic Scientists* about the affair in May 1949 (a full year before the Velikovsky issue heated up), pointed to this metamorphosis:

> Pseudobiological literature of all times is full of books by philosophers, state[s]men, theologians, and cranks who want to replace facts and laws found by the hard work of the active biologists with their own pet ideas or creeds. Lysenko would certainly join this group, which always finds follow-

ers among laymen, if he lived in the Western World, and his doings would remain a curiosity on a library shelf.

However, his case, otherwise completely uninteresting, has become highly significant by the fact that he has succeeded in persuading the high command of the party, including, as it seems, the dictator himself, that his, Lysenko's, line is first, truly Marxian, second, of greatest importance for the economy of his country and full of practical promise, and, third, 100 per cent Soviet Russian.

The allegations against Lysenko were standard after August 1948; new was the emphasis on the role of the state and an analogy with the United States. Goldschmidt insisted that he did not "want to imply that we are in danger of having a Lysenko appear in our midst. But there are different degrees of such things."[59] His worry was that public funding of science, grown so large in the postwar years, might lead to the "planning" of science and take the Americans down a similar path as the Soviets. L. C. Dunn confessed to Goldschmidt that he was "a little disturbed by the association in the last few pages of suppression of freedom in science, such as the 'new line' in Russia, with planning in science generally. I think it would be unfortunate, considering the amount of planning which is required to get public support, were it to raise to suspicion that public support always implies political direction of science."[60] Dobzhansky, however, was inclined to concur that the problem of Lysenko was broader than a Ukrainian agronomist's manias and reached to the heart of the state's relationship to postwar science:

> Lysenkoism may be useful only because it provides a lesson. Whether we like it or not, the days of the independent scientist and of independent science are about over. The more important science becomes in the lives of individuals and of nations, the more it will need popular support and will have to submit to social control. But the forms and techniques of this support and control have not yet been devised and tested. The problem is a new one. The Soviet rulers have tried a solution, but their solution has resulted in lysenkoism, and thus proved to be a dismal failure.[61]

As Dunn was well aware, however, concerns such as Goldschmidt's and Dobzhansky's were not completely fantastical. Postwar America generated a more aggressive and ideological federal policing apparatus, first with the beefing up of the House Un-American Activities Committee

(HUAC)—which investigated the "loyalty" of broad swaths of academics, entertainers, journalists, and other elites—and then its later expansion in the Senate under the auspices of the junior senator from Wisconsin, whose name would become an eponym for this historical period: Joseph McCarthy. Faced with this emergent loyalty apparatus, some scientists feared for their disciplines.[62]

There was a definite correlation between opposition to anti-Communism, resistance to Lysenko, and the uproar over Velikovsky. Dunn was investigated by HUAC in 1950, and in 1953 Ruth Shipley of the State Department denied him an application for a passport because of his alleged Communism; Dobzhansky also was viscerally opposed to loyalty policing.[63] Many of the early principals in the Velikovsky debates were alleged fellow-travelers who had been summoned before HUAC in the late 1940s. Consider the man most often fingered by Velikovsky and his supporters as the ringleader of the opposition: Harlow Shapley, the director of the Harvard College Observatory. Shapley was well known in the years after World War II as one of America's most prominent liberals and was a strong advocate of Soviet-American scientific cooperation (in this he worked closely with Dunn).[64]

Partly as a result of his publicly articulated views and activities (but also due to local dynamics of a Boston congressional race), Shapley was called before HUAC on November 4, 1946, by Congressman John E. Rankin to talk about the records of four organizations he was associated with: the Political Action Committee of the Congress of Industrial Organizations (CIO-PAC), the National Citizens Political Action Committee, the Joint Anti-Fascist Refugee Committee, and the Independent Citizens Committee of the Arts, Sciences and Professions. On November 14 Shapley attended a brief but contentious closed session, which ended when Shapley was berated from the dais and in turn demanded a lawyer, making him among the first in a long line of scientists brought before the new anti-Communist loyalty apparatus. This was only the beginning of Shapley's troubles. In early 1950, anti-Communism received a renewed push: the jury in the second Alger Hiss espionage trial came back on January 21 with a conviction on two counts of perjury (four days later Hiss was sentenced to five years in prison), and then the press reported on the arrest of Klaus Fuchs on February 2 for wartime atomic espionage. A week later, on February 9, 1950, Joseph McCarthy began his extended career as the focal point of federal red-hunting with a speech in Wheeling, West Virginia, where he declared he had a list of 205 card-carrying Communists who were members of the State

Department. He named Harlow Shapley among them—despite the fact that Shapley was not a Party member and he did not work for the State Department. A few weeks later, McCarthy again singled out Shapley by name as one of 57 highly placed Communists. He was cited a month later again in Congress.[65] This was February 1950, *exactly* when Shapley suspended for several months his correspondence with Macmillan about *Worlds in Collision*. He now had substantially bigger things to worry about.

Shapley was aware that a connection—however vague—existed between McCarthyism and the Velikovsky matter, and observed in a letter to Ted Thackrey (the editor of the *Daily Compass* who supported Velikovsky on free-speech grounds, much to the consternation of Harlow Shapley; they ended their friendship over Velikovsky) that the linkage had been noted by other American scientists:

> In the annual address to an important scientific foundation, a distinguished American psychologist on Saturday bemoaned the rather black future, and obvious decadence of our times. We have failed completely in our scientific teaching, he stated, or the "Worlds in Collision" atrocity would not have caught on the way it has. It seemed to him that Dr. V. and Senator McCarthy are symbols of something dire and distressful. But I do not worry about it. Time has curative properties.[66]

Opposition to anti-Communism and McCarthyism remained a substantial part of Shapley's public writings into the 1950s, even as he became silent about Velikovsky.[67] Velikovsky's supporters, however, continued to make the link. The pro-Velikovsky article in the July 1950 issue of *Newsweek* that publicized the boycott and linked it to Shapley, for example, made a point of identifying him as a "board member of the fellow-traveling National Council of the Arts, Sciences, and Professions," an invocation of McCarthyist language.[68] Shapley was not the only critic of Velikovsky who seemed to be simultaneously caught up in the loyalty apparatus. Edward Condon—director of the National Bureau of Standards, who had reviewed *Worlds in Collision* for the *New Republic*—was also fresh from a bout with HUAC. (Coincidentally, his case was discussed in the January issue of *Harper's* that featured the Larrabee article.)

For these liberal scientists and many like them, Velikovsky served as a proxy to demonstrate that their anti-Lysenkoism represented more than a form of anti-Communist propaganda. (In Britain, criticism of Velikovsky and, in muted fashion, Lysenko were both undertaken by J. B. S. Haldane, who had just left the Communist Party.) In the early 1950s, the critics were all *perceived* by Velikovsky and his supporters as representatives of the Left, and not without reason. In her 1952 critique of Velikovsky, Payne-Gaposchkin made a seemingly unconnected aside to denounce "these days of loyalty oaths"; scientists had to hold "to principles, not to dogmas; to respect for evidence—all the evidence, not merely such as fulfils his expectations, and respect for those formulations that embody the evidence."[69] Resisting Velikovsky, for her, was a stab against McCarthyism. Even Velikovsky observed this, noting that Condon, Haldane, and one or two others were "all left-wingers."[70] (The political valence of opposition to Velikovsky would shift somewhat by the 1960s.) This is more than a coincidence. Those on the Left, like Dunn and Shapley and Haldane, came to publicly criticize Lysenko because they believed the Soviet agronomist was deeply mistaken on matters of fact and that Soviet support for him was ruinous for science. But they also recognized that criticism of Lysenko was, as Dobzhansky had observed, a psychological victory for right-wing anti-Communism. So, in a moment of heightened tension over the status of science in American culture, and with a particular fear of perceived crackpots entrenching themselves, along came Velikovsky with a seemingly rightist, pro-religious doctrine that they could slam as pseudoscience. The critics could reaffirm their liberal bona fides under the guise of protecting science. The pump had been primed in 1948, as Hermann Muller observed: "When we criticize the shocking treatment accorded scientists in Nazi Germany and which is now being given them in the USSR, we must also exert ourselves to prevent the same thing from happening in our own midst."[71]

Many shared the perception that there was some lesson learned from Lysenko that was being applied to Velikovsky. To be absolutely clear: there was no suggestion that Velikovsky would become "Lysenko for astronomy," that the federal government would endorse Velikovsky and criminalize uniformitarian, Newtonian, or other traditional modes of astronomical research. No one believed that. The lesson was rather the one drawn out above: it concerned *tactics* when confronting crackpots. In the overheated air of 1950, amid the rising tide of anti-Communism, some American scientists felt that it was simply a tactical error to hope that crackpots would fade away due to lack of attention: charlatans had demonstrated that they were dangerous. And it was not worth trying to engage with Velikovskianism as a scientific problem and refute it claim by claim. The Americans had tried *that* with Lysenko, and look what happened! At this moment when the scientific method seemed vulnerable and the autonomy of scientists was challenged by the state, astronomers reacted convulsively to an article in *Harper's* and a popular book with a conflagration unthinkable only five years earlier.

The early letters to both *Harper's* and Macmillan often bristled with invocations of either anti-Communism or Soviet developments, in the service of pointing out the necessity for freedom of thought for the continued development of science. The Michigan astronomer Dean McLaughlin, for example, one of the more incensed and verbose of the letter writers, informed Macmillan's president in May 1950 that *Worlds in Collision* "is a serious threat to education and, I believe, to the democratic principle itself. . . . No amount of lying will alter the truth,—but lying can alter the willingness of a people to accept the truth."[72] This point was extended, with a subtle nod toward Lysenkoism, in a mimeographed statement that he had handed out to his astronomy students and included in the letter to George Brett:

> Science may *appear* to have no connection with political freedom. . . . Scientists are too busy with *positive measures*, such as productive research and the dissemination of the results of the search for truth, to take time out to refute every crackpot notion that gets into print. . . . If the democratic process were applied *ideally*, and if enough people were to accept the claims in the article as truth, then publicly supported schools and universities could be depopulated of competent faculties, whose places could be taken by quacks and political appointees. Granted that the chance of this is very small, nevertheless the imagined situation has a modern precedent. Something very similar *did happen* (on purely political grounds) in a European country during several years preceding the second world war. It has happened in other countries since the end of that war.[73]

Otto Struve, the noted astronomer from the University of Chicago, told *Harper's* much the same thing. "I am afraid that you have rendered a disservice to America and to science by publishing this altogether unsound article. In doing so you have furnished ammunition to our enemies in the Soviet Union and elsewhere; they are trying to prove that American scientists confuse religious issues with matters of science," he wrote, "and are misleading the people by presenting to them false theories for the purpose of creating an idealistic outlook based upon faith rather than cold logic. When I read your article I was in the process of collecting material to expose and refute some of these Russian arguments, but I am beginning to wonder whether there is not a grain of truth in their criticism."[74]

Some of the analogies between Lysenko and Velikovsky were rather crass, as in the statement in the Dallas *News* that "perhaps *Worlds in Collision* is the current Soviet dogma in regard to paleontology, anthropology, and cosmology, a new science which may best be called Sov-myth," which happens to be the only right-wing attack on Velikovsky I have been able to locate.[75] More reflective were the observations of Martin Gardner, who spent the following decades writing about people he labeled "crackpots." For example, in an article published in late 1950 referring to Wilhelm Reich's "orgone research" program (to which we will return in chapter 5), Gardner pointed to a common feature of scientists' tactics when faced with what they considered pseudoscience:

> The reader may wonder why a competent scientist does not publish a detailed refutation of Reich's absurd biological speculations. The answer is that the informed scientist doesn't care, and would, in fact, damage his reputation by taking the time to undertake such a thankless task.[4] For the same reasons, scarcely a single classic in the field of modern scientific curiosa has prompted an adequate reply. The one exception is the work of the Russian geneticist, Lysenko, unimportant in itself, but with a wider significance because it strengthens a cultural paranoia, and dramatically highlights the conflict between a relatively free and a rigidly controlled science.

Footnote 4 in the above passage leads directly to Immanuel Velikovsky: "Although there obviously is no sharp line separating competent from incompetent research, and there are occasions when a scientific 'orthodoxy' may delay the acceptance of novel views, the fact remains that the distance between the work of competent scientists and the speculations of a Voliva [the prominent Flat Earth theorist] or Velikovsky is so great that a qualitative difference emerges which justifies the label of 'pseudo-science.'"[76]

Over the following decades, numerous other books and articles discussing approaches to "unorthodox science" would point to the responses to Lysenko (after 1948) and Velikovsky as distinctly aggressive in tone, and opponents of Velikovsky would point to the actions of his 1970s followers as "Lysenkoist" in their alleged reliance on bullying and insinuation.[77] (No one mentioned the several occasions in which Velikovsky discussed vernalization and other Michurinist ideas favorably—all dated after 1948.)[78]

Yet many of these critics, such as the noted philosopher of science Michael Polanyi, recognized a difference between Lysenko and Velikovsky. While in the Soviet Union science had been under "attack on a broad front," he noted in 1967, there was still some analogy to the "more personal challenge to the foundations of science . . . from Dr. Velikovsky's highly unorthodox book."[79] The perceived similarity between this reaction to Velikovsky and the American response to Lysenkoism may be more than accidental: the anti-Velikovskians adopted (and exceeded, in the case of the threatened boycott) some tactics of the post-1948 anti-Lysenkoists in the hopes of nipping this in the bud before Velikovsky became entrenched in the popular consciousness. The fact that they miscalculated and only made Velikovsky *more* visible and popular is beside the point.

Interestingly, Velikovsky and his adherents turned that lesson around, claiming they were persecuted in the same fashion as Lysenko's victims, and in much the same way tried to appropriate the mantle of Galileo and Giordano Bruno. Pointing to the boycott campaign and the firing of Gordon Atwater and James Putnam, Velikovsky wrote to Ted Thackrey: "A campaign of distortion and calumny was carried against me and my book here, but there [in the Soviet Union] I would have been silenced if my views contradicted the accepted dogmas. And is it only a coincidence that Shapley and his friends in political thinking tried to use the Lisenko [*sic*] methods on me?"[80] If the Soviet bugbear was the analogy in one circumstance, the American bugbear assumed the same role in another, as M. Abramovich, the president of Liberty Electronics in New York City, wrote to the editor of *Scientific Monthly* in 1951: "I wonder whether McCarthyism had entered the World of science as well as that of politics. I refer to the article of Prof. Laurence J. Lafleur attacking Dr. Velikovsky and his theory."[81] But however applicable some perceived the invocation of McCarthy, it did not compare to Lysenko. Comments on the dictator of Soviet genetics are scattered around Velikovsky's archives and publications by his followers. Consider, for example, the 1979 observations of Leroy Ellenberger, then a vigorous supporter of Velikovsky:

> Thus, having been inadvertently associated with one of the 20th century's greatest scientific frauds, left-wing scientists were sensitized and quick to attack any suspicious-looking idea in order to shore up their reputations as respectable, proper-thinking men of science. To vulnerable astronomers, the advent of *Worlds in Collision* could not have been a more welcome instrument to assist their redemption.[82]

Aside from partisan sniping, the fundamental problem endured: science after the Second World War had rapidly achieved a striking salience in public life, and thus was exposed to a new and (to scientists) surprising degree of attention from the state and the public. This was visible to those outside the scientific establishment as well, such as Eric Larrabee, of all people, the man who introduced America to Velikovsky in *Harper's*. "Scientists as a class, like nearly every other in contemporary America, are prone to exaggerate the degree to which they are persecuted, ignoring the existence of their own prestige in order to visualize themselves as underdogs," he wrote (by invitation) in *Science* in 1953. "The vast admiration that science actually enjoys is not only more widely shared than the antipathies against it, it is partly responsible for them." Larrabee studiously avoided mentioning Velikovsky, but that did not stop others from bringing him up, sparking a follow-up discussion in a later issue.[83] Commentators both at the time and since have pointed out that worries over public legitimacy, which carried concerns of prestige as well as funding, were partly behind the surprising vigor by which alleged "pseudoscience" was policed in this postwar conjuncture.[84] In the post-1948 moment, the dominant metaphor among many styles of argumentation was Lysenko, and it proved surprisingly flexible.

The existence and persistence of this Lysenko trope demonstrates how multifaceted the affair became in its American resonances. There was, on the one hand, the pre-1948 lesson that one needed to engage perceived crackpots and dismiss them on scientific grounds; and, on the other hand, the post-1948 lesson that organized campaigns of persecution of scholars on behalf of a particular doctrine (whether Michurinism or uniformitarian astronomy) automatically rendered that dominant doctrine pseudoscientific. Velikovsky and his followers read the lesson through that latter prism; Payne-Gaposchkin criticized Velikovsky through the former. In either event, it was clear that no one wanted to be in *Lysenko's* position—everyone was a Vavilov in the American inflection of the Lysenko affair.

LEARNING THE WRONG LESSON?

Velikovsky was primarily attacked by physical scientists, and Lysenko by life scientists, at least in the late 1940s and early 1950s. (In the 1960s and 1970s, biologists would also join in the Velikovsky fracas.) The clear resonances between the two controversies, however, demonstrates something significant about the pseudoscience wars as they developed: one of the functions of labeling doctrines "pseudosciences" was to provide a sense of unity for the sciences as a whole, and as a result the cross-fertilization between the controversies for some of the actors becomes more understandable. Dunn and Dobzhansky, for their part, seemed oblivious to the Velikovsky debates, but it is fairly certain that they would have deplored the promiscuity of the Lysenko metaphor. Rather than viewing the Lysenko affair as a template for how pseudoscience should be combated in the future, they saw Lysenkoism as *sui generis*, not to be generalized to other controversial claims (with the partial exception of creationism), or even claims of scientific persecution. To do so would be a mockery of the seriousness of Lysenkoism itself.

Dobzhansky was especially concerned about a particular false generalization. As he saw it in later years, it was an error even to consider Lysenkoism as pseudoscience, not because its claims possessed a chance to become respectable science, but because Lysenko's tactics were so nakedly repressive that even *pseudoscience* endowed them with too much dignity. In the three folders in his personal archives that he labeled "crackpot files," Dobzhansky only included claims of cures for cancer and the like.[85] Lysenkoism was too grave to be dismissed as mere crackpottery. For his part, Michael Lerner (like Dobzhansky a Russian émigré and who similarly took the predations of Lysenkoism personally) fumed at Nobel Laureate William Shockley's claim—in the face of opposition to his plans to gather sperm from geniuses and breed a superior type of human—that he was besieged by a "Lysenko syndrome." Lerner, who had translated Zhores Medvedev's classic exposé of Lysenko (published in 1968) and collaborated with Dunn and Dobzhansky to salvage Soviet genetics, vehemently objected to this as a trivialization of the dead and a complete misunderstanding of the nature of Lysenkoism.[86] Yet this was precisely the same move Velikovsky had already made.

If the rise of Lysenko to unquestioned prominence with Stalin's endorsement in August 1948 set the stage for the morality drama of the pseudoscience wars, Lysenko's fluctuations in power and his eventual collapse did nothing to disrupt the play from continuing on to its conclusion. As mentioned earlier, Lysenko's position was a little wobbly right before Stalin's death, when criticism was allowed to be published against him in the Soviet *Botanical Journal*, but then Khrushchev's patronage resuscitated his career. It seemed fated, when that patron was ousted in 1964 in the wake of economic and foreign-policy disasters (not least the Cuban Missile Crisis), that the client would fall as well. The Academy of Sciences dispatched a commission to investigate Lysenko's farms, and they uncovered gross mismanagement and fraudulent accounting. The agronomist was removed from some leadership positions as genetics began to be fully rehabilitated after an almost twenty-year underground existence.[87] (Lysenko was not, however, stripped of his membership in the Academy.) Dobzhansky and Dunn seemed not to know what to make of this transformation. In a 1965 history of genetics, Dunn grouped Lysenko with the Nazi eugenicists and then hurried on to avoid further discussion, while Dobzhansky minimized Lysenko's long-term damage and threw up his hands at explaining the irrationality that had beset his homeland for decades.[88]

But the fact that a doctrine conceived of by American scientists as the arch-pseudoscience had been dethroned and that rationality resurged triumphant did not change the dynamics of the Velikovsky debates. The Lysenko affair was relevant at the moment of publication of *Worlds in Collision*—it framed the context—but it was not genetically linked to Velikovsky's doctrines. The similarities were not structural and not philosophical; they were similarities *interpreted* by the actors, because Lysenkoism was the historical exemplar closest to hand. And because the astronomers mobilized in such a vigorous fashion, they ran the risk of repressing free discourse in a manner analogous to what they had condemned in the Soviet Union in 1948—or the McCarthyists they excoriated at home. Think what you might about Velikovsky's arguments, a book boycott is not an invitation to open discourse, and neither was arranging for "proper" reviews of a translated scientific document. It is in this specific sense that the American reaction to the Lysenko affair became a "playbook" for the Velikovsky affair: in the debates about the borders between legitimate science and crank doctrines, the defense of Soviet genetics between 1946 and 1949 offered a host of lessons for the combatants. Which they chose to pick up depended on their point of view, but they all saw Lysenkoism as indicative of something vitally important and staked their claims in analogy to it.

4 · Experiments in Rehabilitation

"To this day I have not seen a single defense of Velikovsky's theory by a competent scientist," wrote geochemist Harrison Brown in 1955. "All the defenses of Velikovsky's views that have come to my attention have been written by nonscientists unfamiliar with the facts. Many of the defenders appear to believe that scientific arguments can be won with rhetoric or by taking a public opinion poll and assessing how many persons are 'for' and how many are 'against' the theory."[1] But these issues could not be resolved by democratic methods, even if the trigger for them had been an absence of democracy in Soviet genetics. In the wake of Lysenko's triumph and the American reaction, the scientific community was simply unwilling to listen to Velikovsky, no matter how popular he was. But he was also not about to retire from the fray. Convinced that he had found a core truth about humanity's (and the solar system's) past, he needed to make scholars pay attention. The solution was to temporarily lay down arms and begin an effort at diplomacy. He would start to engage scientists on their own terms. He wanted to become respectable, to have his theories rehabilitated.

After the transfer of *Worlds in Collision* from Macmillan to Doubleday in June 1950, much of the public controversy—and certainly its temperature—simmered down. It was still referred to in occasional articles, as one would expect for the publishing sensation of the year, but those scientists who got vocally upset seemed to be assuaged. Now might be a good time to privately approach them, without the fanfare of *Harper's* and *Collier's*. At the end of the summer, David Arons, an admirer of Velikovsky's and publicity director for the Gimbel Brothers department store in Philadelphia, sent copies of *Worlds in Collision* to a roster of scientist members of the American Philosophical Society, America's oldest and most venerable learned society, founded before the nation itself by Benjamin Franklin.

Convinced that the problem with the scientific reaction to Velikovsky was that his opponents had simply not read the book (as several of them cheerily admitted), Arons proposed that now they take the opportunity and respond to Velikovsky's evidence and propositions.

The venture did not go according to his sunny projections. Many of the scientists did read it, but not a single one (of those whose letters were retained in Velikovsky's personal files) claimed to be converted. The reaction was unanimous across a broad diversity of disciplines. The most detailed and comprehensive response came from George Gaylord Simpson, the dean of American paleontology, another architect of the neo-Darwinian modern synthesis (along with Dobzhansky), and the curator of paleontology at the American Museum of Natural History—the very institution that had fired Gordon Atwater only months earlier for his support of Velikovsky. Simpson would not shirk his duty. He read the book Arons sent to him, writing in response that he had already read *Worlds in Collision* when it appeared, but "have not previously expressed an opinion on this book outside of personal conversation, because the book would have been better ignored and I hoped the furore would die down." There was simply nothing in the book worth salvaging. The book was "utterly worthless to the point of being nonsensical. It reveals ignorance of the facts in fields it purports to cover and even, despite the numerous references, of the really pertinent and reliable literature. It further reveals inability to reason clearly and ignorance of the essentials of scientific method," he noted. "Every year a number of pseudo-scientific works or dissertations on the lunatic fringe of learning are written and a few get published. We are used to these, and they are usually ignored because anyone with serious work to do cannot spend his time in unnecessary consideration of foolishness." He included a small sampling of the errors he found. As for the campaign of suppression against it, he thought Arons had misunderstood that as well. The only reason why anyone had paid attention to this particular book was because of the way Macmillan had promoted it, and this made it "impossible for all those capable of seeing the worthlessness of the book to remain wholly silent, and of course some of them spoke out if only from a sense of duty."[2]

This noblesse oblige, the claim by scientists to police their own domain, was repeated in many of the letters. A dominant refrain—expressed by correspondents such as Ira Bowen (director of the Mount Wilson and Palomar observatories), Joel Stebbins (Lick Observatory), Eliot Blackwelder (professor of geology at Stanford University), and Frederick Seitz (physicist at the University of Illinois at Urbana-Champaign, later president of Rockefeller University)—was that had the book been published by any press but Macmillan, the leading scientific press of the period, scientists would have taken no notice of it.[3] Eminent Harvard psychologist B. F. Skinner put the point most succinctly: "As I understand it, the main objection of scientists was to the fact that the Macmillan Company classified and promoted the book as a scientific document. This it certainly is not."[4] With such unanimity, across so many disciplines, was there any way for Velikovsky to emerge from infamy into respectability?

There was some hope. Walter S. Adams of Caltech and Mount Wilson Observatory, and one of the astronomers to whom Velikovsky had fruitlessly written in the mid-1940s, sent Velikovsky an unsolicited letter that indicated openness to at least limited engagement. "I differ from the critics whom you mention in having definitely read your book. Its impression upon me has been mixed," he wrote in July 1950, a month after Macmillan walked away from the book. "I cannot help feeling that you have overestimated the value of this [mythological] material as evidence. Primitive peoples in small countries, with little or no means of outside communication, are, like children, prone to exaggeration." But that was the history; the science was more problematic. Velikovsky's invocation of electromagnetic forces to explain the celestial mechanics of Venus's erratic movement Adams considered "a distinctly *ad hoc* or even fantastic hypothesis.... It is scarcely profitable to interpret one miracle by another miracle, and from the scientific point some of your hypotheses would require nothing short of miracles for their fulfilment."[5]

Velikovsky, who characteristically reacted to any criticisms of his views with a flurry of aggressive argumentation and a circling of epistemological wagons, responded with cordiality and gentility. Although he disagreed with Adams's opinion of the value of the literary heritage of humanity as a source of scientific evidence, and insisted on incorporating electrical forces into the movements of planets, he was touched. "It was the first letter of an astronomer in this country who read and sincerely debated the problems of my book," he wrote back, two weeks later. "For this, I thank you."[6]

Given the way arguments over Velikovsky's ideas blazed in 1950 and then reignited into a second inferno from the mid-1960s into the 1970s, one might imagine that Velikovsky spent the lull in hostilities polishing his knives for combat. He did not. Instead, he undertook a variety of moves to rehabilitate his theories in the eyes of the scientific community, *not* the broader public who had been the audience for *Worlds in Collision* (in whose eyes he was not besmirched, after all). Although Velikovsky certainly did not explicitly label his actions in the 1950s as a strategy for legitimation, we can with hindsight discern three distinct approaches to persuading scientists to take his arguments seriously: seeking the endorsement of respected scholars; finding new evidence to bolster his historical reconstruction; and emphasizing confirmed predictions of astrophysical findings from *Worlds in Collision*. It was improbable that this venture would succeed in rehabilitating him, but it was not impossible, as other "discredited" fields managed to use strikingly similar approaches at precisely this moment to save themselves from disrepute.

REBIRTHING THE WELL-BORN SCIENCE

In the mid-1940s, Immanuel Velikovsky drafted a peculiar manuscript, a diversion from his constant labor on his *magnum opus*. He was concerned with the effects of climate on people, and specifically the issue of heat: "The Jewish national movement brings the Jews, or at least those of them who will probably escape assimilation, back to Palestine. From the point of view of racial biology there is danger in this transplantation." In this relatively short typescript, he argued that extended exposure to heat inhibited the development of advanced civilizations (he passed over the problem of the Egyptians, the Babylonians, the Mesoamericans, and other contrary examples) and suggested the necessity for the Jewish emigrants to structure their lives so they could cool down. Velikovsky never developed this argument further, and it remains a curiosity among his many projects. The title, however, is striking: "The Climatic Influence upon the Intellect of Races with Special Consideration of the Eugenic Question of the Cerebral Development of the Jews in Palestine."[7] The word "eugenic" stands out.

It is unclear how much Velikovsky knew of eugenics and its history, and I do not turn to this episode because it helps explain him. Rather, the postwar development of this discipline is highly instructive about the rehabilitation of a discredited field, and thus illuminates the contemporary context of debates over the boundaries of what should and should not count as "science," controversies that were much larger, though parallel to, Velikovsky's. Eugenics—a term coined by Francis Galton in 1883, over two decades before the word "gene" itself—was the applied science of heredity to questions of human breeding, whether attempting to breed out deleterious diseases or traits (like Huntington's chorea) or to "improve" the stock of the race (variously understood to mean either a specific race or the human race). Eugenics cropped up in the previous chapter, since one of Lysenko's arguments against genetics was that it went hand in hand with its more suspect cousin, and thus debates over the scientific status of eugenics—although raising different sets of issues than Velikovskianism—formed part of a larger complex, a different front of the pseudoscience wars. A brief exploration of postwar eugenics will thus show us the variegated terrain of demarcation, and the many ways in which another contemporary field tried similar approaches to rehabilitation even though its characterization as a pseudoscience stemmed from different causes.

One of the major differences from Velikovsky was that, by 1945, eugenics already had a history—and a rather checkered one, to say the least. Although the field emerged in Britain, it spread globally in the early twentieth century, achieving particular popularity among activists in the United States and in Germany—who communicated eagerly with one another—both as an elite discipline to expand understanding of human evolution, and as a popular movement to pass legislation regulating or restricting marriage and reproductive choices.[8] In 1920s America, eugenics was wrapped with xenophobia and racism, culminating in the 1927 Supreme Court decision *Buck v. Bell*, which affirmed the constitutionality of involuntary sterilization for eugenic purposes; and in 1930s Germany, the National Socialist party employed eugenic language and policies to aggressively cull the "unfit," and those practices saturated the Final Solution. This horrifying outcome was knowledge run amok and has earned the epithet "pseudoscience" from numerous historians.[9]

And not just historians. In fact, Nazi racial policies and especially the revelations from the camps in the mid-1940s produced a wave of consternation from certain scientists that eugenics was inherently racist and was clearly a pseudoscience blighting the legitimate science of genetics. George Gaylord Simpson, the same man who chastised Velikovsky, noted in 1949 that "eugenics has deservedly been given a bad name by many sober students in recent years because of the prematurity of some eugenical claims and the stupidity of some of its postulates and enthusiasms of what had nearly become a cult. We are also still far too familiar with some of the supposedly eugenical practices of the Nazis and their like."[10] But perhaps the most vocal of the geneticists opposed to racist invocations of their science and especially its eugenical offshoots was Theodosius Dobzhansky, the same Russian émigré who translated Lysenko in an effort to discredit that other abomination upon his discipline. Dobzhansky beat a constant refrain in his publications from before the war through the 1960s about the "prostitution of biology which was employed by the Nazis as justification for their crimes" and "the pseudoscientific foundation" for race prejudice.[11] It was the duty of biologists to speak out, as a friend of Dobzhansky's wrote to him as early as 1936: "It seems obvious that the prestige of genetics in general may suffer considerable harm unless geneticists take the initiative in clearing the situation by frank discussion."[12] Dobzhansky concurred.

It would appear from the views of geneticists and historians that eugenics was discredited after World War II, becoming, according to historian Daniel Kevles, "virtually a dirty word" in the United States.[13] Many histories of the subject stop at war's end, as if eugenics crumpled instantly.[14] Revulsion was omnipresent in the 1950s, to be sure, but so was eugenics, working hard to rehabilitate its image from far greater handicaps than those faced by Velikovsky and his catastrophism. The odd part of this story is that eugenics succeeded in transforming its image from despised pseudoscience to an essential part of population biology and medical genetics. There were active programs in states like North Carolina that called themselves "eugenic" into the 1950s and beyond, and their advocates insisted that "laws providing for voluntary sterilization in democratic countries bear no resemblance to this German experiment," and that "the ethical principles underlying population policy in democratic countries are entirely different from those of Nazi ideology and hold as axiomatic that individual liberty cannot be curtailed except by general consent."[15] Also, the American Eugenics Society (AES)—the linchpin of the eugenics movement in the United States during the "bad old" racist days of the interwar years—held on to its name until as late as 1973![16] (At this point, it became the Society for the Study of Social Biology.) Clearly, if revulsion had set in against the term, many had not heard the news. What follows is a brief history of how the American Eugenics Society, under the leadership of its highly competent secretary, Frederick Henry Osborn, successfully rehabilitated eugenics. As his collaborator Carl Bajema stated in 1971: "Frederick Osborn has probably done more than any other single individual during the past fifty years to bring about the development of scientific knowledge that can be used by societies to make more rational and humane decisions in the realm of eugenics."[17]

A successful financier, Osborn retired from Wall Street at age forty, in 1928, which was surely great timing. He decided to undertake research activities in human evolution and heredity at the American Museum of Natural History (where his uncle, Henry Fairfield Osborn, was curator of paleontology, as well as a leading eugenicist) in the early 1930s, which was just as surely terrible timing, for this was precisely the moment when the Nazi eugenic apparatus became operational. (And it is true that, in 1937, Osborn did praise some of the Nazi laws—although not the anti-Semitism—as forward thinking.[18]) By 1940, when Osborn published the first edition of his major book, *Preface to Eugenics*, he had already distanced his version of eugenics from the authoritarian path in Germany and his racist predecessors at the AES. "Because many early writers misused the word eugenics, its public connotations are quite out of line with the more conservative interpretations of the scientist," he noted. The solution was not to continue the old policies, but to recruit "scientists and trained and informed leaders of public opinion who will be capable of developing the concepts and methods of eugenics into a unified and non-controversial form."[19] By the second edition of his book, in 1951, he was explicit about the break with the racist past, because "science has not produced evidence to support the claim that any nation is racially superior."[20]

In order to present postwar eugenics as anti-racist, Osborn had to engage in a significant effort of historical revision. The first postwar history of the AES, by Maurice Bigelow in 1946, presented a continuous history of the group with no explicit mention of any link with Nazi researchers— clearly a deliberate omission. (The same holds for Osborn's own 1974 history.[21]) To make such an account credible, Osborn had to be subtle. First, he labeled the ideas of the most notorious racists in the AES, leaders such as Madison Grant and Harry Laughlin, "pseudo-scientific work" and minimized their impact on prewar eugenics.[22] Second, Osborn assured later historians—who relied heavily upon his interpretation—that this transition had begun before the war, when "I was made Treasurer and Secretary of the Society, and as I was far the most active member, things fell into my hands, and as the older directors and members retired or resigned in distress over the new eugenics which was emerging, we replaced them with qualified scientific people from the various related disciplines."[23] Third, after the war, Osborn cut his ties with those individuals, such as Pioneer Fund founder Wickliffe Draper, who were eager to continue the "old" eugenics.[24]

Immediately after the war, Osborn was tapped to head negotiations over nuclear arms control with the Soviet Union at the United Nations, a venture that famously got nowhere.[25] By the time he was able to devote his attention to eugenics, around 1950, he had also established clear bona fides in the Lysenko affair, routinely denouncing Michurinism and vocally defending genetics even before the August 1948 session produced a clampdown on Soviet biology.[26] And then he moved to remake the AES into an organization that would present a credible face to contemporary scientists, deemphasizing the prewar attention to social policy. Perhaps the most important structural move was revamping the board of directors of the AES, replacing socialites with scientists. Osborn was particularly concerned to recruit leading geneticists to lend their endorsement to the goals of the AES. The AES was closely linked to the American Society of Human Genetics, founded in 1948, in that five of the first eight presidents of the latter (Lee Dice, C. P. Oliver, C. Nash Herndon, Sheldon Reed, and Franz Kallmann) served simultaneously on the board of the AES. (One of the three who did not was Hermann Muller, an avid defender of eugenics, who thought his explicit participation would damage the cause of eugenics more than it would help.[27]) By the mid-1960s, he could write proudly that "the Board of Directors of the Society is now wholly scientific, and includes geneticists, demographers and anthropologists of the highest competence."[28] Membership was restricted to invitation only, and those invitations were extended exclusively to professional men and women in fields like population genetics or medicine.[29] By 1964 Osborn could declare with satisfaction that the "Society at present is scientific and not propagandistic and we think this is wise until our scientific backing is more completely developed."[30]

Along with a change of personnel, Osborn also spearheaded a change of content. For example, "in order to overcome the injury done to the word eugenics by excesses of propaganda and by Hitler in Germany, we have for some years been devoting ourselves to the scientific development of information in birth trends and heredity."[31] Data collection was important, but the most successful aspect of this rebranding was the transformation of what used to be called "negative eugenics"—the elimination of the unfit from the breeding population, such as by sterilization—into "medical genetics" by promoting the institutionalization of heredity counseling at medical schools to advise prospective parents about the genetic risks to their offspring.[32] Around this same time, Osborn replaced the older *Eugenical News* with the *Eugenics Quarterly* as the organ of the AES, transforming it into a scientific journal that developed a strong subscriber base among medical and scientific institutions.[33]

By 1965 Osborn could confidently write to Alexander Robertson, the executive director of the Milbank Fund (a major funder of population issues), that "the Society moves into the mainstream of scientific investigation."[34] As an indication of just how complete this transformation was, consider ardent anti-racist Dobzhansky. In his 1952 book on heredity and society, coauthored with L. C. Dunn, he had written that "eugenics has often been perverted to a pseudoscientific justification of social inequity and oppression."[35] Upset by this characterization, Osborn wrote to Dunn: "I did, however, very much want you and Dr. Dobzhansky to know the change that has taken place in eugenics in the last 20 years. . . . It is naturally discouraging, therefore, to have such men as you and Dr. Dobzhansky repeating today criticisms of eugenics which would have been proper 30 years ago but which I believe and hope are no longer justified."[36] Dunn immediately consoled Osborn that while they did indeed oppose a certain kind of eugenics, his own views were known to be quite rational.[37]

Osborn had been trying to recruit Dobzhansky into the AES orbit for years. If such a vocal opponent of "bad" eugenics like Dobzhansky would endorse the society, then surely the new age would have dawned. In 1951 Dobzhansky had reviewed the second edition of Osborn's *Preface to Eugenics* very enthusiastically: "What a far cry from the pseudo-science that has so often utilized the name of eugenics!" (His one criticism was that some of Osborn's statements "lend color to the assertions of Lysenkoists that genetics considers genes unchangeable."[38]) In the late 1960s, Dobzhansky even served as the chairman of the board of directors of the AES, wrote a preface for Osborn's *The Future of Human Heredity*, and confidently declared that "eugenics—incorporating the latest social and scientific theories—has entered a new phase. The pioneer of the new eugenics in the United States is Frederick Osborn, whose book *Preface to eugenics* has opened that new phase."[39] Rehabilitation had been accomplished: eugenics was no longer a dirty word.

Nevertheless, the word was not exactly ideal. It might now be a legitimate term among scientists, but among broader circles it still exuded the sulfurous stench of bigotry. For example, in 1965 an outfit named the American Eugenics Party began producing anti–civil rights propaganda, and several board members (including Dobzhansky) wanted the AES (that is, Osborn) to issue a public statement in the *Eugenics Quarterly* distancing the organization from this material. Osborn was more sanguine about the matter. "It is painfully embarrassing to have this kind of crowd using the name eugenics," he conceded to geneticist Lee Dice. "You may remember that some time back there was a pornographic outfit operating under the name of the Eugenics Publishing Company. They died out eventually, and I hope this recent group will too."[40] (There was a similar problem with the radical right-wing International Association for the Advancement of Ethnology and Eugenics, whom Osborn considered "a group of racists who are prostituting science."[41]) Osborn's preferred solution was to ignore them: "Our general feeling is that we might only give them more publicity if we paid any attention to them."[42] The AES did not issue a statement, but the issue of the term endured.

In 1966, for the second time since the end of the war, members proposed changing the society's name to eliminate the potentially distasteful word. As Osborn reported to his London counterpart, K. Hodson, in the prior debate, the board had been "about evenly divided and I think we might have given up the name 'eugenics' if it had not been for old Dr. Bigelow, a great backer of the Society of the old days, who felt violently that the name should be retained."[43] And so the issue lay fallow until 1969. That was the year when the *Eugenics Quarterly* (formerly *Eugenical News*) changed its name once again to *Social Biology* in order to attract more submissions. Changing the name of the society as well came up again at a board meeting and was reserved until the next session. Osborn thought that "here were more in favor of changing the name than of continuing as at present. The argument for changing was that we would get better acceptance and participation by qualified scientists who cannot forget the misuse of the word eugenics in the past and will feel that a normative title is not appropriate to a scientific approach." Although Osborn had agreed with the dissenters when Bigelow had quashed the earlier motion, his "view today is that, in the light of developments in the last ten years, the argument for continuing the name is strengthened. Over the past ten years the strictly scientific operation of the Society has encouraged leading scientists in related fields to join as members and to accept positions as officers and directors, as the letterhead clearly confirms. In the same period, there has been an increasing recognition that scientists have a responsibility with respect to the social application of their work."[44]

But the success of the rehabilitation had hinged crucially on the participation of respected geneticists, and many of them—in particular Dobzhansky—were in favor of eliminating the offensive word.[45] Carl Bajema, one of Osborn's closest collaborators, suggested the "Society for the Study of Eugenics," which was for Osborn "the first time a name has been suggested which I would be really happy about. It seems to me excellent."[46] Few others agreed. At the annual membership meeting in November 1972, it was proposed and approved to change the name of the American Eugenics Society to the Society for the Study of Social Biology. Of 128 ballots cast in person or by proxy, 94 were in favor of the change, with 15 opposed, 1 abstention, and 18 not marking their ballots. Osborn, deferring to the scientists that the AES depended on for legitimacy, acquiesced, reserving the old name so it could not be appropriated for nefarious ends.[47]

The American Eugenics Society was dead—and lived on. For Osborn, who had led the AES for over forty years, the change was hard. "As you will see by this letter head, the name of the Society has been changed in order to bring it more directly in line with its objectives and the wide-ranging activities of its members. The purposes of the Society of course are not changed," he wrote to the Burdick Foundation in 1973. "The change has my warm endorsement, but I can't help feeling some nostalgia about it. In the forty years of my close association with the Society I have seen it change from an apologist for unjustified prejudices to a Society dedicated to re- search and the dissemination of scientific findings."[48] Because, in the end, rehabilitation had worked. The name "eugenics" was not abandoned as a necessary first stage toward sanitization; it was replaced after that process had just about reached completion, as a capstone of sorts. Just because a field had once been designated a pseudoscience, even one so ghastly and dangerous as the abuse of genetics under Hitler, did not mean that it could not attain acceptability among scientists. Resuscitation depended on a host of factors, however, since no two alleged pseudosciences were the same. Yet the transition was possible in principle and in practice, if you played your cards right.

THE PRINCETON SUMMITS

Immanuel Velikovsky almost certainly knew nothing of Frederick Osborn's activities on behalf of the American Eugenics Society, but he managed to deploy many of the same moves in his quest in the 1950s to inch toward the scientific mainstream, and away from the circle of infamy to which he had been decisively condemned. Just as there are a limited number of ways to attempt to break out of a pincer movement on a battlefield, so there was a limited set of tactics that so-called pseudosciences could use to attain legitimacy, and Osborn and Velikovsky seemed to have hit on some of the same ones. These efforts were not necessarily coordinated, and some of them worked at cross-purposes, but they represented serious attempts to force scientists to reevaluate their stance from the *inside*. Only when this approach collapsed in the early 1960s would Velikovsky and his supporters transform his theories into an attack on the establishment.

The first way to rehabilitation was to seek the endorsement of established scientists, and Velikovsky looked for them locally. He and his wife, Elisheva, had moved to Princeton from New York City in 1952 to be closer to their elder daughter (married to a graduate student in physics at the university), and Velikovsky approached various local scientists from their new home at 78 Hartley Avenue. For example, while he was writing *Earth in Upheaval*, his 1955 book on geology, he came to know Harry Hess, the central figure in geology at Princeton University and one of the most important thinkers in the development of plate tectonics.[49] Velikovsky submitted to Hess (and Hess forwarded to the relevant scholarly committee) a list of proposed experiments to be conducted during the International Geophysical Year of 1957–58, in the hopes that a global search would produce confirmation of his scenarios.[50] Why would Hess engage with a heretic in this way? As he wrote to Velikovsky in 1963:

> We are philosophically miles apart because basically we do not accept each other's form of reasoning—logic. . . . I am not about to be converted to your form of reasoning though it certainly has had successes. . . . I do not know of any specific prediction you made that has since been proven to be false. I suspect the merit lies in that you have a good basic background in the natural sciences and you are quite uninhibited by the prejudices and probability taboos which confine the thinking of most of us.
>
> Whether you are right or wrong I believe you deserve a fair hearing.[51]

Until his death in 1969, Hess continued to associate with Velikovsky and made a point of introducing him to graduate students and insisting that these students engage with Velikovsky's geological arguments (to hone their skills in refuting them). Hess also organized the first Cosmos and Chronos Study and Discussion Group in January 1965, an institution that would later proliferate among campus countercultures nationwide, as a forum to discuss Velikovsky's views with faculty and students. (The acrimony of the first session led Hess to demur on further meetings.[52])

As much as Velikovsky valued Hess's support and friendship, this was not the Princetonian whose endorsement he most coveted. As a 1976 book observed: "Velikovsky conspicuously invokes the name of his onetime neighbor, Dr. Albert Einstein, implying, with little real evidence to back him up, that Dr. Einstein approved his ideas."[53] ("Neighbor" is a bit strong. It is a longish walk between Hartley Avenue and Mercer Street.) The relationship with Einstein was absolutely central to Velikovsky's efforts of rehabilitation. Einstein was to play the role that Dobzhansky performed for Osborn.

We should begin by noting a problem of evidence. A number of letters between Einstein and Velikovsky survive, and they seem to embody the most intellectually important of their communications. Yet they are insufficient to draw a complete picture of the relationship because many of their interactions took place in person, and those conversations contextualize the letters. (Besides, their discussions have been so thoroughly reinterpreted by later supporters of Velikovsky that it is difficult to recover the initial context. For example, Elisheva Velikovsky insisted to Velikovsky's associate Frederic Jueneman a year after Velikovsky's death that "Einstein and Velikovsky were *not friends*; to imply otherwise is harmful because Immanuel's opponents try to imply that Einstein devoted his time to Velikovsky's ideas only out of friendship, rather than genuine interest."[54]) We do have a lengthy account of their relationship, in the form of a narrative manuscript that Velikovsky attempted to publish without success in the 1970s, but that version is entirely from his point of view.[55] We do not know Einstein's full attitude, although even in Velikovsky's rendering there are glimpses of Einstein's characteristically mercurial personality. The story that follows, therefore, is an attempt to contextualize this relationship with an eye to Velikovsky's path to rehabilitation.

Einstein and Velikovsky had first met, as noted in chapter 2, during the editing of the *Scripta universitatis* in Berlin. This proved to be the first of several abortive contacts. Velikovsky wrote to Einstein about the latter's correspondence with Freud on pacifism (published as *Why War?*) in 1939, but there is no record of a response.[56] More relevant for this story, Velikovsky and his elder daughter, Shulamit (a student of physics), traveled down to Princeton to consult with Einstein on July 5, 1946, about the general astronomical argument of what would become *Worlds in Collision*. According to Velikovsky's memoir, "Einstein agreed then to look through a part of the manuscript. At that time he advised me to rework my book so as to make it acceptable to physicists and to save what was valuable in it."[57] Einstein's exact words, in a letter sent three days later, were "Thus it is evident to every rational physicist that these catastrophes could have nothing to do with the planet Venus, and that also the Earth's direction of rotation toward the ecliptic could not experience any considerable change without the entire Earth's crust being fully destroyed."[58] He returned the manuscript to Velikovsky and politely declined further discussion. Velikovsky sent Einstein a copy of *Cosmos without Gravitation* a few months later but received no response until July 1950, when Einstein told Velikovsky to stop sending him material, as he would no longer read any of it.[59] Nevertheless, in January 1951 he sent Einstein a copy of *Worlds in Collision* along with the *Scripta* (perhaps as a reminder of their prior acquaintance), and Einstein answered that Velikovsky had "the stuff to thoroughly disprove even the table of multiplication with your historical-philological methods. Of the applause of the laymen, who have a secret grudge against arithmetic, you can be assured."[60] It seemed this discussion was at an end.

And then Velikovsky moved to Princeton, fewer than a hundred yards from Lake Carnegie, the artificial lake that serves Princetonians for sculling and recreation. Einstein, for fresh air and quiet, enjoyed the occasional boat ride on the lake. One day in late August 1952, the two ran into each other and Velikovsky reintroduced himself. Einstein, apparently forgetting their earlier interactions but recalling the recent controversy over *Worlds in Collision*, said (in Velikovsky's recollection), "Ah, you are the man who brought the planets into disorder," in German, and then stopped smiling.[61] Einstein rushed off in a hurry, but Velikovsky took the chance encounter as an opportunity to write a letter. He wrote in English—a rarity; most of their correspondence and conversation had been and would again be conducted in German—that a "physicist cannot prescribe to an historian what he is allowed to find in the past, even if he finds contradiction between the alleged historical facts and our understanding of natural laws."[62] Einstein, seemingly worried that he had been rude, immediately sent a kind letter that clarified the scientific objections to *Worlds in Collision*:

> Against such precise knowledge [as celestial mechanics] come speculations of the sort as you put forward which cannot be considered by one with a knowledge of the subject. Thus your book must appear to the expert as an attempt to deceive the lay public. I must confess that I myself initially also had this impression. Only subsequently was it clear to me that intentional deception was far from your intention.[63]

Velikovsky hoped to turn this into a longer exchange, sending a very long letter on September 10 that insisted, among other things, that "I have not invented new physical forces or new cosmical forces, as cranks usually do; I have also not contradicted any physical law; I came into conflict with a mechanistic theory that completely coincides with a *selected* group of observations; my book is as strange as the fact that the Earth is a magnet."[64] Einstein did not respond, and the interaction stalled again.

The event that triggered their closer relations was a speech that Velikovsky gave at the Graduate Forum at Princeton University, in which he articulated his vision of the universe and made some predictions about phenomena in the solar system. Einstein did not attend, but the "women" of his household, including his trusted secretary, Helen Dukas, did, and they reported back favorably. Velikovsky was permitted to visit Einstein, and, as Einstein reported to a friend on November 7, 1953, after the visit, "I found him very nice."[65] On March 20, 1954, when Velikovsky was invited to Mercer Street, the discussion addressed Velikovsky's scientific theories but did not get anywhere. Then Velikovsky showed Dukas copies of the correspondence to Macmillan, especially that by Harlow Shapley; Dukas told Einstein, and the latter was captivated. From this point onward, Velikovsky would come every few weeks to visit Einstein until his last visit on April 8, 1955, ten days before the physicist's death.

What happened? Why was Einstein, once guarded and hostile, now interested in the man he had dismissed repeatedly in the past? It is impossible to reach a firm answer, but there are some dominant themes. One, not to be underemphasized, was that they could converse in German. Einstein was very attached to his native tongue and never became fully comfortable in English, but after the horrors of the Holocaust felt even more distressed at using this language with a German. Velikovsky, being an émigré Jew and not German to boot, was safe, and one sees this pattern with many of Einstein's associations in Princeton. That Velikovsky was a Zionist and had lived in Palestine also clearly attracted Einstein, although his relationship with the Zionist movement and the State of Israel was complex.[66]

The clearest attraction, however, is the one implied by the story of Dukas and the Macmillan correspondence: Einstein instinctively wanted to defend someone whose ideas had been suppressed by self-appointed censors. The compulsion was partly autobiographical—his theories had been targeted by Nazi ideologues—but also political, drawing from his strongly held liberal views. According to Velikovsky, when Einstein saw the Shapley letters, he cried: "This is worse than Oppenheimer's case."[67] The fate of J. Robert Oppenheimer—the scientific director of Los Alamos and thus responsible for the design of nuclear weapons in World War II, and after the war the director of the Institute for Advanced Study in Princeton—was much on Einstein's mind. Oppenheimer was due in April

1954 before a subcommittee of the Atomic Energy Commission to defend himself against aspersions to his loyalty. In the transcripts of Einstein's phone conversations with Hanna Fantova, the maps librarian at Princeton and a close friend in his final years, he often mentioned the Oppenheimer case—and also, as we shall see, Velikovsky. When Oppenheimer's clearance was revoked, Einstein was incensed about the expansion of McCarthyism, however ambivalent he was about Oppenheimer personally. On the other hand, as angry as he was about Shapley's letters to Macmillan, he had great respect for Shapley and sympathy with his politics, and he wished that Velikovsky "could enjoy the entire episode from the amusing side."[68] The symmetry between Oppenheimer and Velikovsky broke down.

And thus, for all these reasons and more, he was drawn to Velikovsky. Plus, it was interesting to talk about the man's ideas, however mistaken he might think they were. Einstein agreed to read the manuscript of *Earth in Upheaval*. As he reported to Fantova on January 6, 1954: "I am reading Velikovsky's manuscript; he is a professional revolutionary. I consider him gifted, but uncritical. He was a psychoanalyst and has interpreted some dreams completely crazily [*verrückt*]. He is himself a bit crazy [*verrückt*], but is however right when he says that everything goes evenly in nature. He writes interestingly and his books are worth reading."[69] Einstein's view of the manuscript was similarly positive in tone:

> The proof of "sudden" changes (p. 223 to the end) is quite convincing and meritorious. If you had done nothing else but to gather and present in a clear way this mass of evidence, you would have already a considerable merit. Unfortunately, this valuable accomplishment is impaired by the addition of a physical-astronomical theory to which every expert will react with a smile or with anger—according to his temperament; he notices that you know these things only from hearsay—and do not understand them in the real sense, also things that are elementary to him. . . . To the point, I can say in short: catastrophes *yes*, Venus *no*.[70]

But, on July 14, 1954, Einstein told Fantova something different. "I receive manuscripts sent by all possible and impossible people, it is utterly terrible," he said, and then immediately added: "Velikovsky has also sent me his manuscript, and he is certainly gifted, but crazy."[71]

Craziness aside, Einstein continued to receive Velikovsky, and they continued to discuss physics, life, and the reaction to Velikovsky's publications. Velikovsky reported that Einstein said to him on March 11, 1955, after many months of discussing the manuscript of *Earth in Upheaval* and the German translation of *Worlds in Collision* (later found on his desk when he died) that "I think that it is a great error that the scientists do not read your book—there is much that is important in it." A month later, on April 8, during their last meeting, he began with praise—"I have again read in *Worlds in Collision*. It is a book of immeasurable importance, and scientists should read it"—but then added: "But why do you need to change the theory of evolution and the accepted celestial mechanics?"[72] It is precisely this oscillation—between encouraging Velikovsky and then insistent critique of both the Venus scenario and Velikovsky's advocacy of electromagnetism in celestial mechanics—that makes Einstein's opinions hard to decipher.

Those ambiguities were erased from the narrative after the sad news of Einstein's death hit the Velikovsky household. The acknowledgments of *Earth in Upheaval*, for example, gave prominence to Einstein's views of the manuscript and to the closeness of their relationship. Velikovsky also hinted at an imminent conversion of the century's most famous scientist:

> The late Dr. Albert Einstein, during the last eighteen months of his life (November 1953–April 1955), gave me much of his time and thought. He read several of my manuscripts and supplied them with marginal notes. Of *Earth in Upheaval* he read chapters VIII through XII; he made handwritten comments on this and other manuscripts and spent not a few long afternoons and evenings, often till midnight, discussing and debating with me the implications of my theories. In the last weeks of his life he reread *Worlds in Collision* and read also three files of "memoirs" on that book and its reception, and expressed his thoughts in writing. We started at opposite points; the area of disagreement, as reflected in our correspondence, grew ever smaller, and though at his death (our last meeting was nine [sic: ten] days before his passing) there remained clearly defined points of disagreement, his stand then demonstrated the evolution of his opinion in the space of eighteen months.[73]

By 1963, one of Velikovsky's defenders, Livio Stecchini, transmuted Einstein's evident ambiguity into a statement of endorsement: "The newness of the revolution is evinced by the Einstein-Velikovsky correspondence wherein the former soon accepted as tenable the hypothesis of global catastrophes and, though originally quite opposed, at last became sympathetic even to the hypothesis of a recent origin of Venus as a planet."[74]

Thus it was with considerable consternation that Velikovsky opened the July 1955 issue of *Scientific American* and read the account of Einstein's last interview, conducted by I. Bernard Cohen, a historian of science from Harvard. The piece was presented as Cohen's own narrative of his visit to Einstein, rather than a straight interview, and was peppered with direct quotations. As they discussed controversies in science, Cohen reported that Einstein "mentioned a fairly recent and controversial book, of which he had found the non-scientific part—dealing with comparative mythology and folklore—interesting. 'You know,' he said to me, 'it is not a bad book. No, it really isn't a bad book. The only trouble with it is, it is crazy.' This was followed by a loud burst of laughter." Einstein admitted, in his own words, that "there is no objective test" for when an unorthodox theory should be rejected or accepted (certainly his own had been unorthodox enough), but he did lament the way this book had been protested by scientists. Einstein, Cohen reported, "thought that bringing pressure to bear on a publisher to suppress a book was an evil thing to do. Such a book really could not do any harm, and was therefore not really bad. Left to itself, it would have its moment, public interest would die away and that would be the end of it. The author of such a book might be 'crazy' but not 'bad,' just as the book was not 'bad.' Einstein expressed himself on this point with great passion."[75] The word "crazy" set Velikovsky off.

In an angry letter to Cohen, Velikovsky offered a condensed account of his relationship with Einstein and said he simply did not recognize the man presented in this account. He referred to a letter of March 17, where Einstein "made very clear what he thought of my adversaries and their methods of combatting my book," and added that their last meeting had been on April 8, five days *after* Cohen's interview, at which point he was rereading *Worlds in Collision* and "he said some encouraging sentences—demonstrating the evolution of his opinion in the space of 18 months." Where was *this* Einstein in the profile? "Einstein appears from the portion of your interview dealing with me as unkind and cynical—and these features were very far from him. And certainly he was not two-faced."[76]

Velikovsky sought redress. First he wrote to Otto Nathan, the executor of Einstein's estate, and typically very careful about how Einstein's legacy was invoked, asking him to intervene, lambasting Cohen as a member of "that Harvard Group whom Einstein condemned in his marginal notes."[77] (Nathan responded that since Velikovsky was not identified by name in the interview, most readers would not know who was meant, and that it would only do more damage to call attention to it.[78]) The editor of *Scientific Amer-*

ican, Dennis Flanagan, was even less accommodating and echoed many of the criticisms of Velikovsky's public saturation from the heady days of 1950: "I think your books have done incalculable harm to the public understanding of what science is and what scientists do. There is no danger whatever that your arguments will not be heard; on the contrary they have received huge circulation by scientific standards. Thus I feel that we have no further obligation in this matter."[79] (A brief exchange between Nathan and Cohen was published in the September issue, which referenced this passage but largely concerned how Einstein's words should be used after his death. Velikovsky was not mentioned.[80])

To his death, Velikovsky was convinced that Einstein had not said what Cohen reported, or at least he did not mean it in the sense intended. Yet Einstein was rather partial to identifying Velikovsky as "crazy"—almost every mention in the Fantova phone logs finds the two juxtaposed. For a variety of reasons, Einstein clearly enjoyed Velikovsky's company and lamented the way that Velikovsky had been treated, and just as clearly rejected both the Venus mechanism and electromagnetic astronomy. Aside from its intrinsic interest, the importance of the Velikovsky-Einstein relationship lies in how it was deployed as a source of legitimacy. If Einstein could come close to endorsing him, how could ordinary scientists remain aloof?

But endorsements worked both ways. If Velikovsky were to become closely associated with someone considered too "fringy," it risked jeopardizing any effort at rehabilitation. At the 1952 meeting of the American Philosophical Society, when Karl Darrow read Cecilia Payne-Gaposchkin's renewed attack on Velikovsky in a panel on unorthodoxy in science—and, in a piece of coincident casting, Bernard Cohen presented a general historical overview—a botanist-turned-psychologist named Joseph Banks Rhine was also in the audience. In the late 1920s, Rhine moved to Duke University and began experiments in parapsychology, especially extrasensory perception (ESP), testing students with a series of cards to see whether they could do better than chance in guessing the concealed geometric shapes. In the 1930s, he published a series of papers claiming verified instances of ESP, which sparked a vivid debate in psychological journals, and soon J. B. Rhine's work was labeled an instance of "pathological science" and intensively marginalized.[81] Rhine was in Philadelphia for much the same reason as Velikovsky: to hear his work attacked. In the introduction to the symposium, psychologist Edwin Boring explicitly linked them together:

"Velikovsky may yet do us a service by becoming a paradigm for how a scientific belief is not to be induced, and Rhine, if I may here introduce my own opinion on ESP, may become the paradigm for the waste of energy when boundless enthusiasm is directed upon a poorly formulated problem."[82]

In 1963, after his own attempts at rehabilitation had failed, Rhine approached Velikovsky. "It is not likely there are many people better prepared on the basis of experience to appreciate what you have gone through and, to some extent, how you have felt about it," he wrote. "I think it likely there are some of the more general judgments, viewpoints, and lessons, that come out of a life time of the kind we have both lived that can, with some profit and not without some satisfaction, be shared."[83] Rhine was looking for a sympathetic interlocutor; Velikovsky wanted Rhine as a scientific advocate of his own views. "A problem that I would like to discuss with you," he responded, "is the question of a possible inheritance of the traumatic experiences engendered in the great paroxysms of nature on a global scale."[84] Rhine declined, claiming, "I am myself so unorthodox that I need to be careful lest I handicap any good projects like that by giving it my sponsorship." He wanted Velikovsky to come down for a visit to the parapsychology laboratory but insisted that it not happen under his own sponsorship lest it "only multiply the unorthodoxy we share."[85] (In May 1964 Velikovsky did travel to Durham, North Carolina, for what was apparently a very pleasant visit.[86]) This served as Velikovsky's education, however, in the need to monitor one's associates, one that he would take to heart in the mid-1960s.

FROM PAST TO FUTURE

For now, however, Velikovsky pursued his quest for legitimacy within the scholarly community, and historians were the first subgroup he targeted. Even though his entire interpretation of ancient history was rooted in correlating and comparing texts, he recognized that conventional historians simply refused to consider his reconstructed chronology because of his unconventional dating. The science and the history reinforced each other in a vicious circle: he used the ancient evidence to argue for a Venus event, and then he used that event to synchronize his dating of ancient texts. The ancient historians could simply point to the rejection of the Venus mechanism by establishment scientists and thus discredit the revision of the history.[87] But Velikovsky thought his major contribution was in *history*, not astrophysics, and thus he had to solve the dating problem. If Venus was not scientific enough for the historians, perhaps carbon would be.

Radiocarbon dating was developed by Willard Libby and J. R. Arnold at the University of Chicago in 1949, too late for it to be caught in the sweep of Velikovsky's research for *Worlds in Collision*. But as the tool came to be adopted into archaeological research in the 1950s to date artifacts, Velikovsky was captivated by the idea that this new technique would validate his controversial datings, especially for Egypt.[88] Libby's dating technique involved measuring the quantity of carbon-14 (a heavy isotope of carbon, possessing two more neutrons than conventional carbon-12). As nitrogen in the upper atmosphere is bombarded by cosmic rays, it turns into carbon-14, which is then incorporated in plants and consumed by animals, making every living thing slightly radioactive. As long as the organism is alive, it continues to absorb C-14, but once it dies, uptake stops, and then the total amount of carbon begins to decay, acting as a clock. The half-life of carbon-14 (the time it takes for half of the original quantity of the isotope to disappear) is 5,730 years, and by measuring the radioactivity of ancient wooden artifacts and extrapolating back to the original quantity of C-14, one could determine—assuming the sample is large enough and uncontaminated with later carbon inputs—when the tree that furnished the wood was felled. Since Velikovsky rejected the internal philological and historical evidence most historians used to establish chronology, and historians rejected cosmic catastrophes as global synchronizers, he turned to carbon-14.[89]

The first problem was getting an appropriate sample. He needed a hunk of organic matter, such as a piece of wood or cloth that would then be destroyed in the testing, ideally from the dynasties of the New Kingdom in Egypt. But who was going to give the great heretic a valuable archaeological artifact? Robert Pfeiffer, chair of the Semitic Museum at Harvard, wrote letters to Dows Dunham (of the Museum of Fine Arts in Boston) and William Christopher Hayes (curator of Egyptian art at the Metropolitan Museum in New York), asking for samples on Velikovsky's behalf. "The matter should be settled, if possible, once for all," Pfeiffer wrote to Hayes. "It is so vital for all students of ancient history to have all doubts removed that the application of the radio-carbon test seems to be most desirable."[90] Neither Hayes nor Dunham budged.

So Velikovsky mobilized a higher authority. At his last meeting with Einstein, on April 8, 1955, Velikovsky reported news that one of the astronomical predictions (radio noises from Jupiter, discussed below) was confirmed. Einstein was impressed and told Velikovsky that he would write a letter endorsing any test of his theory that Velikovsky desired. Velikovsky could have asked for any astronomical assay imaginable, but he did not; he wanted a *historical* test in the form of a radiocarbon sample. Einstein died within two weeks, not yet having lobbied on Velikovsky's behalf. Helen Dukas, Einstein's secretary, seconded Pfeiffer's request to Hayes in late May:

> During the course of the last eighteen months Professor Einstein had several discussions with Dr. Velikovsky—with whom he had friendly personal relations—about the latter's work. The last such discussion took place on April 8th. In the course of this conversation Professor Einstein said that he would write you and suggest that you should give Dr. Velikovsky an opportunity to have his theory subjected to a radiocarbon test.
>
> As I was present at this discussion I can assure you that Professor Einstein did intend to write that suggestion to you and but for the lateness of the hour the letter to you would have been written then and there.[91]

Again nothing from the Met, but Velikovsky had some success elsewhere, and "three small pieces of wood from the tomb of Tutankhamen were delivered from the Museum of Cairo to the Museum of the University of Pennsylvania. I [Velikovsky] was behind the delivery but was careful that the wood should never pass through my address."[92]

Elizabeth A. Ralph, who performed radiocarbon testing for the physics department at the University of Pennsylvania, analyzed the samples in March 1964. Recall that Velikovsky's dates were six hundred years more recent than stipulated by conventional chronology for the relevant period, and so he expected that the material would be younger by precisely that amount. In this, he was disappointed: the results were only about two hundred years younger than the conventional dating. It was not the significant confirmation he had hoped for, but at least it was in the right direction. Characteristically, Velikovsky took this as a triumph and wrote happily to Ralph that "now it is clear that the conventional dates for this period, too, are by centuries out of conformity with carbon dates whereas the uncertainty of the method is counted only in decades."[93] He wrote to distinguished ancient historians, such as Torgny Säve-Söderbergh in Sweden, to suggest that his reconstruction be given another look and received a blistering refutation of his interpretation of the radiocarbon data.[94] By 1978, however, in *Ramses II and His Time*, the lack of falsification that happened in Ralph's laboratory had spun in the other direction: "The single test which I succeeded in having performed in 1964 brought a result that vindicated the reconstructed version of history."[95]

At the same time, scores of carbon-14 datings of artifacts from the Near East were being published, and they seemed, within experimental error, to confirm the established chronology.[96] What was Velikovsky to do now, when the radiocarbon that was supposed to rehabilitate him had turned traitor? He had several options, and he deployed them all. First, he claimed that most of the radiocarbon tests that had been conducted remained unpublished, precisely because the results they produced were hundreds of years at variance with conventional chronology. These potential confirmations of Velikovsky had been junked as experimental error, he argued, and generated a bias in the published record.[97] But his more persistent tack was to invoke the Venus catastrophe *itself* as the reason why the data was not more consistent with *Ages in Chaos*. During the catastrophe, fire rained down from the heavens, repeatedly, dropping extraterrestrial hydrocarbons on Earth's surface and exposing the planet to higher doses than usual of cosmic rays. "The intrusion of fossil carbon into the atmosphere," he wrote in a memorandum in his files, "must have disturbed the C_{12}-C_{14} balance in the sense of making any organic material that grew and lived after the catastrophe appear in the carbon test as older and belonging to an earlier age." In any event, "organic material of extraterrestrial origin cannot be dated in the same way as we date carbons of organic material of terrestrial origin."[98] So either one could compare material from various regions that *Velikovsky* thought were produced in the same epoch, as opposed to conventional chronology, and see if those dates correlated; or one could test material from the twentieth and twenty-first dynasties, which was far enough from the catastrophes of the eighth and seventh centuries B.C. to be uncontaminated.[99] That would require many more samples, and Velikovsky was not likely to get his hands on them.

History had failed Velikovsky again, and so he was pushed, once more, into staking his claims for rehabilitation on the stage of science. In December 1962, he got a bit of help. Lloyd Motz, astronomer at Columbia University, and Valentine Bargmann, mathematician at Princeton University and once Einstein's assistant at the Institute for Advanced Study, published a short letter in *Science*, the journal of the American Association for the Advancement of Science (AAAS). Bargmann and Motz mentioned two recent planetary discoveries: the high temperature of Venus and the emission of radio noises from Jupiter (a cold body). Both, they contended, had been predicted by the widely reviled Immanuel Velikovsky. "Although we disagree with Velikovsky's theories," they concluded, "we feel impelled to make this statement to establish Velikovsky's priority of prediction of these two points and to urge, in view of these prognostications, that his other conclusions be objectively re-examined."[100] These were distinguished scholars, in the premiere American scientific journal, calling for Velikovsky's rehabilitation. It was, as one historian has put it, "likely the most notable and serious reference to Velikovsky's work in the reputable scientific literature."[101]

Where did this extraordinary letter come from? Velikovsky had consulted with Motz in the 1940s, while living in New York, about some of his physical calculations, and Motz both liked and respected the man, especially his historical research (although he never accepted the physical picture). Beginning in the late 1950s, Velikovsky approached individual scientists to protest his treatment, Bargmann and Motz among them. As Motz recalled in the late 1980s, "Recognizing the justice of his priority claims we therefore wrote a letter."[102] Philip Abelson, the editor of *Science*, was nonplussed by the submission. On the one hand, there were respected scientists; on the other, there was *Velikovsky*. As he wrote to one of Velikovsky's supporters the following year: "I strained my sense of fair play to accept the letter by Bargmann and Motz, and thought that the books were nicely balanced with the rejoinder of [Poul] Anderson."[103] The latter was a parodic letter that suggested that since *Gulliver's Travels* mentioned two moons of Mars before they were discovered, perhaps Motz and Bargmann would agree that Jonathan Swift should be revisited as well. One good prediction or two did not matter, for "while one bad apple spoils the rest, the accidental presence of one or two good apples does not redeem a spoiled barrelfull."[104] But the Anderson letter was a footnote; the Bargmann and Motz letter was news.

The novelty was not that Velikovsky wanted recognition from scientists—this had been part of his campaign for rehabilitation from 1950 onward. Rather, it was precisely at this moment, the turn of the 1960s, that Velikovsky began to talk and write extensively about *predictions*. This was likely another legacy of Einstein, whose career was catapulted to international stardom due to the successful confirmation of general relativity's prediction of the bending of starlight around the sun, measured by Arthur Stanley Eddington in 1919. In his conversations with Velikovsky, Einstein insisted that it was not enough for a theory to be consistent with the evidence, but a theory must also make predictions of novel phenomena not foreseen by prior accounts. Velikovsky took the lesson to heart, and from the mid-1950s until his death he focused intently on the issue of prediction. For example, he wrote one of his supporters in 1967 to "ask three or four historians of science which four or five concrete predictions in science they regard as the most impressive." He would do it himself, he added, but his name was too inflammatory.[105] We do not know of the responses to this query, but we do know Velikovsky's own favorite predictions from the scenario of *Worlds in Collision*.[106]

The first of these has already been mentioned: the emission of radio noises from Jupiter, which had so impressed Einstein. The key citation here was not from *Worlds in Collision*, but from Velikovsky's October 14, 1953, speech to the Princeton Graduate College Forum, which had precipitated his entry into Einstein's household. Toward the conclusion of that speech, he stated: "In Jupiter and its moons we have a system not unlike the solar family. The planet is cold, yet its gases are in motion. It appears probable to me that it sends out radio noises as do the sun and the stars. I suggest that this be investigated."[107] Velikovsky did ask Einstein, in 1954, to push for measurements of Jupiter's radio spectrum, but the physicist did not act on it. And then, in April 1955, Bernard Burke and Kenneth Franklin of the Carnegie Institution announced that they had by chance detected strong radio signals coming from Jupiter. Burke and Franklin, when contacted by Velikovsky's editor at Doubleday about the confirmation, were tight-lipped. "It is not surprising that an occasional near miss should be found in the large number of wild speculations that Dr. Velikovsky has produced," they responded, "but such a coincidence could never be regarded as a true prediction."[108]

Jupiter was a bonus for Velikovsky; the real issue was Venus. Velikovsky proposed what he called three "crucial tests of the concept," only the first of which was mentioned by Bargmann and Motz: that Venus must be very hot; that it should be wrapped in hydrocarbon (or maybe carbohydrate) gases and dust; and that it must demonstrate, by its motion, that it had been disrupted by collisions, such as the fact that it exhibits retrograde rotation (Venus rotates around its axis in the opposite direction from other planets).[109] The first and the third of these claims were undoubtedly true. The issue with Venus's heat was the mechanism: for Velikovsky, it stemmed from the planet's incandescence, still glowing from the tremendous energy of its ejection from Jupiter; while a series of other mechanisms were proposed by astronomers, including Carl Sagan's runaway greenhouse effect.

But the hydrocarbons—well, here was a real controversy. The tremendous advances of the space age meant that probes could fly by, even land on, the planet, and sample the atmosphere directly. (This would also be a way to test the greenhouse effect.) Given the significance of hydrocarbons in Velikovsky's scenario—they made up the tail of the comet and ignited upon contact with Earth's oxygen, producing the epic conflagrations and also depositing much of our planet's petroleum—this seemed truly a crucial test. Without hydrocarbons, the entire Velikovskian mechanism would be in disrepute, a fact recognized by supporters and detractors.[110]

Eric Larrabee, the author of that first article in *Harper's* that set the controversy off, entered the fray again. In 1963 he produced another *Harper's* piece that claimed that when *Mariner II*, an American Venus probe, passed within twenty-two thousand miles of the planet in December 1962, it showed a surface temperature of about 800°C, and also that the planet was enveloped in fifteen-mile thick hydrocarbon clouds. "[Velikovsky] stands or falls by the evidence. There is no appeal here to the esoteric or occult, or to the antiscientific attitudes which make light of fact," Larrabee wrote. "*Worlds in Collision* is filled with statements of fact: about what the ancient peoples said happened to them, as opposed to what we have come to assume they meant by it; about ancient calendars and clocks and astronomical observations; about the craters of the Moon, the surface and atmosphere of Mars, and—as noted above—the gases and thermal balance of Venus."[111] So were there hydrocarbons after all? Indeed, there was an announcement that hydrocarbons had been discovered, but Carl Sagan would later argue that this was a slip of the tongue at a press conference and not an actual scientific finding: "Neither Mariner 2 nor any subsequent investigation of the Venus atmosphere has found evidence for hydrocarbons or carbohydrates, in gas, liquid, or solid phase."[112] The error was actually more interesting than just a misstatement. As Lewis Kaplan of the Jet Propulsion Laboratory at Caltech (the source of the press conference) wrote to physicist Julian Bigelow from the Institute for Advanced Study in Princeton (who occasionally wrote letters to scientists when Velikovsky dared not), the problem was a worry about how to announce the presence of organic compounds that *had* been found in the atmosphere. The word "organic" implied life to the layperson, and that might spark more trouble than it was worth. "The word 'hydrocarbon' was a mistake," Kaplan explained, "and was used only to avoid the use of 'organic compounds' for obvious reasons. The reaction to even 'hydrocarbon' was much too violent."[113]

Indeed it was, so much so that for over a decade *Mariner II* continued to be cited as evidence of hydrocarbons, and hence Velikovsky's correctness. In the meantime, scientists generally remained detached. The lone mainstream exception demonstrates how much Velikovsky's project to rehabilitate himself had failed. In March 1969, *Science* published an article by William Plummer entitled "Venus Clouds: Test for Hydrocarbons," which documented measurements of Venus's atmosphere in the infrared spectrum between 2.1 and 2.6 microns (but concentrating on 2.3–2.5 microns), the range where one would expect to locate evidence of hydrocarbons—and found nothing. "The presence of condensed hydrocarbons in the clouds of Venus," Plummer concluded, "a prediction regarded by Velikovsky as a crucial test of his concept of the development of the solar system, is not supported by the spectrophotometric evidence."[114] Velikovsky's fans wrote him for his response as soon as they became aware of the article (often with some delay).[115] That was coming, he said. But first he wrote a letter to Plummer "to express to you my appreciation: Whatever the results of the present experiments or future finds, you have subjected one of my conclusions to a test—and this is not a common experience to me."[116] And soon enough Velikovsky's rebuttal appeared (but not in *Science*, despite Velikovsky's best efforts), and he redefined what counted as a "crucial test": "Therefore, the presence in our time of hydrocarbons, even in lower strata, could not be construed as a crucial test. . . . The high, near-incandescent heat of Venus . . . constitutes a crucial test."[117]

Velikovsky dealt with his disappointment in not being confirmed by changing the target. This kind of argumentation—slippery though it is—is not exceptional even in mainstream science. The issue with hydrocarbons would fester for many years, but the publication of Plummer's piece provides us a deeper indication of the fate of Velikovsky's efforts at rehabilitation. Clark Whelton, a Velikovsky supporter associated with the *Village Voice*, asked Plummer in June 1969 why he had bothered to look for hydrocarbons. Plummer answered that he was actually engaged in a survey of molecules in the Venusian atmosphere and published the article on hydrocarbons separately only because of the controversy over Velikovsky. In fact, the original submission bore the title "Velikovsky and the Venus Clouds":

The comments my article received were a general acceptance for publication, but only after "major revision." The reviewer basically attacked Velikovsky, not my report. He declared that, "The works of Velikovsky are a mixture of myth, legends, statements out of context, and wild speculation—all presented with absolute authority. The objective application of physical laws and processes play no part in his conclusions. In short, Velikovsky preaches utter nonsense, and any similarity of his conclusions with the real world is purely coincidental. No scientist should give Velikovsky the semblance of credibility by using his predictions as a motivation for serious inquiry." The reviewer then went on to acknowledge that my experimental work was new, had merit, and deserved publication, but he demanded the elimination of all reference to Velikovsky.

Such purging was not acceptable to me for several reasons, one being that the test for hydrocarbons had been reported separately *because* of the Velikovsky affair. . . . The reviewer who wished to purge my work of Velikovsky's name was insulting the scientific method, since science must not become so dogmatic that its powers of experimental test are withheld from any plausible hypothesis, whatever its source. . . . I compromised only by changing the title to something less controversial.

As for the rebuttal, Plummer seemed unworried: "The case against hydrocarbon clouds on Venus is actually stronger than I have illustrated, since the spectra shown in my article range only between 2.1 and 2.6 microns in wavelength. Hydrocarbon features do not show in *other* parts of the spectrum either, and condensable hydrocarbon gases have not been detected in the Venus atmosphere."[118]

A PARTING OF WAYS

Velikovsky at first courted public opinion with the sensational publication of *Worlds in Collision*, but he deeply wished to be treated as a bona fide scholar. He spent most of the 1950s and the early 1960s attempting to transform his reputation into that of a researcher with interesting hypotheses that needed to be tested and, if possible, refuted. (The fact that he did not think refutation likely is immaterial; most scientists do not believe that their hard-won hypotheses are worthless.) But, unlike Frederick Osborn's American Eugenics Society, Velikovsky failed. He took almost all the same actions as Osborn: soliciting endorsements from eminent scientists (and who was more eminent than Albert Einstein?), attempting to establish research programs to legitimate his theories, and striving to publish in respectable journals. As significant as the failure is the fact that seeking rehabilitation with mainstream scientists, not stoking populist fervor, was Velikovsky's first instinct.

Occasionally, he did get high-level backing for his claims as a legitimate scholar, but not from scientists. For example, Mark Casady, a fan of Velikovsky's who also happened to be a staffer for Congressman Thomas M. Rees of California, persuaded Rees to write to Thomas O. Paine, the director of the National Aeronautics and Space Administration (NASA), to request certain tests that would confirm or refute Velikovsky once and for all.[119] Paine was careful in his responses but demurred on altering the mission of the space program.

The resistance Velikovsky met from academics was more forceful and not easily overcome. As an illustrative final case, consider John Holbrook, in the late 1960s an architect at Yale University, and his attempts to obtain for Velikovsky some kind of affiliation with that institution. He approached Derek J. de Solla Price, the chair of the university's new department of the history of science and medicine. Price's response: "No dice!" Holbrook reported to Velikovsky:

> At this point I asked him what was unscientific about seeing anomalies in accepted scientific theories, exhaustively examining the empirical evidence upon which those theories are supposedly based, constructing new theories which will account for all the evidence, deducing the logical consequences of the new theories, making predictions on the basis of those deductions, and finally requesting that men who are in a position to do so conduct experiments which will verify the validity of those predictions, claiming that such experiments will represent crucial tests of the validity of the new theories themselves, all of which you had done.

Price did not answer that question, but instead asked "why I was attempting to assist you? You had chosen the role of the martyr. You should be allowed to play out that role until the end. Would I have withheld the nails from Christ's limbs when he was hung from the cross? To have done so would have denied him the full measure of his martyrdom. To assist you would deprive you of the very same."[120] Price simply refused to treat Velikovsky in terms of science—there was no answering Holbrook's question. Rehabilitation had utterly failed. And if Velikovsky was not to find allies among the worlds of scholarship and science, he would look for them elsewhere.

5 · Skirmishes on the Edge of Creation

Immanuel Velikovsky was always unwilling to compromise on his claims. He believed that the scenario presented in *Worlds in Collision* was accurate as it stood and that establishment science would eventually have to come to terms with the force of the historical evidence he presented. Yet by the mid-1960s it was already clear, even to him, that this was not going to happen anytime soon. Scientists no longer belittled him as much as they had in 1950, during the controversy over that book's publication, but they replaced it with a kind of frosty isolation. Simply put, he was being ignored.

But even once the establishment demonized something as pseudoscience, there could be a path to orthodoxy of a certain sort. If Velikovsky was now to set up long-term camp on the fringes of science—and there was little chance in the immediate future of being brought in from the cold—then he ran the potentially great risk of seeing his doctrines spin off into absurdist emendations as his growing base of fans and quite vocal supporters wielded his catastrophism in their own battles. He certainly was not going to let that happen, and thus he had to assert his authority to keep the doctrines pure while still insisting on their scientific status. This problem confronts everyone rejected by mainstream researchers.

Consider the case of Henry Morris. In 1961 he published, with John C. Whitcomb Jr., one of postwar America's most culturally significant works about the natural world. It was read by hundreds of thousands, spawned its own research institutes, and remains absolutely rejected by every mainstream biologist and geologist. The book was *The Genesis Flood*, and though it did not spark the instant conflagration of Velikovsky's *Worlds in Collision*, its controversy has burned longer and harder.

One aspect of the method of *The Genesis Flood* bears a direct affinity to

Velikovsky's approach to ancient texts: Morris and Whitcomb argued that "the great Deluge [was] inscribed in the records of the Bible and in the legends of early peoples all over the world."[1] But they were not terribly interested in exploring the myths of the globe. What they wanted was two-fold: first, to attack mainstream geology as beholden to a false assumption of uniformitarianism (that the same forces that shape the surface of Earth today have been, although perhaps to a different degree, the same forces that have always shaped it) and contrast it with an alternative catastrophist picture; and, second, to articulate the specific form of the single great catastrophe that shaped the globe—the Noachian Deluge described in books 6 to 9 of Genesis. This flood explained the layering of fossils in the geologic column, the forms of faults, and mountain building consistent with a young-Earth creationism that required only six to eight thousand years from Adam to the present. The authors did not claim they could prove catastrophism over uniformity, since the "assumptions of historical continuity and scientific naturalism are no more susceptible of genuine scientific *proof* than are Biblical catastrophism and supernaturalism."[2] In short, "neither procedure is scientific, since we are not dealing with present and reproducible phenomena. Both approaches are matters of faith. It is not a scientific decision at all, but a spiritual one."[3] They then cited a litany of geological sources to demonstrate that Earth bore traces of a tremendous hydrologic disaster and to refute both geologists who relied on radioactive dating and conventional arguments, on the one hand, and theologians who contended that the Noachian Deluge was purely a local phenomenon, on the other.

The Genesis Flood relaunched the theory of flood geology into the evangelical mainstream from the state of dormancy in which it lay since the 1925 Scopes trial, and it remains the orthodoxy for scientific creationism (synonymous with "creation science"), surpassing its competitors to become "unquestionably the most influential twentieth-century treatment from any perspective"—at least until the advent of intelligent design (ID), a non-biblical variant of the argument from design that came to dominate the movement's intellectual circles by the late 1990s.[4] Until his death in 2006, Morris was, according to one scholar of the movement, "undoubtedly the most influential leader of modern U.S. creationism" and has been called by another "the founder, patriarch, architect, and chief proselyte of the modern scientific creationism movement."[5] The holder of a PhD in hydraulic engineering from the University of Minnesota and the former chair of the Department of Civil Engineering at Virginia Tech, Morris moved to

California in 1970 at the request of an independent Baptist pastor from San Diego named Tim LaHaye. (LaHaye is much more famous as coauthor of the sensationally best-selling Left Behind series of novels, which chronicle the warfare and strife during the Last Days that follow the Rapture of devout Christians.) Together, they organized Christian Heritage College, a new fundamentalist school where Morris headed up a creationist institute with a full-time scientific staff, research facilities, and an ability to grant graduate degrees. By 1972 this was spun off as the Institute for Creation Research (ICR), which remained "the most influential and prestigious creationist organization" through the 1980s.[6] By the early part of that decade, the ICR had about two dozen PhDs working for it and a sizable budget for research.[7] Unacknowledged by the biological and geological community, but supplied with ample funds and an organizational network from evangelical churches, Morris had turned flood geology into its own establishment, and he enforced his own orthodoxy.

Morris's variant of creation science occupied a central battleground in the pseudoscience wars. Scientific creationism, more than perhaps any other doctrine, has been persistently tagged as a pseudoscience, vilified as crackpot knowledge, and forcefully rejected by the scientific establishment.[8] Even liberal theologians and devout geologists have insisted that flood geology occupies the heart of the enemy camp, "detracting from the gospel of Jesus Christ by adding to it the human foolishness of pseudoscience."[9] Historians and philosophers have lavished tremendous attention on creation science, especially on the quest to teach it in the public schools alongside evolutionary biology, where the label of "pseudoscience" meant that Morris's theories could be excluded from the curriculum. The literature spawned by the sparring between evolution and anti-evolution spans many shelves in any research library—and shows no sign of abating.[10]

One topic that rarely comes up in this commotion is the role of Velikovsky. Insofar as creationism and Velikovsky are mentioned by scholars in tandem, it is to demonstrate that both are "fringe" movements that preach absurd stories about Earth's history.[11] While it would be too much to say that Velikovsky played a central role in the history of creationism, it is equally mistaken to discount the influence both he and the controversy over *Worlds in Collision* exerted on the genesis of Morris's brand of scientific creationism. In what follows, I trace three specific links between Velikovsky and creationism, each of which illuminates what counts for legitimacy on the fringe: the close connections between Velikovsky and George McCready Price, the initial architect of flood geology and the intellectual progenitor of Henry Morris's blockbuster; the role that Velikovsky's encounter with the scientific establishment played as a cautionary tale for flood geologists; and the rise of a Velikovskian strand within flood geology itself, which Morris swiftly purged. All these linkages were suppressed by both the Velikovskians and the creationists, which will bring us back to a broader point: when it is impossible to gain legitimacy from establishment scientists, it becomes absolutely vital to maintain authority in your own camp. As Velikovsky began to attract not just readers and admirers, but also acolytes and disciples, the question of orthodoxy would loom large for him as well.

THE PRICE OF UPHEAVAL

In 1955, just five years after the explosive publication of *Worlds in Collision* and three years after his foray into ancient history in *Ages in Chaos*, Immanuel Velikovsky published what seemed on the surface to be the most un-Velikovskian of books—*Earth in Upheaval*. This book contained, as he claimed in the introduction, no sources from the literary heritage of humanity: "I present here some pages from the book of nature. I have excluded from them all references to ancient literature, traditions, and folklore; and this I have done with intent, so that careless critics cannot decry the entire work as 'tales and legends.' Stones and bones are the only witnesses. Mute as they are, they will testify clearly and unequivocally."[12] What followed were pages upon pages of geological anomalies. "The finding of warm-climate animals and plants in polar regions, coral and palms in the Arctic Circle, presents these alternatives: either these animals and plants lived there at some time in the past or they were brought there by tidal waves," related Velikovsky. "But in both cases one thing is apparent: such changes could not have occurred unless the terrestrial globe veered from its path, either because of a disturbance in the speed of rotation or because of a shift in the astronomical or geographical position of the terrestrial axis."[13] In this presentation—the insistent layering of footnotes and snippets, evidence culled from all over the planet and presented as an indictment of conventional scientific assumptions—we see the hallmarks of the Velikovskian approach, despite the absence of mythological material.

The book is no less riveting than *Worlds in Collision*. "One after the other, scenes of upheaval and devastation have presented themselves to explorers, and almost every new cave opened, mountain thrust explored, undersea canyon investigated, has consistently disclosed the same picture of violence and desolation," Velikovsky wrote. "Under the weight of this evidence two great theories of the nineteenth century have become more and more strained: the theory of uniformity and the theory of evolution built upon it."[14] Velikovsky also dismissed repeated ice ages and the theory of continental drift (then just starting to gain acceptance in the geological community) as explanations of apparent ruptures on Earth's surface and the movement of large boulders; both struck him as uniformitarian.[15] For, in language later echoed by Whitcomb and Morris—although the authors would disagree about the timing of the catastrophic event as being either in Genesis or Exodus—Velikovsky insisted that the principle of uniformity had no place in geology: "Rather than a principle in science, it is a statute of faith."[16] Reprising an argument that had proven highly controversial in 1950, he also claimed that the sudden deaths of Siberian mammoths by freezing and the existence of dead coral at the poles could be explained by a deflection of Earth's axis of rotation. However, the mere occurrence of huge catastrophes did not necessarily imply that the narrative presented in *Worlds in Collision* and *Ages in Chaos* was correct. Since Velikovsky made a point here of not basing his case on ancient testimony, he had no way of dating these catastrophes, either absolutely or relative to each other. Yet he maintained that they took place in the period of 1500–600 B.C., during the hypothesized near collisions of Earth with Venus and Mars.

If Velikovsky were correct, and if these anomalies in Earth's geology indicated that large catastrophes had taken place, then the principle of uniformity would indeed be under severe stress.[17] But Velikovsky's point was not just to establish the reality of catastrophes, but also their relative *recency*. If the mammoths had been suddenly wiped out around three thousand years ago, that had implications for the time scale in which evolution, particularly Darwinian natural selection, could play out. Velikovsky tried to steer clear of young-Earth or old-Earth controversies such as those that had been rocking the creationist community for decades. "I do not see why to a truly religious mind a small and short-lived universe is a better proof of its having been devised by an absolute intelligence," he opined. "Neither do I see how by removing many unsolved problems in geology to very remote ages we contribute to their solution or elucidate their enigmatic nature."[18] Flood geology used the Noachian Deluge to explain fossil data and structural oddities in one fell swoop, thus eliminating the possibility of Darwinian evolution and championing a young Earth; for Velikovsky Darwinian evolution also had to be explained away, but without relying on special creation. Instead, he seized upon the recent findings of radiation genetics (scientists were worrying about the effects of atmospheric radio-activity generated in atomic-bomb tests) to argue that rapid mutations and accelerated development took place as the near approaches of celestial bodies exposed Earth to tremendous doses of cosmic rays, a process he called "cataclysmic evolution."[19]

This was a large argument, and—unlike in *Worlds in Collision*—Velikov-sky conceded that he had received a lot of help along the way, including some from scientists (whom he exempted from any responsibility for the claims in the book). The biggest name in his acknowledgments was Al-bert Einstein, who had read the manuscript, and his presence in *Earth in Upheaval* drew the greatest attention from commentators, as discussed in the preceding chapter. But another name lurked there that had far greater implications for both the form and the content of Velikovsky's argument, a "geologist in California, [who] read an early draft of various chapters of this work. Between this octogenarian, author of several books on geology written from the fundamentalist point of view, and myself, there are some points of agreement and as many of disagreement."[20] The man was George McCready Price.[21]

This is a significant name. Price was born in New Brunswick, Canada, in 1870, and he was twelve when his mother joined the Seventh-Day Ad-ventist church. He attended some courses at Battle Creek College, worked as a bookseller, and eventually undertook a teacher training course, begin-ning his career as a teacher and concluding all his formal education in any area, including science. Inspired by the visions of Ellen Gould White, the prophetic founder of Seventh-Day Adventism, he abandoned his doubts about the recent creation of Earth and dedicated his life to espousing strict creationism: a young Earth, with creation in six literal days. At the time, this was decidedly a minority position among American Christians. He wrote several works articulating his theory that Earth's geological features were products of the Noachian Deluge, and in 1923 published a 726-page college textbook entitled *The New Geology*, which sold over 15,000 copies and solidified his position as the leading figure of flood geology. (He was cited in the Scopes trial by William Jennings Bryan as his authority in ge-ology, and his work illuminates almost every argument of Whitcomb and Morris's *Genesis Flood*.[22]) California rancher Dudley J. Whitney teamed up with Price and flood geologist Byron Nelson to establish the Religion and Science Association in 1935, and after that Price organized a nucleus of Adventists in the Los Angeles area into the Society for the Study of Creation, the Deluge, and Related Science (also known as the Deluge Geology Society). The unpopularity of his denomination left Price somewhat marginalized, yet nevertheless through the 1940s and into the 1950s he remained a major figure in the biblical literalist faction of the creationist community.[23]

Immanuel Velikovsky's first letter reached Price in Loma Linda, California, in September 1951, as he started work on the manuscript of *Earth in Upheaval*. He wrote to Price that he had "profited from reading your textbook of Geology"—that is, *The New Geology*—and he hoped that Price might read a segment of his work-in-progress.[24] Price responded a month later with a rather long letter, declaring that he was "greatly interested in your statement of your literary plans, in following up your amazing impact on American thought in *Worlds in Collision*. And of course I shall be glad to look over the MS." As for Velikovsky's views about the Exodus and catastrophe, those needed no introduction: "As the saying goes, everybody and his brother-in-law had to make at least a bowing acquaintance with your *Worlds in Collision*, when it first appeared, if he wanted to pass as intelligent. My son (living in Los Angeles) sent me a copy; and I was intrigued at the boldness with which you knocked down one after another many of the pet idols of the Occidental world." He also suggested that Velikovsky read other creationists like Alfred Rehwinkel and concluded with a parting shot against uniformitarianism.[25] Here the two could concur. "I agree with you what you say in your letter on the basic problems of geology," Velikovsky responded. "In my book I shall attempt to demonstrate that there were global catastrophes in historical times."[26]

Velikovsky sent Price roughly half of the manuscript of *Earth in Upheaval* on May 29, 1952, the same year he published *Ages in Chaos* (which Price read and quite liked).[27] Within a week, Price had sped through the manuscript and sent Velikovsky a host of comments, corrections, and suggestions. "While I have not always been able to agree with some of the details as to how the thing happened," he wrote, "I have admired the handsome way in which you have demolished Charles Lyell as well as the other Charles, who has been made a demi-god by almost all the civilized world." He enjoyed Velikovsky's attack on the ice ages (a notion that Price preferred to leave in scare quotes to "show our disapproval") and also his dismissal of the revolutionary ideas of Alfred Wegener: "I think you have ticked off the nonsense about continental drift in fine shape. I think you have more patience with it than I could have." In particular, he applauded Velikovsky's penchant for citing older geological works—especially early nineteenth-century catastrophist ones—in preference to more recent studies, which were corrupted by uniformitarian assumptions. Yet Velikovsky surprised Price when quoting material from more recent studies, such as on the geology of Arabia: "All this has been developed (so far as I know) since I did my work on geology. I have not attempted to keep up with modern advances in geology." This was understandable: he was approaching eighty-two years of age and thus "can be only an interested spectator of the procession. But I am still keenly interested in the whole subject; and am glad to see some one else picking up these subjects for further publication."[28] When he received the published *Earth in Upheaval*, Price was eighty-five and ecstatic. "I hope it jolts the self-complacency of the high priests of geology, who have fastened a rigid creed on the students in our colleges and universities, a creed that smothers them like a deadly lewisite," he wrote, using a simile of World War I vintage. "Scarcely a breath of free inquiry in geology has blown across England and America in a hundred years. I hope your book may breathe upon the dead bones of this important science, so that these dead bones may live."[29]

But as much as Price seemed to admire the book when typing his missives to Velikovsky, he took very few of the arguments to heart. For example, in 1954 Price published a slim volume entitled *The Story of the Fossils*, in which he reiterated many of his arguments against uniformitarianism, the dating of fossils, and the tremendous length of time that geologists ascribed to Earth's history. While citing many of the same examples of catastrophic change—such as Velikovsky's favorite about the sudden extinction of the Siberian mammoths—he ridiculed his correspondent's explanation of such events through the tilting of Earth's axis: "This is regarded as scientific nonsense. I do not know of any competent student of these subjects who believes that the poles have ever been situated anywhere else than they are at present."[30] Instead, Price held firm, as always: "It seems almost certain that this sudden change in the earth's climate was part of the cosmic change which the Bible describes in the sixth to ninth chapters of Genesis."[31] A suspicious eye was the way to approach geological data, much like a "coroner holding an inquest," a favorite metaphor with him.[32]

This rejection of one of Velikovsky's central explanatory devices did not prevent Price from endorsing *Earth in Upheaval* when Velikovsky asked. Price "dashed off a few lines which may serve your purpose."[33] Much of the endorsement—down to the coroner metaphor—was drawn from their earlier correspondence, and it ended with a strongly creationist slant:

There is no uniformitarian nonsense in *"Earth in Upheaval."* The author's first book, *"Worlds in Collision,"* created so much controversy that he promised to use "only stones and bones as witnesses" in this present one. Most of the facts he lists have come to light since Lyell's day. Also Velikovsky shows conclusively that Darwin never could have had a hearing if Lyell had not prepared the way and conditioned the world to receive him. Thus if we in modern times want to persist in thinking all these problems through to ultimate truth, we must first deal with Lyell's geology. And when we get straight in the geology of the first Charles, we will have no trouble with the biology of the second.

"*Earth in Upheaval*" is one of the most thought-provoking books of modern times.[34]

Velikovsky suggested that Price send this to some supportive magazines in London and San Francisco, but it is unclear whether the sage of Loma Linda complied.[35]

HOW FLOOD GEOLOGY LOST ITS VELIKOVSKIANISM

After the 1925 trial of schoolteacher John Scopes for contravention of Tennessee's anti-evolution statute—a trial that, it bears recalling, Scopes *lost*, his conviction being overturned later—evolution and creationism entered a kind of stalemate, with anti-evolution laws remaining on the books but generally unenforced.[36] Creationist theories were not taught in the classrooms, but often neither was evolution. Although Price continued to proselytize his ideas, flood geology's prominence in and after that trial elicited criticism from within the influential American Scientific Affiliation (ASA), a Christian organization intended to reconcile science and religion. Notably, in 1950 J. Laurence Kulp, a geochemist from Columbia University, published a prominent attack in the ASA journal on Price's geological notions. "This paper has been negative in character," he noted in conclusion, "because it is believed that this unscientific theory of flood geology has done and will do considerable harm to the strong propagation of the gospel among educated people."[37]

Kulp's writings circulated among those Christians willing to endorse theistic evolution, or perhaps some form of old-Earth creationism, but they did not end the debate over flood geology. If anything, Kulp bifurcated the Christian community into a set that opposed Price's notions and a growing cohort that continued to develop them. In 1951, for example, Alfred Rehwinkel, a prominent Lutheran theologian at Concordia Seminary in St. Louis, published *The Flood in the Light of the Bible, Geology, and Archaeology*, a passionate defense of biblical inerrancy, including the Flood.[38] The book impressed Price enough for him to recommend it to Velikovsky. But Rehwinkel published on the Deluge after Velikovsky's 1950 *Worlds in Collision*, which proved a rather important difference.

At least it did for Henry Morris. Morris came to flood geology in 1943 through reading Price. When he subsequently went to the University of Minnesota to obtain a doctorate in engineering, he specialized in hydraulics and geology so that he could develop Price's ideas further. And then, at an ASA meeting in 1953, he met theology student John C. Whitcomb Jr., and the two began to discuss their mutual interest in flood geology.[39] In 1957 Whitcomb and Morris began collaborating on a book blending Whitcomb's theological expertise and Morris's specialty in hydraulics. The timing was propitious. On October 4, 1957, the Soviet Union shocked the American population—not least the scientific community—by launching *Sputnik*, the first artificial satellite. In the ensuing transformations of science policy, curricula across the nation were overhauled and standardized from the ground up, in hopes of producing a generation of Americans up to the Soviet technological challenge. Biology, of course, was not exempted, and the Biological Sciences Curriculum Study (BSCS) reintroduced evolution into the American schools, breaking the fragile truce that had lasted since Scopes. These reforms not only restructured American biology education; they galvanized and transformed creationism.[40]

Morris and Whitcomb began exchanging drafts of chapters of their proposed book in 1957. They spent roughly two years finishing the manuscript and another year in significant revisions before turning it over to the press for its publication in 1961.[41] While the central influence on *The Genesis Flood* was clearly George McCready Price, especially his mechanism of the Deluge to explain the geologic column, Immanuel Velikovsky lurked in the background. When Morris received Whitcomb's first draft chapters, he was horrified to see how heavily—and explicitly—Whitcomb relied on Price and Velikovsky as authorities for geological anomalies. In a letter of October 7, 1957, Morris chastised Whitcomb for the misstep: "Price and Velikovsky are both considered by scientists generally as crackpots, although no one ever takes the trouble to answer their arguments save by ridicule and summary dismissal."[42] Whitcomb took the lesson to heart, for on January 24, 1959, while commenting on a significantly fuller draft, he noticed that Morris had also sprinkled Velikovsky citations over the text: "Even the references to Velikovsky should be thought through carefully," he shot back, "because his name, like that of G. M. Price, waves a red flag immediately before some people's eyes."[43] Yet, for all their care, they never managed to excise Velikovsky entirely. A careful reading of *The Genesis Flood* finds two citations to Velikovsky in the footnotes, both to *Earth in Upheaval*. The second is a block quotation of Velikovsky's prose in the middle of their text, without direct attribution to the arch-heretic.[44] (For what it is worth, Velikovsky returned the favor, writing to one of his admirers who recommended he read *The Genesis Flood* that he was not impressed "because the intent is so childishly fundamentalist, the book has no scientific value and certainly no impact on sciences."[45])

The reason Whitcomb and Morris were so worried about Velikovsky is clear enough. At first, according to his own account, Morris did not intend for the book to be specifically Christian; he considered it sound science, appropriate for a scientific publisher. Reflecting on this from the 1980s, Morris recalled a different kind of centrality of Velikovsky for his book:

> Dr. [Thomas] Barnes approached the 15 leading high school textbook publishers, told them all about the manuscript, and was expecting them all to compete for the contract to publish it.
>
> *But not one of them would even look at the manuscript!* They said (no doubt remembering the infamous "Velikovsky affair" of the early fifties) that all of their textbooks would be boycotted if they would dare to publish a creationist book. Consequently, Dr. Barnes and I finally turned to a Christian publishing house.[46]

Velikovsky was a cautionary tale, an indication that the scientific establishment was brutally suppressive.[47] His role in framing their catastrophic geology was elided.

The Genesis Flood transformed Whitcomb's and Morris's lives. They became "highly sought-after celebrities, famous among fundamentalists as the Davids who slew the Goliath of evolution."[48] Having introduced a new generation—especially of non-Adventists—to Pricean ideas, the two planned a prequel to their blockbuster hit that would cover the actual creation, but lecturing and administrative duties sapped their time. Whitcomb remained at Grace Theological Seminary through the 1980s, where he presided over the Spanish World Gospel Mission. Morris, on the other hand, was a member of the so-called Team of Ten who broke off from the ASA in 1963 on the grounds that it was too soft on evolution, and he became the patriarch of creation science we met earlier.[49] And in that role, he encountered Velikovskianism in a different guise.

PATTEN'S CHARGE

The problem with safeguarding orthodoxy on the margin is that fringe doctrines have margins of their own, and creationism was no exception. Donald Wesley Patten was born in Conrad, Montana, in 1929. He attended the University of Montana for three years before transferring to the University of Washington, where he obtained a BA in geography in 1952 (with a minor in history), at which point he was shipped out to serve in Korea. Upon his return to Seattle, he founded a successful microfilm service, raised his seven children while faithfully attending the Evergreen Baptist Church, and earned a master's degree in geography in 1962.[50] Patten was the leading representative of the Velikovskian strand of creation science; he published widely and often, and yet he remains completely absent from histories of creationism.[51] The link between his virtual erasure from the record and his Velikovskian inspiration is not accidental: in the 1960s, Patten gained significant visibility in both Morris's circle and Velikovsky's— and was systematically purged from both.

Patten's whole body of work derives from and expands upon his first book, *The Biblical Flood and the Ice Epoch*, published in 1966 by Pacific Meridian (a press founded, owned, and operated by Patten himself, exclusively for the publication of his writings). In contrast to theistic evolutionists or those (like Morris) he dubbed "geocentric catastrophists," Patten was an "astral catastrophist": "Astral catastrophism involves occasions of sudden and overwhelming cataclysmic changes in the conditions of the Earth in a brief and limited time. . . . We do not maintain that the period of crisis referred to in Genesis as the Flood was the first conflict or the last; we only maintain that it was the worst."[52] As the name implies, astral catastrophists attributed the transformations to extraterrestrial causes and saw their attention to mechanism as their greatest virtue. Patten viewed any theory (like Morris's) without a mechanism as simply positing God waving a miraculous wand; this was nothing short of "theomagical."[53]

The entire (admittedly small) school of creationist astral catastrophism centered on Patten.[54] He saw himself as merely a further step in a tradition of catastrophists, including George McCready Price, Alfred Rehwinkel, and Henry M. Morris. But perhaps his greatest influence, and the end of his genealogy of catastrophists, was Immanuel Velikovsky, "the most important figure among the secularly oriented catastrophists."[55] ("Secular" for Patten was identical to "non-Christian.") Patten was no orthodox adherent of Velikovsky's version of Earth history, as the fact that he directed his attention to the Deluge and not to the Exodus already indicates. In fact, he devoted considerable space to offering "friendly and constructive" criticisms of Velikovsky's three major faults: his "deep predisposition to modern humanism," by which Patten meant Freudianism; his lack of "geophysical perspective," in which Patten included the complaint that "although he wrote hundreds of thousands of words about catastrophism, he never produced a single line diagram, not a single illustration, not a singular tabular form"; and his obsession with his critics.[56] But despite these differences, Velikovsky got one major thing right: he realized that the cataclysms on Earth were caused by near approaches of planets. For that insight, as well for his historical work dating the catastrophes and his collection of ancient textual evidence corroborating biblical accounts (which Patten relied upon heavily), Velikovsky was to be admired. But Patten believed his predecessor had fingered the wrong planet, misunderstood the mechanism, and failed to see that the systematic pattern did not require the invocation of electromagnetic forces: "Some writers analyzing these ancient events would suggest celestial mechanics be set aside or challenged. This is a mistake."[57]

While Morris concluded that the ice ages happened after the Deluge (if they happened at all), for Patten the ice age (there was only one) and the Flood "were one and the same catastrophe," caused by a near approach of Mars to Earth.[58] Venus had nothing to do with it. In fact, all of the biblical catastrophes—the Flood, the Tower of Babel, the destruction of Sodom and Gomorrah, the Exodus, the Long Day of Joshua, the Isaiahic catastrophes, and several others—resulted from the same process, as Patten would elaborate in later works into the 1990s, and especially in a 1973 collaboration with Ronald R. Hatch (a physics BA who worked for Boeing) and Loren C. Steinhauer (PhD in aeronautics and astronautics). According to Patten and his coauthors, the ancient orbits of Mars and Earth used to intersect (much as the orbits of Neptune and Pluto do) with a 2:1 resonance, and every 108 years—two cycles of 54 years, alternating between mid-March and late October (as the two elliptical orbits intersected on the same side of the sun)—for 9,200 years, the two planets would approach each other, causing massive catastrophes on both bodies.[59] The Deluge, for example: Mars and Earth came closer to each other on this cycle than they ever had before or have since, and an icy satellite of Mars (dubbed Glacis) entered within Earth's Roche limit (11,000 miles, at which point tidal forces destroy any approaching body, a process known to have created the rings of Saturn) and was utterly shattered, spraying ice particles on both planets. On Earth, these produced both the ice age at the poles and the Deluge in the temperate regions, while on Mars floods cut the famous canals visible on its surface.[60] This near approach was the most catastrophic mutual interaction until the resonance orbits began to unravel in the eighth century B.C. The Exodus, too, was caused by Mars (like clockwork in March, just in time for Passover), leaving Venus to orbit the sun in peace.

As committed as Patten was to a Velikovskian interpretation of solar system history—down to the extensive citations from myth and legend—he also moved in a creationist milieu increasingly dominated by Henry Morris. And Morris, as we have seen, was not keen on citations to Velikovsky. To that end, Patten, too, especially when writing for an audience that included Morris and other figures associated with the Institute for Creation Research (or in publications in its journal), lopped off the Velikovsky references.[61] Those few supporters of Patten who continued to work within the creationist camp also tried to insert some daylight between Velikovsky and Patten. For example, Charles McDowell, in the Symposium on Creation series that Patten took over editing in the early 1970s, declared that Velikovsky was a "neonaturalist," "a thoroughgoing evolutionist who bases evolutionary development upon what he calls cataclysmic evolution." Patten was different, the founder of a "second school." "Patten is dependent upon Velikovsky for much of his specific historical data," McDowell wrote. "Nevertheless, Velikovsky and Patten are widely separated on many specifics and more importantly in ideology."[62] Those differences, and the selective camouflage of citational apparatus, were supposed to keep Patten in the creationist fold.

It didn't work. In Morris's 1984 history of creationism, there was no mention of Donald Wesley Patten, even to be excoriated. It was as though he had never written. But in the 1960s, after Patten's first book had appeared, Morris certainly paid attention to it. In 1968 the *Creation Research Society Quarterly*, a journal largely controlled by Morris and like-minded supporters of the young-Earth, flood-geology picture from *The Genesis Flood*, published a harsh review of Patten's monograph. Signed by four credentialed scientists employed at Bob Jones University, the authors declared themselves "both disappointed and disillusioned. It is neither a book of *Christian* apologetics nor a vast storehouse of scientific information. . . . [T]his effort is definitely a step in the wrong direction."[63] The problem was not just the technical content of the book, but also Patten's resolute attempt to find a naturalistic mechanism for biblical miracles—what Patten saw as his strongest point. For the reviewers, this denied the possibility of miracles altogether.[64] By not explicitly endorsing a young-Earth position, Patten was unable to grasp the central theological message of *The Genesis Flood*: "The physical condition of the world before and during the Flood is not subject to scientific investigation. There can only be speculation about the cosmological conditions operative at that time as far as modern science is concerned. The only way we can *know* what happened is by direct revelation from God."[65] As an aside, in a veiled snub of Velikovsky, the authors noted that the "references cited often represent the work of individuals possessing little scientific knowledge in the area of concern."[66]

Patten was both dismayed and enraged—he believed he had provided a crucial improvement on Morris's framework, and his ideas were shunted aside. He turned for solace to someone he thought would understand him: Velikovsky himself. The two first met at a Velikovskian conference in Portland in August 1972, and Patten capitalized on the personal contact to complain about this very review in terms he thought that his intellectual inspiration would understand. The review "has been widely appealed to to discredit, or try to discredit my work as well as the entire astronomical approach to catastrophism. It was probably drafted by a person named Henry Morris, author of another work on the deluge which is geocentric in philosophy, even though it happens to be signed by four of his colleagues." They were not simply trying to discredit him, however; they had "treated my work like Shapley and his colleagues treated yours. The parallel is just quite remarkable."[67] The analogy did not strike Velikovsky as particularly strong: "I do not find that your critics handled you so badly; actually they unwittingly made you a compliment by saying that you are less fundamentalist than they."[68]

That meeting in Portland was the result of over a decade of Patten trying to get close to Velikovsky. In 1960, while still studying for his master's degree, Patten had written Velikovsky with praise for his books and offered his own thesis: "I have come to a clear conclusion, for instance, that the deluge described in Genesis was tidal in nature, and was synchronized with the cataclysmic appearance of the ice age in time. On this basis, I am prepared to make a substantial number of conclusions and inferences."[69] Velikovsky had no patience for this: the Deluge "was not, in my understanding, a mere tidal wave; the water came from the space and I shall give close details on the origin of this water. It will be written with the same

abundance of quotations from ancient sources as Worlds in Collision is."
To Patten's request that Velikovsky look at his manuscript, the latter declared that he had no time to read a whole book, "but if you prepare for
me a summary on 4–5 pages, I shall gladly examine it and give you my
opinion."[70] Patten sent fifty-seven pages.[71]

Velikovsky did not read them, but that did not stop him from huffily
asserting his priority:

> The subject of the Deluge will be the theme of a book contemplated for over
> two decades—and most probably you have not come to the same conclu-
> sions of the cause and the phenomenon of the flood as I did. I am afraid
> that with my views elaborated on the Flood I will be a destructive critic of
> your paper that though following the same principles as found in "Worlds"
> and "Earth" cannot incorporate my unpublished view of the great upheaval
> known as Deluge. . . . Yet I cannot release my theory in a few short state-
> ments: it is a work of the size of Worlds in Collision, and actually is one of
> the two volumes that I have planned to add to it. If after these discouraging
> for you statements, you wish to send me, at first, a few pages of your thesis,
> I will not see my way of refusing you, though my time is simultaneously
> claimed by several of my unfinished works.[72]

So Patten sent another letter with another typescript, and again Velikovsky begged off, this time on the grounds of demarcation—that Patten was
not scientific because he included too much religion: "The quick glance
through the piece left the impression that you have a theological approach
to the problems of geology; if this should substantiate itself by a careful
examination, I would be entirely out of my sphere: theology is a matter of
faith and credence; science and history of fact and evidence."[73]

Patten, two years into this abortive attempt to set up a relationship, was
undeterred. Having received a nibble from University of Washington Press
about his manuscript, he inquired of Velikovsky about the reception of
the latter's work—which, amazingly, he seemed never to have heard of—
and also inquiring about royalty arrangements.[74] Velikovsky's (drafted but
unsent) response was blistering: "I am returning your typescript. I have
taken two hours from my very busy schedule to read carefully more than
twenty pages of it besides the concluding pages. It will serve no purpose
useful to you or to science in my reading the rest of it. You have no material
for a book, nor for an article."[75] Instead of sending this version, however,
a nameless "secretary"—Velikovsky employed no such person in 1962—

responded in words that sound so much like the man himself that they were certainly dictated, if not written directly, by him: "Although Dr. Velikovsky is not a fundamentalist, he found interest in the books of Price, a fundamentalist. But your brand of fundamentalism is so different from the approach of Dr. Velikovsky that he feels his criticism of your work would not serve any useful purpose. Where arguments are built on faith, they are not given to fruitful debate."[76]

This point about fundamentalism recurs frequently in Velikovsky's letters and writings.[77] Perhaps the oldest mention appeared in a letter to Price: "It is not my intention to prove the fundamentalist creed; but I am pleased to find the Old Testament a truthful document on history and also natural history of the epoch."[78] Velikovsky's supporters, such as Eric Larrabee, likewise wanted to insulate Velikovsky from the charge that the use of biblical texts as *evidence* in scientific arguments implied anything like inerrancy or theology.[79] In fact, as Morris and others keenly noted, Velikovsky's arguments were not like creationists' citations of Genesis: for the latter, these were revelation, telling how the past actually was; for Velikovsky, they were utterances to be interpreted naturalistically. Velikovsky was always careful to keep his distance in order to avoid being tarred with the brush of fundamentalism. For example, when a publisher (Kronos Press) associated with his followers released a book attacking evolution, the author took pains to stress that "Velikovsky himself is *not* a 'fundamentalist.'"[80] Due to all these caveats, it is perhaps no surprise that Patten understood these avowals to mean that Velikovsky was "an atheist and a Freudian."[81]

Patten had by then been through the wringer with Velikovsky. Having self-published his book, he attempted to wheedle out of Velikovsky a list of places that had reviewed *Worlds in Collision* and *Ages in Chaos* (both published by major trade presses) so he could get notice for his own work, and he added the following puzzling mathematics: "You may be interested in our judgment that we are about 75% in agreement with your conclusions and about 25% not in agreement. . . . You may be in harmony with the bulk of our conclusions; yet we anticipate that perhaps $\frac{1}{4}$ to $\frac{1}{3}$ of our conclusions may not be pleasing or acceptable to your thinking."[82] Yet still there was no substantive response from Velikovsky. Patten finally realized that he was getting the cold shoulder: "It would look like you have been somewhat reluctant to correspond or to respond [to] a couple of my last letters sent to you."[83] He took a different tack, noting that he was a pre-millennialist and thus "wholly enthusiastic toward the general Zionist perspective."[84]

Now seven years after the correspondence began, Velikovsky brushed him off again: "I cannot express myself as to your theory, since I have not read nor seen your book; neither do I have time to read anything besides my immediate needs for the manuscript in preparation for print."[85] He still had not read the book a year later.[86] Patten tried once more in 1969, with the same results.[87]

Velikovsky's followers treated Patten no better.[88] Lynn Rose, Velikovsky's closest disciple through the 1970s, tried to block Patten from any Velikovskian outlet. When the British journal *SIS* [Society for Interdisciplinary Studies] *Review*, a pro-Velikovsky organ, proposed publishing a review of one of Patten's works, Rose urged against it:

> My own feeling is that it might be best for a Velikovsky-oriented journal to devote as little attention as possible to Patten. The danger is that Patten and his colleagues might use even the slightest amount of attention to them as an occasion for filling the journal with a protracted exchange, the main result of which would be valuable time and space diverted from Velikovsky's own theories.
>
> Patten and Velikovsky are not in the same league, and I would prefer that they not be mentioned together at all—a maxim that is difficult to formulate without thereby violating it![89]

If they insisted on publishing a review, Rose preferred they use one he wrote himself, dismissing the work as creationist propaganda. (The journal did publish Rose's review of *The Long Day of Joshua*, juxtaposed with a favorable review of the same work by Robert W. Bass, in spring 1980.[90]) Rose was also very upset when he realized that he was on the masthead of another journal (*Kronos*, discussed in the following chapter) that mentioned Patten without unambiguously condemning him.[91]

Rose was only following Velikovsky's lead. According to Velikovsky's associate Alfred De Grazia, when Patten first approached Velikovsky in Portland, the latter blanched and said: "[']You are trying to destroy me, but you will fail in the end!['] So relates Patten and there is no reason to doubt him, especially when he adds that a while later V. returned to him and apologized."[92] Or, as De Grazia quoted Patten: "Velikovsky viewed me as an unwanted protege, not to be encouraged. . . . Often criticized as he was (and many times unfairly), Velikovsky regarded me as yet another critic trying to destroy his work. He was uncomfortable with my evangelical,

Christian faith."[93] The rift between the two was never resolved—in fact, it is almost too much to call it a rift. For there to have been a rupture, there had to have been contact, and that Velikovsky steadfastly refused.

There are two significant lessons in this excursus into Patten's marginalization by Velikovsky and the Velikovskians. First, the symmetry between Morris's and Velikovsky's attitudes to Patten. Both mainstays of their own fringe doctrine, rejected by establishment science, they could not afford to lose focus or open themselves to (further) ridicule by diffusing their message with revisionism and new ideas. As a result, they needed to establish their authority and reject heterodoxy wherever it appeared, *especially* if it was heterodoxy emerging from another point on the fringe that used the same evidence (the Bible) for radically different purposes. If Velikovsky was too "pseudo" for Morris, creationism was the same for Velikovsky.

The second lesson stems from how Velikovsky enforced his own orthodoxy: by using precisely the same tactics that were used against him. Much as Harlow Shapley had ignored Velikovsky's letters begging for support and mutual exchange of ideas, Velikovsky had no time for Patten. If he had been slammed in reviews, then so would Patten be. Finally, Velikovsky and his supporters often raised the (correct) point that many of those who damned *Worlds in Collision* in 1950 were proud of the fact that they had not read the book they were condemning. How did Patten's case measure up on this front? "Patten is a swollen head and his inquiries addressed to me did not recommend him as a candid researcher," Velikovsky wrote in 1969. "But I have not read his book."[94]

RISE OF THE VELIKOVSKIANS

For all the intertwining of the Velikovskian and creationist cases, there was an abiding asymmetry between the two, at least in the 1960s. Shortly after the publication of *The Genesis Flood*, Morris found himself equipped with a well-funded network of people who already agreed on fundamental principles (the Revelation of God) independent of the specifics of flood geology, which made it easy to effectively discipline the movement and keep it on message. In the early twentieth century, before the Pricean orthodoxy emerged, any attempts to enforce coherence often risked suffocating the incipient movement. Making assertions to authority while defending a fledgling doctrine on the fringe was a risky proposition, as Velikovsky soon discovered. Rejecting Patten was a story of enforcing order on the periphery, guarding the borders of one's own domain against incursions. It was a separate matter altogether when Velikovsky had to ensure that his own ranks were pure.

Velikovsky had always had fans, even supporters, but in the early 1960s he began to acquire disciples. Among the first was Ralph Juergens, who in 1961 moved from the Midwest to Hightstown, New Jersey—right outside Princeton—to be closer to Velikovsky, taking a job as assistant editor for technical writings at McGraw-Hill publishers. (By August 1969, he had moved his family to Flagstaff, Arizona, yet remained in constant contact with Velikovsky.[95]) The most influential early disciple was surely Alfred De Grazia, who was professor of government and social theory at New York University when he met Velikovsky. According to his own account, he read *Oedipus and Akhnaton* shortly after its 1960 publication, and, since he lived in Princeton, decided to seek out its author.[96] The two struck up a friendship, and De Grazia became intrigued by both Velikovsky's theories (although later he would develop intellectual differences, along with Juergens just about the only disciple to be tolerated to do so) and the suppression narrative of the events of 1950. He decided to use the journal he founded and edited, *American Behavioral Scientist*, to renew inquiry into Velikovsky's case, devoting the entire September 1963 issue to Velikovsky's biography, catastrophic ideas, and especially its reception by the scientific community.

De Grazia argued in the journal that the mixture of historical methodology with cosmological claims was one of Velikovsky's strongest features. "What has not been appreciated in the course of the conflict is the high degree of involvement of the social and behavioral sciences. The social sciences are the basis of Velikovsky's work; his proficiency in the natural sciences, except medicine, is derived," he wrote. "It is by the use of the methodology of social science and the dates of history that Velikovsky has launched his formidable assault upon the heroes and theories of the classics, astronomy, geology, and historical biology. Yet social scientists have been generally unaware of his work and almost totally disengaged."[97] His journal, he proposed, would change that. Claiming that he only wanted to investigate the Velikovsky affair from a sociological angle, he wrote an analysis of the reception system in science in which he claimed that Velikovsky, and not his opponents, behaved in line with the standard model of scientific reception (reasoned evaluation of opposing theories). According to De Grazia and Livio Stecchini (who penned a piece placing Velikovsky in a lineage of cosmic catastrophists dating back to William Whiston at the end of the seventeenth century), there was one person to blame: Harlow Shapley.[98] This version was reinforced by a history of the Velikovsky affair by Juergens that was entirely (and silently) derived from the manuscript of Velikovsky's own *Stargazers and Gravediggers*.[99]

De Grazia succeeded in refocusing attention on the Velikovsky case. Prefaced with endorsements by respected scientists and scholars that the controversy over Velikovsky (if not his theories) deserved closer scrutiny, the *American Behavioral Scientist* issue generated real interest, even provoking positive comment from such intellectual luminaries as literary scholar Jacques Barzun.[100] De Grazia's venture drew criticism as well. The *Bulletin of the Atomic Scientists*, one of the leading journals about the role of science in public policy, published a scathing review of the special issue by Howard Margolis, who declared it an attack on science itself.[101] Such a prominent drubbing proved to be excellent publicity, and the issue took off, all the more so when De Grazia published it in 1966 as a stand-alone book, *The Velikovsky Affair*, amplified with additional essays by Stecchini and Juergens, the latter updating his story of the controversy to the present, including the Margolis rebuke.[102]

From this point forward, Velikovsky's public visibility grew anew, reaching even greater heights in the 1970s than it had in 1950. That public story is told in the next chapter; here, I emphasize the core of Velikovsky intimates, which eventually grew to a few dozen but was animated by a group of roughly ten (the composition changing over the years). One might think Velikovsky would have been happy about this development, and at times he clearly was, although it did not seem to change what one of his daughters characterized as his cyclical mood swings, "deteriorat[ing] in depressions between periods of productivity and optimism."[103] Frederic Jueneman, an industrial chemist who joined the inner circle in the early 1970s, worried that the very enthusiasm of the group troubled Velikovsky: "Yet, there is a distinctly Freudian aroma in the air, where the 'sons' appear to be hoping against hope that the father-image will die, pass away into oblivion so that the scions can pillage the inheritance and squander the spoils."[104] That may have been so in later days, but when Velikovsky's star began to rise again, he made the most of it to vindicate himself.

It was one thing to have active supporters who would devote time and energy to his cause—writing letters to the editor against his critics, filing his correspondence, engaging skeptics in debate—it was something else entirely to have a dedicated organization, and this is what Velikovsky saw growing around him in the late 1960s. Bruce Mainwaring, an affluent businessman interested in Velikovsky's ideas, proposed the creation of such an organization in 1968: "Therefore, I feel that a foundation could be organized, sponsored by individuals like myself, which would be in a financial position to support and direct the efforts of scientists who are willing to take an objective view."[105] Soon the Foundation for Studies of Modern Science (FOSMOS) was established, with Mainwaring as the treasurer. The board of FOSMOS (including Juergens, De Grazia, Mainwaring, and other stalwarts) set as its chief task obtaining new archaeological evidence that would confirm the historical reconstruction proposed in Velikovsky's books. They attempted to sponsor a dig at El Arish in North Sinai—then occupied by Israel but now in Egypt—where Velikovsky was convinced that ruins would be found confirming his identification of the Hyksos with the Amalekites, but the organization bungled its applications for permits and the expedition foundered.[106]

One collapsed expedition would have been easy to absorb, but FOSMOS had a thornier issue to deal with, and this one was chronic: the precise relation of Immanuel Velikovsky (the person) to this organization entirely dedicated to Velikovskianism (the ideas). Although it is hard to reconstruct an objective account, it is clear that this struggle crippled FOSMOS profoundly. Here is where concern for enforcement of orthodoxy—discipline in the ranks—became most visible. After a soul-searching discussion at a board meeting after the early collapse of the El Arish project, Juergens wrote to Velikovsky on behalf of the board that "we have insisted among ourselves that FOSMOS is to go its own independent way, guided, as it were, only in certain projects by your recommendations."[107] This was supposedly what Velikovsky wanted; he feared that he would be blamed for actions taken by FOSMOS. Or, as De Grazia put it less charitably in a journal entry of November 30, 1968: "He warns against everything to be ready to be proven a prophet should things go badly. He cannot let go of any power over things or people, but plays upon every means of entrapping and embroiling them, sucking them in and pushing them off as he feels the one way or the other in his succession of mobilizing-for-action and trust-nobody moods."[108] From this point, things only got worse.

"I wish to dissociate myself from the activities of this Foundation," Velikovsky declared the day after De Grazia's journal entry. Although he had been flattered that FOSMOS wanted to support his theories, he was worried about fallout. The issue of dissociation, as he noted in the next sentence, was a bit tricky: "Since I have no official status in it, no further steps are necessary; however, the Foundation would need to abstain in all its activities from using my name and mentioning my work in its various efforts like solicitation for membership, advertisement of a journal, application for funds, etc." The fund-raising issue was the most crucial, since he worried that "a money drive of such nature would jeopardize my name and position."[109] Velikovsky also denied to the group custody of his archive—a prize that De Grazia in particular was eager to claim, since the documents there would provide a wealth of information about the Velikovskian project. In exchange, Velikovsky would *consider* writing some articles for a FOSMOS journal.[110]

Demoralization set in. Even Juergens was disconsolate at this rebuke: "Simply put, it is that no matter what we do, or how we try, our efforts never quite measure up to your expectations."[111] Characteristically pulling back, Velikovsky much later offered what amounted to an apology of sorts.[112] Yet he remained resolute that while FOSMOS had to adhere to the strict orthodoxy of his theories, it was not permitted to claim that it had any official connection to him. It was not an official army, but it was not to pursue battles on its own initiative either. Less than a year after the blow-up, Velikovsky wrote Mainwaring with a proposal that persisted in blurring the boundaries of authority:

> Should the Foundation over which I exercise no control, develop into a useful instrument, its two tasks would be [a] testing various claims and implications of my work and [b] preparing young researchers for original work.
>
> Such researchers, working directly under me, assist me in bringing my work closer to completion; and equally important, they are taught to develop original ideas of their own in any of the manifold directions to which they may feel impelled.[113]

Such grants were in fact disbursed, although not always through FOSMOS sponsorship. Perhaps the most surprising thing about the whole nonofficial official supportive independent organization is that it took until 1972 to liquidate it.[114]

THE GREAT PURGE

The disputes over FOSMOS were about *how* to support Velikovsky's cosmic catastrophism, and they foundered over disagreements about what would be the most effective way to do that while leaving Velikovsky uncommitted. There was a deeper concern, however, with what counted *as* supporting Velikovsky in the first place and in particular whether one could combine Velikovsky's ideas with those of others. Could alliances be formed in the pseudoscience wars, and if so, who had the right to make them? If Patten had been locked out of the circle for contaminating Velikovskianism with creationism, what one might term the "Stephanos affair" was more serious: how to cut out a perceived cancer at the center.

The whole business began with a pamphlet written by Ted Lasar, an enthusiast for Wilhelm Reich, one of the great figures in the pseudoscience wars. The Austrian-born Reich, like Velikovsky, was trained as a psychoanalyst, but he broke with Freud and published a series of books that took an even more sexualized approach to neurosis than Freud himself—for example, arguing in *The Function of the Orgasm* (1927) that all neuroses stemmed from the inability to achieve satisfactory climax. Controversial touch therapy and semi-nude counseling ensued. Again like Velikovsky, Reich moved to the United States in 1939, in his case to avoid Nazi persecution as Austria was folded into Hitler's empire. At this point, Reich formulated a cosmic theory of sexuality in which the universe was permeated by a blue substance, orgone—responsible for the color of both the sky and electric discharges—which enabled the interconnectivity of the cosmos as well as sexual orgasms, even sustaining life itself. He developed boxes he called "orgone accumulators" to harvest atmospheric orgone, and he sold them to those who believed his claims that they helped cure cancer and other ailments, simultaneously producing more orgasms. (This notion has been parodied in numerous films, including Woody Allen's *Sleeper* and *Barbarella*, starring Jane Fonda.) Reich's career took a turn for the worse as the U.S. Food and Drug Administration turned its gaze on him in 1954. After an exhaustive investigation, they declared that orgone did not exist and ordered him to stop shipping the boxes. Reich defied the injunction and was convicted in 1956 to federal prison in Lewisburg, Pennsylvania. He died less than a year later, at age sixty-five, a martyr to his own cause.[115]

Velikovsky evidently never commented upon Reich's notions until Ted Lasar, an orgone partisan from the New York area, began a correspondence in 1960. In the first letters, Lasar stuck to more strictly Velikovskian topics, presenting a vision of a radiocarbon conspiracy against verifying the reconstructed chronology that was, if anything, more radical than Velikovsky's own:

> You are like a building inspector who has just found out that the Empire State Building must come down; it is unsafe. The careers and books of cele-

brated men are in danger of being reduced to meaninglessness and there is a cruelty and sadness in it. They will resist perhaps even at the risk, when all is said and done, of doing so lunatic a thing as refusing to take a piece of wood from some museum into the laboratory and finding out how old it is, falsifying results or declaring them to be invalid on some pretext or other.[116]

So far, so good. Trouble began when Lasar wrote in September 1960 about his own efforts to combine Velikovsky's cosmic catastrophism with Reich's orgone. Velikovsky was resolutely opposed, saying he did not know Reich's writings, "but what I know of his therapeutic procedure makes me feel that it is sick, sicker than the patients and their diseases."[117] Lasar defended Reich in a response, but Velikovsky cut off contact.[118]

Lasar continued to pursue his notions. In 1969 a pamphlet put out under his auspices came to Velikovsky's attention, describing a Reichian innovation in healing. The fourth paragraph of the pamphlet states:

> Out of Reich's methods and the new discoveries it has become possible to devise a new therapy based upon the realities of human existence. This new therapy, called "Earth Therapy" for reasons that will become apparent as time goes on, is based upon the fundamental principle that armor originated in recent historical times as a result of the collisions of the earth with other planetary bodies. These collisions are described in the book Worlds in Collision by Immanuel Velikovsky.[119]

Velikovsky was horrified: "Much of this material was put together in a way that an unaware reader could believe that I am the author."[120] The fact that it asked for donations made things even worse—and surely was responsible for some of Velikovsky's skittishness about FOSMOS as well.

Before Lasar's pamphlet entered the Velikovskian universe, Robert Stephanos had all the credentials of an excellent disciple. He became devoted to the cause in his college days, before the publication of *The Velikovsky Affair* and the journals of the 1970s had swelled the ranks, and he fought and was martyred for Velikovsky's sake. He earned his spurs in 1966, when he arranged for the Rittenhouse Club, a noted Philadelphia lecture group, to invite Velikovsky to give a talk. Unfortunately, the venue for the Rittenhouse lecture was to have been the Franklin Institute, Philadelphia's venerable science museum. Robert Neathery, vice president of the Franklin Institute, would have none of it and canceled the lecture lest the institute be seen as endorsing Velikovsky. What followed was a revival in miniature of the dismissal of Gordon Atwater at the American Museum of Natural History in 1950: increased publicity for Velikovsky, public condemnation of the Franklin Institute as an elitist establishment of suppressive know-nothings, and the valorization of Stephanos. Stephanos set up the first formal Velikovskian college campus group outside Princeton, dubbed "Cosmos and Chronos" after the original, at Temple University, where he was taking graduate courses.[121] He was soon put in charge of developing the proliferating Cosmos and Chronos groups through mailings of materials and coordination of events.

Stephanos was quite taken with Lasar's ideas and had sent the "Earth Therapy" pamphlet to several of these groups. To describe Velikovsky's reaction as outrage would be an understatement: "My favorite creation—Cosmos and Chronos groups was made to a channel of spreading a feebleminded and vicious material that can appear as written by myself, never signed by its author, and soliciting funds, offering a 'sexual' Earth therapy not without fee."[122] Warner Sizemore, a pastor who for two decades was among the most loyal of Velikovsky's supporters, at first defended Stephanos, but when the extent of Velikovsky's ire became manifest, assured the catastrophist that none of the inner circle would have any contact with Stephanos, nor would Stephanos be a part of any future Velikovskian projects.[123]

Velikovsky found out about these mailings, which happened within the "independent" Cosmos and Chronos organization, through the very closeness of the network he had built. In early 1969 a high school student named Eddie Schorr, who had been corresponding with Velikovsky for several years about his interest in the *Ages in Chaos* project, became one of the first FOSMOS grantees to be brought on scholarship to work with the master in Princeton, fact-checking articles and organizing files.[124] Apparently, working closely with such an intense and psychologically astute individual as Velikovsky led to Schorr feeling raw and troubled, and he confided in Stephanos, with whom he was staying. Stephanos consoled him by pointing to Velikovsky's genius and also told him that there were some other doctrines that could be combined with cosmic catastrophism to assist with such psychological distress, at which point he mentioned Earth Therapy.[125] Schorr told Velikovsky, and the latter sprung into action, informing Schorr that "I told him [Stephanos] that in my life I have not seen a case of worse treachery.... I told him that his function as President of Cosmos and Chronos is terminated and this was decided earlier because he has no contact with a campus as a student or teacher, but now he is removed because I lost faith in him."[126]

Stephanos also told Schorr that he had sent a copy of the pamphlet to Mary Buckalew, an English professor at North Texas State University who devoted herself wholeheartedly to Velikovsky after his lecture at that school in 1968: "Your visit has changed profoundly the lives of many people in this city. My own life, given a new direction when I first discovered your works, has now real purpose."[127] (She began preparing an index of his books, which was never completed.) Velikovsky was in constant phone and mail contact with Buckalew, of whom he was genuinely fond, and he plumbed the extent of Stephanos's treachery. At first she minimized the issue: "I dismissed the sheet with a laugh, thinking that Bob had included it as a joke. . . . Sir, I did not have the impression that Stephanos sent The Pamphlet in your name. On the contrary, he stressed that evening on the telephone that you did not approve of the material, nor would you approve of his sending it to me."[128] Within two days, and after a phone call with Velikovsky, she changed her story: "Sir, I have come to realize now that his very guilt lay in distributing questionable materials while still operating as your right-hand man in matters of corresponding with your many disciples. . . . So that even though he dissociated you from the contents of That Pamphlet, his very position of being in your trust and high regard made acceptable and influential what on its own would never have been."[129]

Velikovsky began to put his house in order. Stephanos was cut off and ostracized—not just from Velikovsky, but from everyone in the inner circle. Even Patten had more contact with the group than Stephanos did after 1969.[130] And in a mirror of the FOSMOS conflict, Velikovsky began to assume control of the Cosmos and Chronos group that was supposed to be independent. "There is a great need to re-organize and develop the Cosmos and Chronos study and discussion groups," he wrote to Albert Burgstahler, professor of chemistry at the University of Kansas. "Stephanos who was in charge betrayed my confidence by sending out to the groups some crazy, almost vicious literature . . . ; the amalgam of my work with organon and sex, leaving the impression that I am behind all this, is very damaging now I take over the charge myself, until the time when a new committee consisting of representatives (faculty members or also bright students) from many campuses should be construed."[131] The only problem was that Velikovsky had very few contacts on college campuses that he could trust. Cosmos and Chronos groups continued to exist, but since Velikovsky did not have the time to devote to their coordination and could not trust others to maintain orthodoxy, they failed to become redoubts of a movement for establishing Velikovsky's legitimacy.

Depending on one's point of view, the 1960s were either a period of tremendous setbacks in Velikovsky's battle for transformation of the scientific establishment or one of triumph. He had not managed to interest mainstream scientists in his work, but he had acquired a coterie of people who were committed both to his theories and to him personally—although the tension between those two remained a persistent, unresolved problem. There were, after all, advantages to being the loner heretic. Once he had disciples, he acquired the problem of enforcing orthodoxy. And without the well-heeled organization that enabled Henry Morris to so successfully police his own ranks of scientific creationists, Velikovsky found himself torn between becoming popularized and becoming vulgarized.

6 · Strangest Bedfellows

Harold Urey disliked Immanuel Velikovsky. Winner of the 1934 Nobel Prize in Chemistry for his 1932 discovery of deuterium (a heavy isotope of hydrogen), Urey had long been a leading statesman of science, and in the 1960s he became involved with the rapidly expanding space program. This attention to matters extraterrestrial, as well as his visible position among the luminaries of American science, brought him to the attention of the Velikovskians. It was not attention that Urey relished.

"I am sorry to see that you have gotten mixed up in the Velikovsky case. Velikovsky was a charlatan," he wrote to Velikovsky's close associate Alfred De Grazia in 1964, in response to an approach by the latter decrying the sharply negative review in the *Bulletin of the Atomic Scientists* of De Grazia's special Velikovsky issue of the *American Behavioral Scientist*. "I am terribly concerned at present about the lack of control in scientific publication. Science has always been aristocratic. Not everyone could get his ideas published in effective journals. . . . Today anyone can publish anything. . . . [T]here is often so much noise that one cannot hear the signals."[1] Intimates of Velikovsky's inner circle did not keep secrets from their polestar, and De Grazia shared the letter with Velikovsky. Velikovsky, predictably, was infuriated, and he penned a hostile letter in De Grazia's name that he urged be sent to Urey, to put the elitist obscurantist in his place with what sounded almost like epistolary blackmail: "Would Velikovsky be less forgiving he would have made public the documents that would ruin completely and for ever several of his chief adversaries, a number of luminaries some of which are fading in their own lifetime. The late Einstein urged him to do so and to publish this documentary. The provocation of the BAS and of your letter may compel him to do exactly this."[2] De Grazia, at this moment a cooler head and never one to take kindly to ghostwriting (even from Velikovsky), opted for a milder tone and picked up on Urey's choice of language: "Your kind of scientific aristocracy is precisely why your subsequent claims are laughable: if there is any villainous theme in the history of science, it is the continuing attempt to deny a voice in the organs of science to iconoclasts, outsiders, and just plain *kleine Menschen*."[3]

This is a familiar theme. In the 1950s, in the wake of the first battles surrounding *Worlds in Collision*, the hostile reviews by scientists, and then the public exposure and interpretation of a fledgling boycott campaign against Macmillan, Velikovskian rhetoric centered on Galileo against the Church, which in itself was a variant of David against Goliath. Velikovsky spoke alone, a voice of reason against hidebound dogmatism and privilege. The Urey exchange lay on the cusp of something new. This was no longer just David against Goliath. Velikovsky was gathering allies in the pseudoscience wars—or, to put it more accurately, allies came to him without his doing much at all to recruit them (and often a good deal to drive them away). Amid the tumults that rocked American culture in the late 1960s, an insurgency on behalf of cosmic catastrophism was shaken loose—and once free to fight, it would stake its claims through exploiting new venues of publication and publicity, from campus activism to novel Velikovskian journals. Through these new forums, Velikovsky became, as one of his critics put it in 1977, "the grand curmudgeon of anti-establishment science."[4]

Urey already knew this. "Velikovsky is a most remarkable phenomenon of the last 20 years," he wrote in 1967 to University of Kansas chemist (and Velikovskian) Albert Burgstahler in response to an earlier missive. "If someone of this kind should turn up in science once a year I think it would wreck science completely."[5] Urey had a hard enough time with just one. At first he thought the best approach, as he responded to young Eddie Schorr, was to stay mum: "I do not believe that Velikovsky has any scientific basis for anything that he has written. I believe it is a mistake to pay any attention to him at all. Scientific predictions are only important if based upon logical arguments of some kind."[6] But it became harder and harder to simply ignore Velikovsky.

For one, Velikovskian enthusiasts—usually much younger and less established than Burgstahler—began to deploy new ambush tactics behind the front lines of conflict. For example, several telephoned to solicit Urey's views of how his research on moon rocks for the space program would confirm Velikovsky's scenario for near collisions with Venus and Mars.[7] Stephen Talbott, editor of the pro-Velikovsky journal *Pensée* (these journals themselves being a new feature of the 1970s), was one such caller. When Urey refused to speak on the record about Velikovsky—the call was being taped—Talbott assured him that he would not publish the conversation but wished to hear Urey's arguments off the record. Urey echoed his earlier views.[8] Now Talbott had him. Despite his promises to keep the conversation private, he wrote Urey a threatening letter demanding a retraction and an apology for Velikovsky. "I am not a scientist, but it does not take a scientist to recognize the elementary rules of argument and fairplay. Velikovsky must be disproven by a discussion of the facts, not by a general attack on his person or a misstatement of his views," he fulminated. "If indeed you have changed your mind about Velikovsky being a 'charlatan' or a 'fraud,' then it seems the only honest course of action would be to publicly disavow your earlier charges."[9] Urey retreated to his earlier position of silence. There was just no dealing with such people.

Silence, after all, had been the scientific community's stance toward Velikovsky after Macmillan had abandoned *Worlds in Collision*, and the war had then settled into a hostile truce. But in the 1970s one could no longer pretend that Velikovsky did not exist, even though he had not published a book since *Oedipus and Akhnaton* in 1960. Velikovsky had broken out everywhere. In 1971 Murray Gell-Mann, who had received the Nobel Prize in Physics two years earlier, viewed the burgeoning movement with alarm: "We are seeing among educated people a resurgence of superstition, extraordinary interest in astrology, palmistry and Velikovsky; there is a surge of rejection of rationality, going far beyond natural science and engineering."[10] Velikovsky in post-1968 America was no longer "fringe": he was one of the most popular authors read by college students.

The story of Velikovsky's final decade, then, is one of triumph—after a fashion. Velikovsky had always craved acceptance by the scientific community, but the scientific community never came to him. The counterculture did, and that is where he found willing soldiers. "The counterculture," of course, was not one thing, being comprised of a diverse array of peaceniks, New Age spiritual seekers, Black Power activists, the drug-addled, the musically hip, and those who just refused to keep on keeping on: "a culture so radically disaffiliated from the mainstream assumptions of our society that it scarcely looks to many as a culture at all, but takes on the alarming appearance of a barbaric intrusion."[11] This characterization came from Theodore Roszak, the writer whose 1969 book, *The Making of a Counter Culture*, gave the phenomenon its name. But Roszak also imbued it with his own agenda: a critique of scientific and technical objectivity.

Many scientists saw the excitement surrounding Velikovsky's theories as more evidence of the same: "It is as if many Velikovskyites are saying 'Nya! Science isn't so great—look at all the things it can't explain!['] ... Many Velikovskyites, like many others who have no experience in research, betray a basic hostility to science per se."[12]

This interpretation is misleading. There were many ways to be hostile to science in the 1970s—getting stoned instead of going to class, joining a radical pastoral commune, or bombing a computer center—but it is not obvious that the detailed study of orbital paths, geological formations, ancient inscriptions, and the latest reports from Soviet Venus landers were among their number. Being interested in Velikovsky meant being interested *in* science, just science of a different sort. Rejected decisively by the establishment, Immanuel Velikovsky entered the 1960s endowed with a set of supporters, and by the 1970s he had acquired, in a fit of absentmindedness, a counter-establishment: his books assigned in college courses, peer-reviewed journals dedicated to his theories, and countless invitations to address packed lecture halls. It seemed to his acolytes that finally, at long last, the time for recognition of Velikovsky's essential correctness had come.

In this chapter, Velikovsky's story takes two different paths. First, I will examine how the growing swath of Velikovsky fans perceived the scientific "establishment," a fighting word for the counterculture. Scientists were again feeling embattled in a sense just as real as (if not more so) the events that provoked the spasmodic reaction of 1950. The barbarians had come through the gates, they were sitting in classrooms, and they could vote. By 1972 Alvin Weinberg, director of the Oak Ridge National Laboratory, was very concerned: "Today, however, one wonders whether science can afford the loss in public confidence that the Velikovsky incident [of 1950] cost it. The republic of science can be destroyed more surely by withdrawal of public support for science than by intrusion of the public into its workings."[13]

But there was a second part to the story. At the very height of his popularity in the mid-1970s, Velikovsky began to withdraw from his engagement with the counterculture. Youth had appropriated him for its own reasons—not his—and Velikovsky neither liked nor trusted these camp followers of cosmic catastrophism. Earlier, as we have seen, Velikovsky had a hard time disciplining his own disciples to keep him from being classed with the "lunatic fringe." By the late 1970s, Velikovskianism had entered a phase of involution, peeling away from notoriety and into its own set of concerns defined by the whims of Immanuel Velikovsky himself. He retreated to his redoubt at 78 Hartley Avenue in Princeton, where he strove to enforce the same controls over publication that he had earlier deplored when voiced by Harold Urey.

VELIKOVSKY 101

It was difficult to attend college in the 1970s without being somewhat aware of Immanuel Velikovsky and his revolutionary theories. The extent of his popularity is hard to measure, but there is no question that, starting in the late 1960s, when one tallies up the letters from fans across the country, the tremendous sales of his books (especially *Worlds in Collision*) in college bookstores, and the numerous invitations to lecture, Velikovsky was becoming something of a phenomenon—even, one might say, a celebrity.[14] (Peter Fonda, the easy-riding poster child of the counterculture, once name-checked Velikovsky in an interview.[15]) This kind of fame was sharply distinct from the infamy that surrounded him in earlier decades. That notoriety still clung to him among older scientists, but the young found something strongly appealing.

Velikovsky had changed his tactics from diplomatic overtures to established scientists to trying to recruit the young, who were less likely to be indoctrinated into uniformitarian dogma. Velikovsky declared in 1969 that a decade earlier "I evaluated my resources and concluded that I should not spread myself on all fronts but dedicate my efforts to the goal of reaching the young generation—college students and young professors."[16] Velikovsky was very concerned with youth, and he particularly enjoyed the contrast of his own aging frame with the boundless energy of his fans. Writing in the late 1970s, he gloried that "I, an octogenarian, stride with the young of mind. There is no cult of Velikovsky; there is only the cult of scientific and historical truth."[17] He had made this same point to George McCready Price (himself a generation older than Velikovsky) in 1956, and even as far back as 1951, at the venerable age of fifty-six: "The young generation—as I learned when I lectured on *Worlds in Collision* in the Harkness Auditorium before a large audience of Columbia University students—is well able to face and deliberate on a new and unorthodox theory. The scholars who have taught and written and published not only have a vested interest in orthodox theories, but they are for the most part psychologically incapable of relearning."[18]

And so Velikovsky looked to the students. He continued to promote his campus organization, Cosmos and Chronos, despite the scandal with Stephanos. By 1967 the fledgling clubs received a four-page mimeographed newsletter from the Princeton chapter of the "Campus Study Groups in Interdisciplinary Synthesis."[19] These intermittently produced newsletters mentioned recent pro-Velikovsky publications, confirmations of predictions, and his impressive roster of upcoming talks. Between 1964 and 1969, by his own count, he had lectured at sixty college campuses, seeding Cosmos and Chronos groups along the way. On April 27, 1966, Velikovsky spoke at Yale University on "The Pyramids," and on January 24, 1968, he lectured at the Towne School of Civil and Mechanical Engineering at the University of Pennsylvania on "A Changing View of the Universe." But the peak of his tour of the Ivy League must have been "My Star Witnesses," presented by invitation before the Society of Engineers and Scientists of Harvard University, the very headquarters of enemy forces, on February 17, 1972.[20] (Harlow Shapley was still alive; he died on October 20 that same year.) Elitist about many things, Velikovsky was assiduously democratic when it came to speaking about his theories. He accepted an invitation from the Forum for Free Speech at Swarthmore, and he did not shun San Fernando Valley State College or the University of North Texas. Occasionally, he even spoke at high schools.

"The new generation on campuses—in this country—is definitely following the heretic; the professors find themselves before unbelieving audiences," Velikovsky crowed. "My visits to campuses are triumphs. And more recently some large universities re-evaluate the entire situation; thus I was selected to address the Honors Day Convocation (June 3 [1967]) at the Washington University, St. Louis, over a two-times Nobel Prize winner (Lynus [sic] Pauling)."[21] On April 14, 1970, the first Earth Day, Velikovsky achieved top billing at the Parsons School of Design with the talk "Is the Earth an Optimal Place to Live?" Stewart Brand, the editor of the *Whole Earth Catalog* and fixture of the counterculture, played backup.[22] Fitting for a man who claimed to have predicted the great discoveries of the space age, on August 14, 1972, Velikovsky spoke at the National Aeronautics and Space Administration (NASA) Biotechnology and Planetary Biology Division at the Ames Research Center at Moffett Field, California, and he returned to NASA—this time to the Langley Research Center in Virginia—on December 10, 1973, to share his "New View of Man and the Universe in Light of the Space Age." The appeal spread northward. At McMaster University in Hamilton, Ontario, in 1974, Velikovsky drew a crowd of 1,100, and he received an honorary doctorate of arts and sciences in spring 1974 from the University of Lethbridge in Alberta, accompanied by a Velikovskian conference.[23]

The change was in the audience, not in Velikovsky: his claims remained almost identical to the position sketched out in 1950 in his blockbuster book. Why were people lining up for Velikovsky? And why *now*? Velikovsky served as a middle ground for people of all political persuasions. He was an underdog in an age that had ceased to trust scientists (capturing the Left), but he also promoted deeper study of the Bible (seducing the Right) in a decade whose best-selling work was Hal Lindsey's *Late Great Planet Earth* (1970), an application of biblical eschatology to Cold War geopolitics. Velikovsky was anti-establishment but not New Left, and thus shared affinities with strands of the counterculture that have dimmed in our memory today.[24] To a speaker at the 1974 Lethbridge conference, Velikovsky was the choice of a new generation:

> The veil of amnesia has been lifted, the result is the awakening of consciousness, whether the apocalyptic agent is perceived to be an extra-terrestrial jostling, or biospheric poisoning, atomic weaponry overkill, or overpopulation; or whether one has experienced the disintegration of his world view by chemical inducement—a magical mushroom or the fabled LSD. The generation of the *Whole Earth Catalogue* has experienced the catastrophe and, consistent with Dr. Velikovsky's amnesia theory, they no longer itch to re-enact the primordial paroxysm that heralded our present age—the bomb has gone off![25]

This view fits nicely with Roszak's anti-technocracy interpretation of the youth movement and Gell-Mann's fears about a dawn of obscurantism, but the Velikovskians themselves did not share it. De Grazia noted some of the irony in that Velikovsky "could easily be fit (noone [*sic*] knowing his character) into the mold of anti-authoritarian ideas and leadership exceedingly popular among those in that era, town, and age-group."[26] Could be, but only with a bit of Procrustean hacking and stretching. Velikovsky deplored the student rebellion, and his politics remained conservative.[27]

For Velikovsky and the inner circle, the youthful exuberance for his doctrines was both flattering and a bit of an embarrassment. His core of local followers continued to present Velikovsky as a scientist and propagandized for catastrophism in somber tones. Lynn Rose, a professor of philosophy from Buffalo, and C. J. Ransom, who held a PhD in plasma physics, attempted in 1974 to assemble a petition of credentialed scientists in Velikovsky's favor, and other followers sensed a change in scientists' attitudes to the interdisciplinary synthesis.[28] Walter Kaufmann, the renowned Nietzsche scholar in the Princeton University philosophy department, was a huge fan and urged his friend into more scholarship, not public propagandizing or sniping with critics.[29] If youths were following Velikovsky en masse, the Velikovskians wanted it to be because he was correct, not because he was rejected by the "establishment."

Given that the main constituency for Velikovsky among college students were those interested in science, excited by flights of the imagination, and enthusiastic readers of paperbacks, one might think that another group that appealed to the same constituency would be equally engaged with Velikovsky: science-fiction authors.[30] One would be wrong. Among the most persistent and hostile critics of Velikovsky across his entire career were the luminaries of science fiction. The celebrated Polish science-fiction master Stanislaw Lem classed *Worlds in Collision* among "fake-science books" and characterized it and its like as "excrements of the mind."[31] American authors were no kinder. A single example will suffice, although one could extend the list much further.

The leading anti-Velikovskian among the sci-fi crowd was the grand titan of the genre: Isaac Asimov himself. Although he was clearly aware of *Worlds in Collision* in 1950 (the year before his own *Foundation* appeared as a stand-alone book), and other science-fiction writers (like L. Sprague de Camp) had joined the chorus of hostile reviewers, Asimov waited until 1969 to review the book. Partly this was because the paperback release gave him an excuse to revisit the topic, partly because Velikovskians had written him angry letters decrying his dismissal of Velikovsky's reconstructed chronology in his own book on the Bible, and partly because of the mood of the times. "There is always something pleasant about seeing any portion of the 'Establishment' come a cropper, and the Scientific Establishment in particular," he noted. "Scientists, these days, are so influential, so far out of the ordinary clay, so supreme in their self-confidence, and (to put it in a nub) so 'smarty-pants' that it is a particular pleasure to see them stub their toes and go flat on their faces."[32] His review raised physical facts (such as the existence of long-lasting stalagmites in limestone caves) that seemed inconsistent with Velikovsky's catastrophic scenarios.

As Asimov continued to comment on Velikovsky, a backlash emerged, not least because of his association with and defense of Carl Sagan, whom the Velikovskians cast as Shapley's successor, the new commanding general of the anti-Velikovskian forces. Among Asimov's many sins was his denial of the suppression thesis about the Macmillan boycott: "Wrong though the reaction of some astronomers was, there was no attempt made to suppress the book as a book; merely to withdraw from it any official label as 'scientific.'"[33] Suppression was not the issue, and Velikovsky was no Galileo. "For every Galileo who was right there were a thousand crackpots who were wrong," he wrote to Frederic Jueneman in 1972. "If Velikovsky were to be right, it wouldn't be because the establishment opposes him. If that were all, every idiot would be right."[34] Nothing, however, seemed to dissuade the supporters, and by 1975 he had had it, ending a correspondence abruptly: "[Velikovsky] is a cult-leader and his followers are cultists and I am against him and them, and I am not going to change just to keep you as a fan."[35] Invoking Velikovsky often generated such an angry response.

And surely this was part of his growing appeal: bringing Velikovsky up in class enraged science faculty. As Chris Sherrerd, a marginal member of the inner circle, wrote to Velikovsky in 1968: "I suspect that much of the support you are finding on college campuses is mot[i]vated not so much epistemologically but rather socially: as part of a general revolt of today's youth against 'the establishment.'"[36] Support for Velikovsky concentrated among the lay public, humanists and social scientists, and, quite interestingly, scientists working for private industry.[37] Yet explaining support of Velikovsky by invoking "anti-establishment" sympathies is no explanation at all. In the 1970s, *everyone* was opposed to the establishment. As historian Bruce Schulman has observed: "Richard Nixon hated the establishment. He loathed the prep school and private club set, the opera-goers and intellectuals, the northeastern Ivy League elite."[38] When the president of the United States can claim anti-establishment credentials, we need a more nuanced framework. The point was not opposition to an establishment, but what the establishment signified to those who opposed it.

For many who cheered Velikovsky, resentment was only part of it. People backed Velikovsky because they thought he was *right*. A pro-Velikovsky article stated clearly in 1968 that his resurgence "is due to one circumstance that the Scientific Establishment did not foresee when it all but unanimously dismissed Velikovsky as a crank and mocked his theories as ridiculous. With the accumulation of new knowledge, especially that gathered in the last decade by space probes"—such as the temperature of Venus, the radio noises of Jupiter, and especially the disputed existence of hydrocarbons in the Venusian atmosphere—"Velikovsky's picture of the solar system has proved to be more accurate on many important points than the theories embraced by the Establishment."[39] A combination of excitement about new astrophysical discoveries, a chafing at the bonds of authority, and the widespread distribution of Velikovsky's works in paperback changed the climate.

Perhaps nowhere is this more visible than in the rise of college courses dedicated to exploring Velikovsky's work. Much as he had long predicted and fervently desired, *Worlds in Collision* became required reading in colleges across the United States and Canada. Velikovsky's first major entry into a college classroom was in a course he taught himself at the New School for Social Research in New York City in fall 1964, at the invitation of its dean, Horace Kallen, a supporter of Velikovsky's since the heady clashes of 1950. Called "The Changing View of the Universe and Man's Past," the course consisted of six sessions on Tuesday nights from 6:20 to 8:00 in the evening, with a fee of $3.50 for a single class or $18 for the whole series.[40] Kallen had hoped that presenting Velikovsky in this dignified environment would dim the sensationalism of his views and spark a real scientific conversation, but he confessed his disappointment at the course's deterioration into a cult of personality: "I had hoped that your New School lectures might be a positive step in this direction; but they seem to have focussed more attention on Velikovsky than Velikovskyism."[41]

Of course, most courses on Velikovsky were not taught by the master himself, but by others who either admired or despised his work. The latter was arguably more common. For example, W. C. Straka, an assistant professor of astronomy at Boston University, taught a course called "Science and Anti-Science in Astronomy," where he assigned *Worlds in Collision* in order to debunk it. As Straka saw it, the only possible place for Velikovsky in a university was as an *exemplum malum*: "From the standpoint of a valid or usefuly [sic] hypothesis, Velikovsky's work merits no further consideration. But the situation is useful as an illustration of the conflict of science and anti-science, with Velikovsky clearly in the latter category. It is as such I use him in my course."[42]

More intriguing were the courses that defended Velikovsky's theories. Given the outsider status of Velikovskianism, it is not surprising to learn that many of the instructors were adjuncts or others who had to fight to get their courses listed. In 1971 C. J. Ransom struggled to get a course on Velikovsky's theories accepted at Texas Christian University in Fort

Worth, Texas. This night course proved very popular, with an enrollment of twenty-nine students. "Overall the students agree with your theory," Ransom wrote to Velikovsky. "Most of the discussion concerns details, and no one seems opposed to the total concept."[43] Three days later he added: "To the young people, the theory seems quite logical and some do not understand why there is so much controversy."[44]

While giving talks about Velikovsky's theories at campuses in preparation for this course, Ransom came to a realization: "The more discussions I have, the more apparent it becomes that opponents do not know the theory and are unable to refute it when confronted with the facts. This points to a great need for personal contact where there is the availability of immediate rebuttal."[45] Courses were an ideal mechanism to address youths. Lynn Rose used some Velikovskian material in his philosophy of science class at SUNY-Buffalo in 1971, and by 1973 he was teaching courses entirely devoted to *Worlds in Collision*.[46] Rose was tenured; he could do as he wished. Others had to exploit the makeshift experiments of the age of Aquarius, such as the proliferation of "free universities" that paralleled established institutions of higher education. There were courses on Velikovsky at the Free University at the University of Pennsylvania, the University of Connecticut Free University, and even at the Medical College of Virginia.[47] Mary Buckalew (the recipient of the mailings from Stephanos) taught courses in the English department at North Texas State University with stunning results: "The students *love* it! They respond to my enthusiasm and conviction, of course; but they are also eager for alternatives to the training given them in the scientific disciplines. I have brought in this way hundreds of young minds to your work."[48]

But the story was not all optimism. A scheduled course at the University of Alabama was canceled at the last minute because of controversy over an advertisement the teacher posted for the class.[49] A course at Penn State was also scrapped, with the argument that "students at the freshman and sophomore level can't judge what is correct or incorrect reasoning. We feel they should only be taught material that is correct beyond any doubt."[50] (The wrangling about the meaning of that statement occupied many pages of appeal and protest. Although the course was never reinstated, the professor was eventually granted tenure.) Things ended poorly for Buckalew as well. She became an increasingly erratic correspondent as she fretted about Marxist radicalism on the campus. By 1975 the final letters from her stored in Velikovsky's archive explicate her fears of a United Nations take-over of the United States, Masonic conspiracies, and other fever dreams from the fringes of the American Right.[51] The house of counterculture had many rooms, and Velikovskians dwelled in several of them.

CONSCIENTIOUS OBJECTORS, INDIANS, AND OTHER ALIENS

It became difficult to keep on message. If in the late 1960s Velikovsky could mobilize swiftly and with extreme prejudice against possible Reichian contaminations of his views, by the early 1970s there were too many different trends, such a potpourri of speculations and fantasies that he could not enforce orthodoxy on all fronts. Instead of merely attempting to confirm or defend his writings, this new generation took their basic correctness for granted and then strove to build new structures on the edifice of his interpretations of ancient myth. One might think of this as the birth of "applied Velikovskianism."

Applied sciences take fundamental theories and focus them on specific, real-world problems. The Velikovskians were no different, and the questions they confronted armed with Venus and the Ipuwer papyrus were problems of their time. Perhaps foremost among them was the Vietnam War. Consider Curtiss Hoffman, who began corresponding with Velikovsky in the early 1960s in connection with a high school science fair project he put together on *Worlds in Collision*. Velikovsky encouraged Hoffman's interest in antiquity. (Many of those who gravitated to Velikovsky before graduating high school were drawn to the reconstructed chronology.) Hoffman joined the ranks of those college students excited about Velikovsky and did quite well for himself. He attended Brandeis University and then enrolled in graduate studies in ancient history at Yale.

And then, in 1968, he encountered the draft board. Hoffman had no wish to be sent to fight in the jungles of Southeast Asia. He turned to his mentor and asked for a letter to the board, arguing that his adherence to Velikovsky's doctrines undergirded his pacifism and that he should be considered a conscientious objector. Velikovsky complied, writing that "Curtiss Hoffman became convinced in the truth of this concept and realizing that conflicts between races and nations and even individuals are rooted in subconscious racial memories of traumatic events on a global scale, and, thus early in his life, became a conscientious objector to all military solutions of international conflicts."[52] Hoffman was ecstatic. "If anything can move a draft board, this should," he wrote. "As you probably

know, most conscientious objector claims are supposed to be based on 'religious training or belief.' I think that in my case I can point to your ideas
as a source for my belief that man is not a tabula rasa and carries within
him, not as an imprint of society necessarily, certain basic patterns—of
which the memory of catastrophism is one, if indeed the most striking
example."[53]

Velikovsky heard nothing from Hoffman after this exchange and wrote a
year later to find out whether Hoffman had in fact been drafted. He hadn't;
he had failed the physical. At this point, however, Hoffman had advanced
in his graduate studies and no longer found Velikovsky's account of the ancient world persuasive. "As you are a psychiatrist, you must have realized
that my reaction to your work was not entirely based on reason, coming as
it did at a time when my emotions were coming to the fore and I was struggling to control them," he wrote back. Velikovskianism had been a kind of
therapy, but the therapist had strayed from the path, becoming "attached
to his theory and spend[ing] more time devising arguments to attack his
opponents than to further refinement, development and publication of his
own work." Velikovsky was now "doing a disservice to himself," and Hoffman wanted nothing more to do with him.[54]

For one who tackled another signal problem of the 1970s—racial
justice—Velikovsky was also the solution. Vine Deloria Jr. was one of the
most controversial activists for Native American causes in the 1970s and
continued to publish broadly until his death in 2005. Sometime after the
appearance of his first book in 1969—*Custer Died for Your Sins: An Indian
Manifesto*—he discovered the works of Immanuel Velikovsky. In 1973 he
published *God Is Red: A Native View of Religion*, which critiqued Judeo-
Christian ideologies built upon Genesis and instead argued for a more
polyvalent theology grounded in American myths. His appropriation of
Velikovsky was twofold: to bolster his critique of biblical miracles, to be
sure; but primarily because if Velikovsky were right and myths referred to
actual events, then Native legends would prove essential to recovering humanity's ancient past. "Hitherto we have had our oral traditions debased,
our religious myths and legends downgraded, and our perceptions of law
and social reality derided as superstitious fictions of savages," he wrote to
Velikovsky in 1977. "Yet when these legends are placed within the framework of a planetary history which coordinates the separate traditions of
human societies within a joint framework connected by catastrophes
of planetary significance, a great many Indian traditions take on a new
historical importance."[55] After meeting Velikovsky at the 1974 McMaster

symposium, Deloria became a devoted correspondent and active defender of *Worlds in Collision*, at one point calling Velikovsky "perhaps the greatest brain that our race has produced."[56]

Neither conscientious objector claims nor justice for the American Indian had been imaginable paths for Velikovsky when he started down his path to catastrophism. What he did foresee was someone taking his claims as evidence for another cosmic theory to which he objected. Ted Lasar and Wilhelm Reich were one matter—those he could handle internally. But what could he do about Erich von Däniken? For in this Swiss hotelier-turned-author, he encountered an applied Velikovskian who not only used *Worlds in Collision* as evidence for a completely different theory, but one who rivaled, even exceeded, Velikovsky himself in popularity among the American counterculture. This was a parallel front in the pseudoscience wars, and one that threatened his own efforts.

Today von Däniken is substantially more recognizable than Velikovsky. This is in part because he is still publishing—his most recent book appeared in 2010, although with markedly reduced sales from his heyday in the 1970s (when he sold in the millions)—and probably more because his arguments and scenarios for ancient history have continued to be appropriated in science-fiction blockbuster movies (*Indiana Jones and the Kingdom of the Crystal Skull*, *Alien vs. Predator*, and *Transformers: Revenge of the Fallen*, to name three recent ones). The link with fantasy should not be too surprising. On the copyright page of von Däniken's huge international bestseller, *Chariots of the Gods?: Unsolved Mysteries of the Past*—published in German in 1968 and soon translated into multiple languages—von Däniken stated: "This is a work of fiction." One simply cannot imagine Velikovsky reissuing *Worlds in Collision* with such a disclaimer (or, for that matter, with a question mark at the end of a title). The arguments of the two authors diverge. Von Däniken claimed in this work that the wonders of ancient civilization were the products of "gods"—visitors from other worlds—who came to Earth in antiquity, interbred with the almost simian humanoids, bequeathed civilization to their progeny, and introduced technological advances (pyramids, cities on mountaintops). But there are also similarities, and von Däniken certainly owed a debt of method to his precursor. He too read the Bible and mythology in his quest for ancient astronauts: "The almost uniform texts can stem only from facts, *i.e.*, from prehistoric events. They related what was actually there to see." Von Däniken even saw the two approaches as compatible: after the Venus catastrophe, perhaps the extent of planetary destruction witnessed from space prompted the Martians to visit Earth.[57]

Von Däniken's theories skyrocketed in popularity, often among the same set of college enthusiasts who devoured the collected oeuvre of Immanuel Velikovsky. As both journalistic and debunking accounts lumped the two writers together, purist campus Velikovskians feared conceptual slippage.[58] One fan from North Texas State wrote shortly before being drafted for a tour in Vietnam with worries that Velikovsky might agree with ancient astronaut claims, and he wished to be assured that this was not so.[59] Velikovsky received many such queries, and he generally replied in the negative, but not unambiguously. Usually, he would note that any account of UFO sightings and contemporary alien visitations was "a view which does not find any credence with me," or that he regarded "the stories about visitors from other worlds in our days as utterly nonsensical."[60] There is some conceptual sleight of hand here: Velikovsky placed doubt on claims of aliens visiting Earth *in our times*. In his unpublished writings, on the other hand, Velikovsky prefigured views similar to those von Däniken would later espouse. For example, he wrote as early as August 1950 to Charles Jacobs, professor of English at the University of Bridgeport, in atypically stilted language:

> However, it is in the realm of the possible that, what we hope to attain in a not too far futures, could have been achieved by dwellers of another planets a few thousand years earlier. Therefore I would not, without further inquiry, regard as utterly impossible a visit of intelligent beings from the outer space. A group of titans descending on Mount Hermon (and unable to leave the earth) is described in an old Hebrew legend, which elaborates on Genesis 6:1–4.[61]

He continued to hold to this interpretation even *after* von Däniken. In his unpublished manuscript on the catastrophes of Genesis, Velikovsky referred to this same "Nephilim" story as "a literary relic dealing with a visit of intelligent beings from another planet," noting that "the extraterrestrial visitors made their landing as if in advance knowledge of the impending catastrophe of the Deluge."[62] No wonder that Lewis Greenberg—then an assistant professor of art history at Franklin and Marshall College and soon to become a close acolyte of Velikovsky and editor in chief of the leading pro-Velikovsky journal (*Kronos*)—conflated Velikovsky and von Däniken in a 1971 letter shortly after their first meeting.[63]

In public, however, Velikovsky and the Velikovskians were careful to put significant distance between their own claims and those of the godly charioteers. They insisted that Velikovsky was more careful with evidence, while von Däniken admitted to massaging facts to suit his narrative. Even critics, like science-fiction writer Ben Bova, agreed that "there is no question of fraud, or of winking at known facts, in Velikovsky's case."[64] But for the Velikovskians, the major reason why the two men were different, despite their easy-to-confuse assonant foreign names, was that "astronomers studiously ignore[d] this immensely popular unorthodox theory" of von Däniken's, but they did not (according to this letter writer) ignore Velikovsky.[65]

With *Chariots of the Gods?* selling four million copies, and public funding for archaeological research collapsing while the costs of establishing digs at foreign sites ballooned, professionals felt there was cause for concern. The problem—as it had been for those confronted by Velikovsky in 1950 and by Lysenko in 1948—was what to do about it. At the 1978 meeting of the American Anthropological Association, a resolution was introduced about whether the organization should condemn von Däniken as a "misrepresentation of science," as well as racist (implying the world's non-white populations could not have built civilizations on their own) and just plain wrong. The resolution was overwhelmingly defeated—not because the archaeologists agreed with von Däniken, but because they felt that overt hostility would stoke the countercultural flames instead of dousing them.[66] In this, the Velikovsky example loomed large for them, much as it had for the creationists.

COUNTER-ESTABLISHMENT SCIENCE, IN PRINT AND IN PUBLIC

Ben Bova, who as editor of *Analog* science-fiction magazine believed that the enthusiasm for Velikovsky was news and thus should be addressed, privately fumed against the Velikovskians. "Sometimes it's not your enemies that hurt you, it's your friends. The only thing more tedious, sententious and lacking in physical proof than Velikovsky's own writings are the writings of many of those who attempt to support his thesis," he wrote to Lynn Rose in 1974. "I'm not interested in counting alleged errors in articles either by or for Velikovsky. I am interested in physical evidence either for or against his ideas, the kind of evidence that one uses to decide the validity of any other physical theory."[67] And that evidence was sorely lacking: "Velikovsky's ideas hold about as much water as a well-worn piece of cheesecloth. They're the result of trying to find one sweeping explanation for every strange and wonderful event that confronts us; this is a syndrome that's very common in science fiction."[68] Nonetheless, in the spirit of fairness, he spent two years negotiating with Velikovsky's inner circle to get him to write a piece for *Analog* to address negative articles that had appeared there. After countless stipulations about copyright, billing on the cover, space constraints, and more, Bova called the whole thing off in 1975. "This hardly seems like the attitude of a man who wants to use rational discourse to convince skeptics," he told Frederic Jueneman. "He's acting like a petulant child."[69]

There was a time when Velikovsky would have leapt at the chance to be published in a broad-circulation magazine like *Analog*. Behind his rebuff of Bova and in his demands for top billing was a recognition that the times had changed. The 1970s saw the emergence of dedicated journals that promoted Velikovskianism, packed with articles bristling with footnotes, equations, and new arguments.[70] Though Velikovsky had not been accepted into the establishment, he now found himself with a full-blown counter-establishment. This was a recent development. In the late 1960s, he felt so locked out of print venues for his ideas—aside from his books, of course, which continued to sell—that he even took special pains with an undergraduate magazine.

In 1967 *Yale Scientific Magazine*, "operated by undergraduates with complete editorial freedom" from the elite educational institution in New Haven, published a special issue focusing on a dispassionate scientific discussion of one aspect of Velikovsky's theories: the issues surrounding Venus, including recent discoveries from space probes. The editor, John W. Crowley, insisted that the magazine "does not pretend either to vindicate or to demolish Velikovsky's ideas in this issue; we seek only to present a paradigm for further discussion by avoiding the abusive tone" of prior discourse.[71] The centerpiece of the issue was Velikovsky's article "Venus—A Youthful Planet," which had been written in 1963 in reaction to the Bargmann and Motz letter in *Science* and submitted to the *Proceedings of the American Philosophical Society* by Princeton University geologist Harry Hess. The dispute over whether to publish it almost ruptured the editorial board of that journal, so a separate panel was established to decide upon the fate of the piece. In January 1964 Velikovsky was informed that the article had been rejected, and it was likewise rejected by the *Bulletin of the Atomic Scientists*, where Velikovsky had subsequently submitted it (again via Hess) in reaction to the negative review by Howard Margolis that had prompted De Grazia's letter to Harold Urey. In the end, *Yale Scientific Magazine* was to be its home.[72] It was followed by critiques from University of Kansas chemist Albert Burgstahler and Columbia astronomer Lloyd Motz, rebutted at length by Velikovsky. The issue concluded with a letter from Horace Kallen, who applauded the venture, only regretting that he was unable to submit a contribution by the deadline.[73] In 1967 this was the best that Velikovsky could do; within five years, the situation had utterly changed.

The first Velikovskian journal initially had nothing to do with Velikovsky. It was called *Pensée* and was officially published in Portland, Oregon, through the Student Academic Freedom Forum of Lewis and Clark College. It is very difficult to reconstruct the early history of the journal. According to a press release for a Velikovsky symposium hosted by *Pensée* at that college in 1972, it was founded in 1966 by David Talbott, then an undergraduate at Portland State University.[74] No issues seem to survive from those early years. In the winter of 1970–71, the journal was reactivated under the editorship of Stephen Talbott, a graduate of Wheaton College (where he had edited the school paper). Judging from these early issues, *Pensée* began as a rather typical student journal in those countercultural days, with opinion battles pro and con on issues like Vietnam (June 1971), local environmental activism (November 1971), and abortion (January 1972). Each issue began with an amusing series of sarcastic commentary on national and local issues—often with a conservative bent—and signed pieces hailed from undergraduate and graduate students across the Portland region. Only rarely did Stephen (who characterized himself in his byline as "an on-and-off-again student") choose to pen a piece, as he did in June 1971 arguing that "The Population Crisis Is a Put-On," a view that skewed slightly to the right in the wake of Paul Ehrlich's *The Population Bomb*.[75] The journal had nothing to do with science, and not even a hint of Velikovsky—no Venus or Egyptian king lists or Harlow Shapley. In May 1972, that all changed. Stephen's brother David Talbott suddenly appeared as the magazine's publisher (earlier it had been Robert G. Wallenstein) and Stephen remained the editor, but the journal's entire emphasis shifted. They launched a series entitled "Immanuel Velikovsky Reconsidered," which hoped to examine the debates over cosmic catastrophism, including some contributions from Immanuel Velikovsky himself.[76] It was a fateful decision: the circulation of the magazine spiked, its content became entirely dedicated to Velikovsky, and by 1974 the editorial board was populated by the inner circle, including Ralph Juergens, William Mullen, and C. J. Ransom.

The counterculture gave birth to *Pensée*, Velikovsky's theories gave it a mission, and then *Pensée* returned the favor by bringing Velikovskianism to the counterculture. *Pensée* provided a forum for his supporters (and some critics) to discuss their thoughts, to puzzle through problems in the chronology of the Middle Kingdom or the orbital damping of Venus—in short, to build a community. Circulation boomed: for the two and a half years of its Velikovskian adventure, *Pensée* had an annual circulation of ten to twenty thousand, but the first issue in the Velikovsky series was reprinted twice with a total run of seventy-five thousand copies.[77] Not bad for a fly-by-night operation in Portland. Submissions flooded in, and the Talbott brothers (principally Stephen) had to develop a system to filter out the good from the bad. They borrowed one from the academic establishment, one that has often been held up as differentiating "real science" from "crackpot works": peer review. Every submission to *Pensée*—with the important exception of the many writings by Velikovsky—was reviewed, often by the inner circle. Not surprisingly, many of the critical anti-Velikovsky pieces were rejected or returned for revisions, usually based on logical flaws or poor grasp of the empirical data.[78] (Several were published upon revision.) If peer review serves as a metric to differentiate science from pseudoscience, then *Pensée* was on the side of the angels. People read it and came to think of Velikovsky as more than just a fun read—this might be the birth of a new science.

Velikovsky had for years been trying to get a hearing before a committee of scientists, an organized panel of diverse experts in Assyriology, astrophysics, planetary science, and history. In 1966 he approached the American Association for the Advancement of Science (AAAS) and got no response.[79] Yet around the same time as the phenomenal success of the first Velikovskian journal, certain members of the academic community thought, for the first time since the uproar of 1950, that they should confront Velikovsky directly. There is no question that student interest in *Worlds in Collision* and his other books motivated these establishment scientists, and in the late 1960s Cornell astronomer Carl Sagan and a few others informally suggested that perhaps it was time to refute Velikovsky so that students would not be led astray by one-sided endorsements of cosmic catastrophism.[80] But although Sagan put together a symposium at the AAAS meeting in 1969 on UFOs, there was no movement on the Velikovsky issue.

That changed in 1973, when Ivan King, the chair of the astronomy section of the AAAS, and Owen Gingerich, chair of the history of science section, decided to sponsor a symposium on Velikovsky for the 1974 annual meeting, scheduled for February. At first, Velikovsky and his circle were excited: the time had come for a public apology from the scientific community and a recognition of Velikovsky's correct predictions. But in November 1973, in response to a query from Stephen Talbott, Ivan King offered this explanation of the invitation:

> Thus the presentation of this symposium does not in any way imply that Velikovsky's ideas are any more acceptable to scientists than they have been in the past. What we do recognize, however, is that his ideas continue to attract a large following. In its role of attempting to build a bridge between science and the public, the AAAS does not wish to turn its back on an influential movement whose tenets appear to be destructive of some of the basic principles of science as we know it. . . . This is not a debate on the correctness of Velikovsky's view of the planetary system; none of us in the scientific establishment believes that such a debate would be remotely justified at a serious scientific meeting.[81]

"Disappointment" is too weak a word to describe Velikovsky's reaction to this statement. King had visited him personally in Princeton to persuade him to attend, and he would not take these words at face value. "Since I cannot imagine that you acted in bad faith," he wrote, "I must assume that you were urgently approached by a number of 'guardians of the dogma,' who in the past have made themselves known by their vituperations, not arguments, and thus will be known in the history of science as obscurant."[82] The conspiracy against Velikovsky lived. He would go to San Francisco and meet it head-on.

The February 1974 AAAS meeting is, next to the 1950 affair, the single most discussed episode of Velikovsky's career, the climactic Battle of the Bulge of the pseudoscience wars. Everyone who has written about it has been eager to declare one side—the scientists or Velikovsky—the victor.[83] My goal here is somewhat different: to focus on the interpretations attached to this event, especially by the Velikovskians. For 1974 was not just the year of the AAAS meeting; it was an *annus mirabilis* of four separate Velikovsky symposia, of which San Francisco was just the kickoff. Putting the controversial first symposium in context also highlights how the

AAAS event was not so much an attempt to staunch the countercultural current on which Velikovsky was borne, but more a stab at joining it in order to shape the flow, specifically to ameliorate the distrust of science.

If that was the goal, it didn't work. The symposium took place at the St. Francis Hotel on February 25, 1974, before a tremendous audience (estimates vary between five hundred and the room's capacity of fifteen hundred, with the actual number likely closer to the latter) and lasted for a total of seven hours—four hours in the morning and an additional three in the evening. The participants—sociologist Norman W. Storer, astronomer Carl Sagan, physicist J. Derral Mulholland, statistician and amateur Assyriologist Peter Huber, physicist Irving Michelson, and Velikovsky himself—were all supposed to speak within defined time periods and then address some questions from the audience. Sagan and Velikovsky both went significantly over time (necessitating the evening session), although Velikovsky more so, invoking the argument that only he and Michelson were speaking on his behalf, and that most of the audience came to see him anyway (which happened to be true).[84] (Michelson broke with Velikovsky in July 1974, publishing a refutation of the energy requirements for the Venus scenario in Ben Bova's *Analog* and sending Velikovsky a wounded farewell letter.[85]) The anti-Velikovsky papers were later published as a volume by Cornell University Press after negotiations broke down over the length of rebuttals Velikovsky wished to include. He and Michelson published their presentations in *Pensée*.[86]

After the fact the discussion was reduced to a head-on confrontation between the seventy-nine-year-old catastrophist and America's most popular astronomer, Carl Sagan. Those looking for sharp verbal fireworks were mostly disappointed. Velikovsky spoke at great length and eloquence about his concepts, which he refused to call a "theory": "My work is first a reconstruction, not a theory; it is built upon studying the human testimony as preserved in the heritage of all ancient civilizations. All these civilizations, in texts bequeathed beginning with the time man learned to write, tell in various forms the very same narrative that the trained eye of a psychoanalyst could not but recognize as so many variants of the same theme."[87] After making some suggestions for the *Viking* probe to Mars based on his conclusions, he ended his speech by hurling down a gauntlet: "None of my critics can erase the magnetosphere, nobody can stop the noises of Jupiter, nobody can cool off Venus, and nobody can change a single sentence in my books."[88] Applause ensued, and so did questions, which Velikovsky addressed at length, his responses to single inquiries at times occupying whole pages of the transcripts and wandering across the range of human erudition (but not always answering the question).

Sagan's paper was also far too long for a twenty- to thirty-minute presentation, and he skimmed through the paper, reading selected segments here and there. He did display moments of sparkling frustration, as in response to a question about the radio noises on Jupiter: "There is bound to be some residual magnetism everywhere. There is bound to be, just as in the Earth's oxidizing atmosphere there are today hydrocarbons. Methane is one part per million of the Earth's atmosphere. That has nothing to do with manna. It has nothing to do with any of this."[89] Likewise Velikovsky, in the midst of a voluminous "short comment" on Norman Storer's presentation, had one of his several rhetorical triumphs: "But neutral is not objective. You cannot be objective between evil and the victim of evil, neutral between the behavior of science—how it was and how it started from 1950 and continued till today, almost till today, till yesterday, better let us say."[90] But the net result was more confusion than enlightenment. Both the Velikovskians and their critics had unanimous views of who won—they just were different unanimous views. From these accounts, we should note the intense effort both sides took in spinning the AAAS symposium, confirming the hunch that the event was more about public relations and propaganda (for both sides) than coming to a scientific evaluation of Velikovsky's theories.[91]

Norman Storer, whose opening paper at the symposium was decidedly lackluster, found the case more interesting in retrospect—not because of Velikovsky, but because of the audience. "My private opinion is that the old guy is quite out of his tree, and I am much more negative about him after having seen him in action than I was before the San Francisco meeting," he wrote later. "But the interesting thing is his following—who are they, what structural circumstances might account for their 'faith,' and what sustains them in the face of overwhelming evidence to the contrary? Questions like this suggest a rich lode of sociological material for someone who wants to dig into 'em."[92] But there was to be no extensive study of Velikovsky's countercultural following, nor any further attempts to confront Velikovsky directly in a forum on the scientists' own turf. Dennis Rawlins, a Fellow of the Royal Astronomical Society, deftly noted the catch-22: "If one simply ignores the crank, this is 'close-mindedness' or 'arrogance.' If one then instead agrees to meet him in debate, this is billed as showing that he is a serious scholar. (For why else would the lordly establishment agree even to discuss him?) Irksome *either* way."[93] So the 1974 experiment was never repeated. It had been neither success nor failure. It raised the visibility of scientific opposition, but it had resolved nothing.

The Velikovskians, however, were on a roll. Lewis and Clark College, in Portland, hosted the first Velikovskian conference on August 16–18, 1972, under the prodding of *Pensée* and its editors. But 1974 was different: studded with panels and discussions of Velikovsky, all of them adulatory. In May, Velikovsky received his honorary degree from Lethbridge University in Canada, and then traveled to McMaster University in Hamilton, Ontario, for an oversubscribed conference on June 17–19 on "Velikovsky and the Recent History of the Solar System." He had some time to rest before traveling to Duquesne University in Pittsburgh for a History Forum on October 27–November 2, and immediately shuttling to a November 2 session of the Philosophy of Science Association meeting at Notre Dame on "Velikovsky and the Politics of Science," where the speakers were three Velikovskians (the man himself, Lynn Rose, and Rose's fellow Buffalo professor Antoinette M. Patterson) juxtaposed with one critical physicist (Michael W. Friedlander of Washington University in St. Louis).[94] Papers from all of these venues were published shortly afterward, indicating that the San Francisco meeting had done little to dampen enthusiasm for the author of *Worlds in Collision* and his theories.

And when those papers appeared in 1974, they usually appeared in *Pensée*, which had become—next to Velikovsky himself—the single most important clearinghouse for information about his concepts. Not imagining that they could contact the man in Princeton, people wrote to the journal, which they assumed had a direct connection to Velikovsky. That connection was in fact rather tenuous, and the tensions surrounding just how much control Velikovsky did or should have over the contents of *Pensée* would soon come to a head. The journal was not like FOSMOS, entirely staffed by individuals with whom he had close personal relationships. The Talbott brothers were not in the inner circle, and the contributors for *Pensée* included critics of Velikovsky's theories. When Velikovsky attempted to assert control over it, he shattered the entire venture.

THE LION IN WINTER

Why was *Pensée* so popular? To judge from the correspondence of contributors and editors (the subscribers are, sadly, impossible to trace fully), the primary reason was that this seemed to be *real*, to look like actual scholarship, actual science. It had peer reports, boasted a large readership, and featured detailed discussions on many sides of an issue. Given how it behaved, this couldn't be pseudoscience. True, many of the citations were to other articles in *Pensée*, but that was because establishment journals had unfairly frozen out these debates. According to Lewis Greenberg, writing to Velikovsky in 1974: "The credibility of *Pensée* depends upon its ability to remain positively inclined to your work without seeming to be too one-sided or partial."[95]

Not everyone thought the situation was clear. For example, while it was obviously beneficial for the journal not to appear as a propaganda organ, perhaps some skewing was permissible. "I do not fully agree that our effort is scholarship only and not propaganda, if the definition and not connotation of propaganda is used," C. J. Ransom wrote to Stephen Talbott in 1975. "We do not need to propagate discussion of the other side since they have their own propaganda machines. However, a certain amount of mixing would act as a catalyst to encouraging discussion of Velikovsky's works."[96] Talbott, in a slightly earlier letter to Lynn Rose discussing the same issues, disagreed, and his response is worth quoting at some length:

> Considerations centering on the overall balance among journals do not by themselves suggest any particular balance for *Pensee*. . . . [B]ut we have to deal realistically with the fact that, for example, the nearly universal opinion among "experts" is that available data do not allow for a recent Velikovskian episode on the lunar surface. If our more informed readers see us failing to interact with the weight of conventional opinion on the subject, they can only discount us. After all, there exists every sort of wierd [*sic*] publication and society, surviving merrily on in isolation, while boldly "challenging" accepted viewpoints in their widely unread pronouncements. Nobody bothers with them. The reason *Pensee* has achieved what it has is that it went straight into the scientific community (read: community of conventional thinkers) with its bold challenges, seeking in every way possible to avoid the isolation that normally would befall such an effort. That meant involving conventional antagonists.

The problem, he continued, was partially the perennial editorial conundrum of how to fill an issue: "It is my opinion that, up until now, and even now, we could not put together a presentable series of issues consisting solely of contributions by Velikovskians. The scholarship is simply not there; the result would look anemic. There are too few Velikovskian researchers, and not in enough fields."[97]

Velikovsky had actually been intimately involved with the editing of the journal for some time. For example, Thomas Ferté, of the humanities department at the Oregon College of Education, was incensed with Stephen Talbott as early as 1972 for allowing Velikovsky, he claimed, to edit his submission for *Pensée* for both content and style; he threatened legal action if the piece was published.[98] Likewise, an article in *Science* on the AAAS meeting attributed the rise of Velikovskianism to *Pensée* and noted that while there was no financial connection between the two, "there is a kind of symbiotic relationship—he is good for circulation and circulation is good for him—and Velikovsky has, on occasion, exerted editorial influence." When interviewed for the piece, Velikovsky told the author that he had at one point given the editors an "ultimatum" when he wished to respond to a critical article in the same issue it appeared, as opposed to waiting for the next issue: "I said if they didn't do so, I would never write for them again."[99] *Pensée* backed down.

Ironically, the very success that prompted Velikovsky and his cohort to exert stronger control over the journal also induced the Talbotts to assert their autonomy. Relations soured among the principals, and in January 1975 Greenberg called Stephen Talbott "an inflexible, arrogant egomaniac who employs his editorial position as a dictator wields political power."[100] Still, in direct letters to the Talbotts, the tone remained civil. Ransom worriedly wrote to Stephen that if he did not continue to grant Velikovsky editorial say over the journal's contents, he might cut all connections to the journal.[101] Talbott, in the same long letter to Rose quoted above, stood firm. "The 'proposed break' with Velikovsky, I trust you realize, is not anything I am proposing, but rather he has threatened. I shall work to avoid it—though, indeed, I'm not quite sure what there is to break," he wrote. "Nothing would sadden me more than to see Velikovsky advertise some kind of 'break' with Pensee, for surely nothing would so effectively negate the gains that have been made, and provide raw material for ridicule by his opponents." But Talbott insisted that *Pensée* was much more than a mouthpiece for Velikovsky, and that he did not believe "that the consideration of any single man's work is a sufficient base upon which to operate a journal." Furthermore, "it would surely be suicidal for us to commit ourselves *editorially* to the truth of his work. . . . *Pensee* cannot be '100% pro-Velikovsky'—or pro-Velikovsky at all, editorially. The commitment which 'has already been made' is—just as you remark—that 'Velikovsky was worth reconsidering'—neither more nor less. To commit ourselves further would be to remove ourselves from the ranks of truth-seeking journals to those of the axe-grinding journals."[102] It proved suicidal, however, to fail to take Velikovsky's threats (always communicated through third parties) seriously. After ten pro-Velikovsky issues, the catastrophist withdrew.[103] *Pensée* lay fallow in 1975, publishing no issues that year.

New Velikovskian publishing projects leapt into the breach. In 1974 there appeared a single issue of *Chiron: Journal of Interdisciplinary Studies*, run out of the Oregon College of Education by the same Thomas Ferté who had stormed out of *Pensée* in 1972. This journal was interested principally in the humanistic aspects of catastrophism and looked forward to compiling a special issue "on theoretical psychology (archetypes, collective amnesia, neo-Freudianism, three-brains hypothesis, etc.)" and also sought papers "on the work of Joseph Campbell, Cyrus Gordon, and Nikos Kazantzakis."[104] Lewis Greenberg attributed its speedy demise "to some rather inexplicable behavior on the part of Ferte [sic] who has failed to send out the first issue to many subscribers and chooses not to communicate with anyone since early October from his secluded retreat in Pocatello[,] Idaho."[105] Greenberg and Warner Sizemore seized the initiative and moved the journal east to Glassboro State College in New Jersey—where history professor Robert Hewsen announced the opening of a Center for Velikovskian Studies "as a focal point for the collection and dissemination of information relevant to the work of Dr. Immanuel Velikovsky."[106] Contrary to his previous statement in favor of *Pensée*'s objectivity, in late 1974 Greenberg assured Velikovsky that the new journal would toe the party line: "I will promise you this right now. The new journal will have its doors solidly barred to any would-be critics. We are not interested in the open forum posture presently assumed by PENSEE. I should like to go on record with you personally on that account."[107] The new journal was *Kronos* and ran from spring 1975 until 1988.

Kronos was a smaller affair than *Pensée* by an order of magnitude: its circulation peaked at 2,400 in its second year, settled to roughly 1,500 by its tenth year. It also had subscribers in twenty-four foreign countries in 1980, while competing with the British "revisionist" journal *SIS Review*. In the first issue, the editorial preface declared: "Thus we present KRONOS, a journal of interdisciplinary synthesis, whose initial contents are dedicated to *Immanuel Velikovsky*—progenitor and inspirational force for the ideas contained herein." But, careful to avoid a repeat of the FOSMOS controversy and risk angering Velikovsky by attributing to him positions he did not hold, they immediately added a footnote: "The views expressed by the authors in this journal are their own and do not necessarily reflect editorial opinion nor that of Dr. Velikovsky."[108] (They also added, later in the issue: "*KRONOS*, an independent journal, is in no way affiliated with *Chiron* or *Pensee* or the Student Academic Freedom Forum."[109]) Velikovsky's unhappiness with the Talbotts did not imply he would be kinder to Sizemore and Greenberg. He received the first issue after returning from a visit to the hospital for some medical treatments (he was eighty-one), and he wrote Greenberg a letter trashing *Kronos* and offering some suggestions. Greenberg acceded to all of them.[110] He was not about to risk a confrontation à la *Pensée*.

Back in Oregon, *Pensée* was about to fold. Stephen Talbott professed that he had "no great personal investment of any sort in *Pensee*. I have periodically thought of resigning the editorship, and if *Pensee* should cease functioning, I would probably feel more relief than anything else."[111] The investment may not have been personal, but it was certainly financial, and *Pensée* was running out of money. Stephen Talbott moved to sell off the *Velikovsky Reconsidered* edited volume—culled from pieces in *Pensee* and a huge seller on college campuses—and also offered to sell the extensive Velikovskian mailing list to the fledgling *Kronos*.[112]

C. J. Ransom was incensed: "Pensee, in the form of Steve and David, is interested in making money. They see your [Velikovsky's] work as a vehicle for this. They view Kronos as competition, and would like to take over from Kronos the publication of Velikovsky related publications if Pensee is ever out of debt."[113] David Talbott was perplexed by this reaction, as he wrote in a letter for the record to his brother in April 1976: "As a non-profit, tax-exempt organization *Pensee* never was and never will be in a position to acquire or disperse [sic] the 'profits' which Ransom seems to fear we are hauling off by the suitcase." In fact, the brothers had bankrolled *Pensée* in its early years, beginning the Velikovsky series "with no initial capitalization, but rather with a debt over $10,000. Indeed, $65,000 would not seem an unreasonable amount of capital to get *Pensee* from point zero to the status and circulation it achieved."[114]

As with so many internal conflicts among the Velikovskians, it would be impossible to tease apart accusations and counter-accusations in a reasonable space, even if there was complete documentation from all sides (which there is not). Suffice it to say that the Talbotts moved from an alleged promise to give the two mailing lists to Velikovsky gratis to selling them at $400 apiece. Ransom grudgingly purchased them out of his own pocket in 1977 to promote *Kronos*'s future growth.[115] And so *Pensée* shut up shop for good. Harvard astronomer Donald Menzel penned a gloating letter to Stephen expressing his delight in the news that *Pensée* was no more, since the "magazine is, in my opinion, detrimental to the best interests of science."[116] A bewildered Frederic Jueneman wrote Menzel to see whether this was a mistake or a prank. No such luck, Menzel retorted: "*Pensée*, from its inception, has been primarily devoted to the glorification of Velikovsky, one of the greatest Cranks of modern times. . . . It is the magazine, *Pensée*, which is irresponsible."[117] And he would dance on its grave while he could.

Meanwhile, *Kronos* plowed ahead, accepting and publishing articles from the inner circle and well outside it on questions of planetary atmospheres, moon craters, and the exact dating of various Babylonian conflicts. Sizemore was pleased: "We now have an organ—a powerful organ— that will allow no distortion of your work to go unanswered."[118] But only if people read it. The editorial board attempted to place advertisements for *Kronos* in major science periodicals, but those journals had for several years refused to print any material promoting Velikovsky's books or his ancillary organs. To Ransom, conspiracy was to blame. "The most reasonable explanation appears to relate to pressure from someone who does not wish the public to have access to documented evidence of irrationality in the scientific community," he wrote to the sales director of *Science News*. "This leads one to suspect that you are bowing to outside pressure."[119]

There was indeed pressure affecting the conduct of *Kronos*, but it was pressure of a decidedly inside kind. In 1971 a biochemistry professor with an active sideline in biblical scholarship named Donovan A. Courville from Loma Linda University (a Seventh-Day Adventist institution), published an exposition of biblical chronology that engaged critically with Velikovsky's arguments from *Ages in Chaos*. While he respected Velikovsky's diagnosis of the problems with the conventional dating, he believed the compression of centuries in Egyptian chronology simply would not work: "The writer[,] however, is convinced that there is a more credible and more convincing manner in which this shortening is to be attained than that proposed by Velikovsky, which may have been a large factor in incurring the wrath of the archaeologists and historians."[120] Courville, understandably, wanted to continue the conversation and viewed first *Pensée* and then *Kronos* as likely forums. He wrote an essay for submission to *Kronos*, arguing that "the evidence he has given to support the construction for this later era [in *Peoples of the Sea* (1977)] bears no resemblance to the potency of the evidence in support of the earlier era." The essay was returned, and Courville was fed up. "It was because Pensee would not publish articles in refutation of the points on which Velikovsky's chronology was not acceptable that I gave up on Pensee," Courville noted. "I had understood that Kronos was initiated on a different basis which would present both sides of this problem of chronology."[121]

And so we return to the issue of Velikovsky's control over the journal and the tension between allowing people to debate his views—letting them live in the conflict of scholarship—and having them remain faithful, even if the project died in the process. The very insurgent quality that had drawn the counterculture to Velikovsky made them want to push further, question his conclusions, edge closer to finding the *real* truth, and he was no more going to allow that now than he had for *Pensée* in the early 1970s or Ted Lasar and Donald Patten in the 1960s. He realized the tension and tried to get Rose and others to publish some of their critiques of the establishment in venues besides *Kronos*, because it was important to have such views "presented outside of the ghetto."[122]

Yet he was unwilling to take steps to expand. For example, a man named Jerry Rosenthal inherited a bit of money in 1977 and, having come to admire Velikovsky's theories and his tenacity in defending them, proposed donating some of those funds toward an expansion of Velikovsky's audience. The problem, as Rosenthal saw it, was not too little scientific research, but a biased older generation blind to the potential of Velikovsky's system. With better public relations, they could win over the young, the scientists and decision makers of the next generation. "Your primary emphasis, as I talk with you," he wrote to Velikovsky, "is toward research and print media publication. This reaches only a small fraction of the public today as movies, TV, radio, even lectures influence many more people. Even young scientists today, because your books and theory are virtually blacklisted, cannot easily be introduced to you." But with the right medium, Velikovsky's reach could be extended dramatically. "Young people have a strong desire to know the facts. They are misled and feel empty with the pseudo-answers of the establishment," he continued. "Some spring off into religious cults or into escapist philosophy; some remain in the establishment knowingly frustrated. There is also a large group of people interested in space, science fiction, and the sciences that would be potential customers of a Velikovsky media event. Young people must be made aware before they get a vested interest in the existing system."[123] The fact that he had to alert Velikovsky about this at the end of a decade that had begun with an approach to youth culture shows how far the involution had progressed.

Rosenthal proposed bankrolling a documentary or a TV series. This was not a new idea: the Canadian Broadcasting Corporation had produced Henry Zemel's *Velikovsky: Bonds of the Past* in 1972, and the British Broadcasting Corporation released *Horizon: Worlds in Collision* that same year. (Velikovskianism even went beyond television: in September 1979 modernist composer Philip Glass contacted Velikovsky with a proposal to write an opera based on *Oedipus and Akhnaton* as a sequel to *Einstein on the Beach* and *Gandhi*—eventually released in 1983 as *Akhnaten*, with not a trace of Velikovsky aside from the Oedipal motifs.[124]) Velikovsky suggested Rosenthal give $4,000 to *Kronos* instead. The latter was unhappy. "I specifically stated that my continued support was dependent upon the expansion of the readership and the necessary broadening of the base for whom *Kronos* could be a useful journal," he complained. "There is a mass market for these ideas; Close Encounters, Star Wars, etc. all prove the mass appeal that popularly packaged Velikovsky would have. Talbott had 35,000 copies printed—over ten times what you might reach, and Sagan reaches millions!" If Velikovsky were not more cooperative, then Rosenthal would no longer give money to ventures like a special *Kronos* issue on the AAAS meeting, which he considered "an ego-boosting rejoinder to Sagan."[125] Velikovsky might be "a great scientist and researcher, but is a failure at leading a revolutionary movement. Both his and your conservatism have failed those of us who need strong leadership," he wrote to Greenberg. He suggested that the "movement must be simplified, digested and regurgitated for the masses. They will then do the work you are trying to do—apply pressure on the scientists, universities, Congress. Then the power, money, success, research grants, books, television shows, expeditions and fame will be given to you."[126] Velikovsky, Greenberg, and the *Kronos* set would not cooperate, and so Rosenthal walked away, bewildered at the man who had captivated hundreds of thousands with his writings but now seemed content to converse only with a faithful band of a few dozen.

The core battalion continued to be riddled with sudden purges, along the lines of what happened to Robert Stephanos in 1969. (In fact, Eddie Schorr, a principal in the move against Stephanos, was exiled even more brutally in 1977 by Greenberg and Jueneman, with Velikovsky's explicit approval.[127]) Meanwhile *Kronos* labored along, as Alfred De Grazia saw it, "essentially and in many details under V.'s thumb until his death, performing very much the function of *Imago* for Freud."[128] In the wake of his publication of *Peoples of the Sea* and *Ramses II and His Time*, both of which extended the story of *Ages of Chaos* further and sparked controversy among his followers, Velikovsky demanded (according to De Grazia's journal entry of December 27, 1978) to have Peter James removed "from the editorial board of *Kronos* in three months, or else he would give them no further material of his own to print. . . . Then, says Sizemore, V. reconsidered and told them that he didn't mean what he said. . . . It is not the first time that V. has come perilously close to practicing the behavior of his enemies."[129]

James's defection to revisionism was only a further fragmentation in a movement that was almost impossible to hold together, even had Velikovsky possessed extraordinary managerial skills. For just as there was no unified counterculture, but rather disparate movements pulling in different directions at the established bastions of authority, so there was no prototypical Velikovskian. Each of these individuals had come to Velikovsky for different reasons and with different agendas, and none could possibly replicate the master's own devotion to (and understanding of) the cause. Since the high point of 1974, with the whirlwind tour of confrontation and celebration, Velikovsky had retrenched in his advancing age, seemingly bewildered at what had become of his original flash of insight in 1940.[130]

It was a symbol of the fraying edges of his movement when Ralph Juergens, among his oldest and most loyal collaborators, died on November 3, 1979, a day after his fifty-fifth birthday. This was a deep-cutting blow. Juergens had been one of the first foot soldiers in the revived pseudoscience wars; he could always be trusted—and now he was gone. Velikovsky and his family had barely had time to process the shock before November 17, 1979, two weeks later, when Immanuel Velikovsky died in his home in Princeton, New Jersey.

Conclusion: Pseudoscience in Our Time

"No one has disproved *Worlds in Collision*," said Immanuel Velikovsky in the fall of 1979. "They cannot just *say* it is wrong, that it cannot be. They will have to *show* that it is mistaken."[1] Shortly after speaking these words, before they made it to the printed page, Velikovsky died in Princeton of heart failure. This was his last interview. As befitted a man who engendered so much notoriety over the decades of his public career, obituaries of Velikovsky popped up in newspapers nationwide: the *Boston Globe*, the *Los Angeles Times*, the *Washington Post*, the *New York Times*, and others narrated the arc of his career and the circumstances of his death at age eighty-four.[2] Most of these derived from a single obituary flashed across the Associated Press wire service, and they were riddled with mistakes. His wife's name was not Elizabeth. He had not studied medicine at the University of Berlin. The facts of his life were melting away at the very moment of his passing.

Velikovsky's lived presence—even if only on the printed page—had always been crucial to the waging of the pseudoscience wars, on whichever front line combatants happened to be. In 1950 he had unwittingly triggered hostilities by publishing *Worlds in Collision* with Macmillan, striding confidently into what happened to be a demilitarized zone still on a hair trigger from the aftermath of Lysenko's rise to dominance in the Soviet Union in 1948. As scientists of many stripes, but principally astronomers, leveled their artillery at Macmillan—the convoy that carried this intruder into the realm of American science—what had been marketed as a popular book aiming to reconcile science and religion was transformed into the manifesto of a new Galileo facing the Inquisition, at least among certain members of the intellectual elite and legions of readers. Active hostilities subsided within a year, and Velikovsky began diplomatic overtures to members of the scientific elite (like Einstein) bearing tokens in their language (radiocarbon assays, predictions about Venus), hoping to gain some modicum of acceptance. He was rebuffed, while new forces—the Velikovskians—centered around his homestead in Princeton, avidly proposed their own scientific and historical inquiries to validate his theories. In the mid-1960s, however, the uneasy truce was shattered as America's college students formed multiple insurgencies against what they derided as "Establishment science" with Velikovsky as their rallying cry. Fractures within the movement soon began to tear it apart, as individuals pursued Velikovskianism for varied ends, and the nineteenth-century visitor from Vitebsk refused to fight on these terms. In November 1979, the invading general of the pseudoscience wars was gone, and his forces began to disperse.

But not immediately. His ideas lived on for a while, continuing to grace the pages of magazines and journals eager to proselytize, refute, or simply discuss cosmic catastrophism. Before drawing some general conclusions about the place of science in our world today, how scientists attempt to solve the problem of dealing with outsiders, what thinking is like on the fringe, and what the Velikovsky case can tell us about all these issues, it is instructive to see what happened to all the sound and fury after the furnace that powered the dynamo, Immanuel Velikovsky, burned out.

The quiet was not apparent to either his supporters or his opponents in the aftermath of his funeral. Doubleday, Velikovsky's publisher since the Macmillan debacle of 1950, continued to advertise the now-deceased catastrophist's theories as "proven," eliciting a blistering broadside that attacked the press for its venality.[3] At almost exactly the same moment, a lingering debate over Velikovsky's cosmology surfaced in the most unlikely place: the letters section of *Physics Today*, a publication of the American Institute of Physics. Velikovsky's elder daughter, Shulamit Kogan, published a letter criticizing the calculations in Sagan's famous rejoinder to Velikovsky. Two separate counter-salvos and counter-counter-salvos ensued, combining recalculations of the relevant equations and general bemusement at the persistence of this controversy.[4] Then the hubbub quieted down. The situation was much the same in other domains, such as geology, where Velikovsky had been repeatedly endorsed by various popular writers through the 1960s and 1970s. While some Velikovsky-style catastrophist arguments (massive recent impacts, for example) are still made occasionally within geology or geography, their influence is slight.[5] A stray biblical scholar might endorse Velikovsky's chronology, but in general the cannons went silent.[6]

Not so among Velikovsky's closest supporters. The Velikovskian journals that had begun in the 1970s in the wake of *Pensée*'s initial success did not vanish with his death. *Kronos*, under the leadership of Lewis Greenberg and monitored by Lynn E. Rose—the philosopher who remained a stalwart defender of the original Velikovskian orthodoxy as articulated in *Worlds in Collision*—limped on until 1988.[7] *Aeon: A Journal of Myth, Science, and Ancient History* took over that year and persisted until September 2006. *The Velikovskian: A Journal of Myth, History and Science* first appeared in 1993 and ran until at least 2005 (it has proven impossible to determine whether it is now defunct or merely on hiatus). The most successful Velikovskian journal survived in part by breaking off from the American tradition: the British *SIS Review*, founded in 1974, changed its name to *Chronology and Catastrophism Review* in 1986 and lives on to this day.[8] Many of the same characters, such as Lewis Greenberg and David Talbott, populate the mastheads of these journals, and they were joined by Charles Ginenthal, publisher of *The Velikovskian*, who is the most significant orthodox Velikovskian to become active in the movement *after* Velikovsky's death. In 1990 he published *Carl Sagan & Immanuel Velikovsky* (with a second edition in 1995), a monograph devoted entirely to refuting Sagan's 1974 AAAS presentation, and he also organized a massive 1996 edited volume to continue assailing those who attacked Velikovsky.[9]

There was plenty to defend, for Velikovsky's theories faced more than the slings and arrows of debunkers, although those continued to hail down as before, sometimes with new evidence and better-articulated arguments.[10] More serious was continued fragmentation within the tight orthodoxy that Velikovsky had maintained with such vigor. The first significant schism concerned Velikovsky's reconstruction of ancient chronology. This conflict had been brewing for some time. In April 1979 a summit of British, Dutch, and Swiss Velikovskians gathered at Heathrow, England, to outline a European wing of a research program to complement the American undertaking at *Kronos*. In the minutes, the group noted a strong degree of consensus, *except* with respect to how Velikovsky chose to complete his chronology in the 1970s. Here the British demurred: "At the same time[,] however, obstacles are met when dealing with the revised chronology as found in 'Peoples of the Sea' and 'Ramses II and his time.' The chronological structure of 'Ages in Chaos' they feel is sound, but the rest of the revision does not square with the evidence."[11] The reinterpretation initiated by Peter James in the mid-1970s about how to complete the *Ages in Chaos* timeline only deepened, and in the 1980s even the British movement fragmented further as James broke with his former collaborator David Rohl.[12] At the periphery, the hold of the center was too weak.[13]

The center could not maintain itself in the United States, either, although the fragmentation of the American disciples into different camps of cosmic catastrophism—more or less at variance with the picture consistently built up in Velikovsky's writings—waited until after the old man's death.[14] The issues here did not concern religious appropriations of Velikovsky's arguments to predict the apocalypse; that was neither new nor terribly troubling to the core of disciples.[15] Rather, important members of the inner circle started to peel off and defend their own views about how to relate ancient mythology to the structure of the solar system. One of these, Frederic Jueneman—an industrial analytic chemist who wrote a series of Velikovskian pieces in the 1970s for *Industrial Research*—had always maintained his independence, as he wrote to Carl Sagan: "I consider Velikovsky a personal friend; however I am *not* a so-called Velikovskyite in the sense that his work is all encompassing—far from it."[16] In his 1975 collection of articles, however, he toed very close to an orthodox line on almost every issue, and the book functioned as high polemic.[17] Yet in 1995 he produced another volume of short pieces on his unorthodox scientific views, and he had indeed drifted quite far from the central concerns of *Worlds in Collision*.[18]

The greater defection was surely Alfred De Grazia's. After coming to Velikovsky's defense with his 1963 special issue of the *American Behavioral Scientist*, the essays of which were collected and then updated in the popular book *The Velikovsky Affair* (1966), De Grazia maintained a close relationship with Velikovsky and was able (and willing) to assert a greater degree of independence than the rest of the inner circle. Yet he remained a stalwart defender of Velikovsky and promoted his doctrines both in the United States and abroad. That changed with the master's death, as De Grazia rapidly published his own series of books on "quantavolution" (also called "revolutionary primevalogy"), which sifted through ancient legends and scientific anomalies to produce a different account of the history of the solar system.[19] Unlike the British, De Grazia not only questioned the historical reconstruction—he diverged on the astronomical picture as well.

As was perhaps inevitable, Velikovsky's arguments were appropriated and revised by those further removed from the devotees who had frequented Hartley Avenue in Princeton. David Talbott presents an interesting case in point. The publisher of *Pensée*, Talbott experienced some estrangement after he and his brother Stephen formally folded the journal in 1976. Velikovsky had encouraged David to continue his research project about Saturn myths from around the world, and Doubleday published it in 1980.[20] *The Saturn Myth* argued that at one point in Earth's recent history (during the mythical "Golden Age"), Saturn hovered over the North Pole, which explained the global ubiquity of tropes associated with the ringed planet and its eponymous god. Talbott shied away from presenting a detailed cosmological scenario for that arrangement and how it changed to our present one.

He remained affiliated with certain Velikovskians for the next decade or so, building on acquaintances formed at the banner conferences of the 1970s. In 1974 Talbott met Wallace Thornhill, an Australian who had begun postgraduate training in physics, partially "inspired by Immanuel Velikovsky," and who concluded that academia would not tolerate his unconventional ideas. The two renewed their association in 1994 and 1996 at similar conferences in Portland and began to collaborate on a physical worldview they called "The Electric Universe," replacing conventional gravitational and nuclear forces with a plasma-and-electricity alternative.[21] Although in one of their joint publications they attempted to minimize Velikovsky's influence to the purely biographical—the name does not appear in the index, for example—both the arguments (an electric cosmos) and the citations to Velikovskian disciples C. J. Ransom, Ralph Juergens (their only reference to any Velikovskian journal, in this case *Kronos*), and Earl Milton indicate that this project retains more than a whiff of its origin.[22]

As alternative cosmologies proliferated rather quietly, criticism emerged from within the community in a more explosive fashion. In the late 1970s, a young man named C. Leroy Ellenberger began writing pointed letters to various editors who had published anti-Velikovsky articles. He did so on his own initiative, not yet having met Velikovsky. That soon changed, and Ellenberger in 1979 became an editor at *Kronos*, continuing his spirited defense of the sage of Princeton. Although he knew Velikovsky for only a short while, he was no less embroiled than any of the veterans in internal disputes, and those tensions strained to the breaking point in the 1980s. The first rupture concerned De Grazia, who had attempted to enroll Ellenberger for his new project. "If this book is a preview of the work for which Velikovsky had to die so the real work could go on, as you said to me at the Capitol Hilton in early 1980, then it's a grave disappointment," Ellenberger wrote in a sharp letter in 1982. "There are serious problems we catastrophists have to solve before we have a strong claim to the attention of mainstream scientists. I simply do not see CHAOS AND CREATION helping catastrophism gain credibility with any scientifically literate reader."[23]

The point about scientific literacy is crucial, for over the first years of the new decade, Ellenberger, once as doctrinaire as Velikovsky himself in defending *Worlds in Collision*, began to experience doubts in terms of how to fit the Velikovskian picture with the geological and climatic evidence gathered since 1950. In 1984, encouraged by editor Lewis Greenberg, he published an article in *Kronos* entitled "Still Facing Many Problems"—a reference to the final section of *Worlds in Collision*, "Facing Many Problems," which outlined a research program to confirm Velikovsky's cosmic catastrophism—documenting several serious physical objections to Velikovsky's cosmology. "Over the past four years I have come to appreciate that, even if Velikovsky were right, there are good physical reasons why astronomers and other scientists have opposed him so tenaciously," Ellenberger wrote. "Whether the key events described in *Worlds in Collision* can happen or did happen is a key issue."[24] The more he learned about the Greenland ice cores, which indicated that no global catastrophe occurred at the time when Venus supposedly threatened this planet, the more skeptical he became.[25] In a 1986 letter in *Skeptical Inquirer*, long a bête noire for Velikovskians and others on the fringe, he published what amounted to his resignation: "The less one knows about science, the more plausible Velikovsky's scenario appears, especially when most of the discussion is hand-waving. Conversely, the more knowledgeable the reader, the easier it is to see that Velikovsky's entire physical scenario is untenable. But unless a critic explains *why* something is wrong, the rejection is more *ex cathedra* than a credible refutation."[26] Ellenberger's exile from the community was by then well under way. He broke off relations with Lynn Rose in 1983 when the latter refused to emend errors that Ellenberger had found in the posthumous manuscript of *Stargazers and Gravediggers*, Velikovsky's memoir of the controversies of 1950; and irreconcilable scientific and doctrinal differences precluded any cordial relations developing with Charles Ginenthal from their first interactions in 1984. (Both accounts are according to Ellenberger.[27]) He resigned from *Kronos* in 1986 and stopped filling back-issue orders in November 1987. Since then, he has remained a prolific critic of Velikovsky's ideas.

These schisms and debates had ramifications far beyond interpersonal drama and personality clashes (which, of course, they also were). Consider, for example, the role of catastrophic ideas within the sciences today. Velikovsky had leveled two major attacks at the dominant attitude of uniformitarianism, the notion that the forces that work at present (such as erosion) are the same forces that worked in the past, with roughly the same intensity. His first claim was that catastrophes had wracked Earth; the second was that these happened in the *recent* past and were witnessed and recorded by humanity. The second claim is still rejected, but the first is not. There are several significant theories in contemporary science, such as the proposal that the Chicxulub Crater off Mexico's Yucatán Peninsula indicates a massive meteor impact that initiated the extinction of the dinosaurs, which are quite prominently entertained (although not wholly uncontroversial). Geologists routinely discuss meteor impacts and other catastrophic changes.

Most of the orthodox Velikovskian literature produced after his death claims that, although Velikovsky was vilified for questioning uniformitarianism, his ideas about catastrophism have since become mainstream—although without granting due credit to their progenitor or offering an apology for his treatment. But do these "neo-catastrophist" theories owe their origin to Velikovsky? Here the evidence is murkier. British astronomers Victor Clube and Bill Napier proposed in their 1982 book *The Cosmic Serpent* an argument for "coherent catastrophism": that the Earth had indeed been threatened by comets diverted into the solar system by the outer planets, and that some evidence for this can be drawn from ancient documents. They insisted, however, that superficial resemblance aside, the "Velikovsky thesis was therefore not so much wrong as hopelessly misguided."[28] More than that, they maintained that Velikovsky had so alienated the astronomical community that "an objective reappraisal of the historical evidence has been rendered almost impossible in America by a particularly myopic scientific lobby."[29] A neo-catastrophist book based on dendrochronology (the analysis of tree rings) also claims to find evidence for catastrophes in human history and notes that "Velikovsky was almost certainly correct in his assertion that ancient texts hold clues to catastrophic events in the relatively recent past," but his "more lunatic ideas had the effect of scaring off most scientists from having anything to do with him or his ideas or the issue of recent bombardment events." The author, Mike Baillie, proposes instead to "go back to Velikovsky and delete all the physically impossible text about Venus and Mars passing close to the earth."[30] That position is more generous than most; in a 2007 book full of geological analysis of ancient myths, Velikovsky is so forgotten that he is not cited at all, even for the purposes of refutation.[31]

The balkanization of Velikovsky's followers into rival factions after his death reminds us about how hard it is to maintain orthodoxy on the fringe. Repeatedly in the preceding chapters, we have seen how Velikovsky and others (like the creationists) relegated to the fringes of science worked to preserve the integrity of their convictions and to avoid schisms that would weaken their movements for an alternative scientific picture. If they were unable to remove the taint of pseudoscience in the eyes of mainstream scientists, at least they could continue their research and wait for another opportunity. As a consequence, we observe highly contentious and vituperative recriminations on the fringe between members of ostensibly the same camp. One might be tempted to say: "Well, they are crazy, so of course they will go after each other's throats this way." But they are not crazy, and demonizing them once again for this behavior prevents us from seeing the deeper pattern.

"Pseudoscience" as a concept is built around the notion of imitation, or, to use a literary term, mimesis.[32] According to establishment scientists who throw the accusation around, pseudoscience is that which is not science but *resembles* or *mimics* it; it has the trappings but not the essence of science. This is partly why simple demarcation criteria fail us; as soon one such standard (such as falsifiability) is erected, the alleged pseudoscientists adapt to meet it. But there is a deeper level of mimesis *within* the community banished to the fringe. Again and again, Velikovsky and his allies used the term "pseudoscience" and its grammatical permutations. At the 1974 AAAS meeting, for example, Velikovsky called uniformitarianism a "pseudo-scientific statute of faith, elevated to a fundamental principle."[33] This cognizance of the pseudoscience label goes back to the beginning. In 1950 Velikovsky wrote historian of science George Sarton of the controversy around *Worlds in Collision*: "How many wrong theories and pseudoscientific books and science fiction are printed year in and year out and do not rob the scientists of their peace of mind?"[34] In 1961 he even called himself a "pseudoscientist" in a letter to Claude Schaeffer, although he was careful to put the term in ironic scare quotes to reflect the accusation back on those who would attribute it to him.[35]

What are we to make of this, as well as of Velikovsky's efforts to deny "legitimacy" to Reichian orgone theory or Patten's astral creationism, or Ellenberger's early attacks on De Grazia, or Rose and Ginenthal's ostracizing of Ellenberger? Quite a bit, I believe. One of the characteristics associated with "pseudosciences" is their mimicry of establishment science, and one of the chief activities of the mainstream scientific community is *the process of demarcation itself*. That is, scientists routinely castigate other doctrines as pseudoscientific, and it stands to reason that those on the fringe would adopt a penchant for demarcation. Not only would they call other competing fringe doctrines pseudoscientific; they would also call *establishment* science so. The mimesis of the fringe goes all the way down, and this is one of the central lessons of examining the development of these doctrines historically. This aspect, not coincidentally, was identified by Richard Hoftstadter as a characteristic of the paranoid style: "A fundamental paradox of the paranoid style is the imitation of the enemy. The enemy, for example, may be the cosmopolitan intellectual, but the paranoid will outdo him in the apparatus of scholarship, even of pedantry."[36] But this is no paradox; it is a central feature of the category (and the process of categorizing) itself.

Meanwhile, the disputes within the community of post-Velikovsky Velikovskians continued to burn, albeit less brightly, and by now have almost burned themselves out—not completely extinguished, to be sure, but the embers glow but faintly these days.[37] After roughly 1985, the public prominence of Velikovsky's catastrophist doctrines collapsed. Where anxiety about Velikovsky used to crop up in popular science journals or newspaper editorials, his became a name recalled only dimly, and then finally not remembered at all. He is not, however, entirely gone. In 2004 and 2006, for example, two separate monographs appeared written by individuals unrelated to the original Velikovsky cohort of the 1960s and 1970s, the former devoting an extensive chapter to articulating a rather orthodox version of Velikovsky's theories, maintaining that they had been confirmed by the discoveries of the space age, and the latter a straightforward endorsement of Velikovsky that could have been written, with little change, in 1967.[38]

Velikovsky also lives on, perhaps unsurprisingly, in the more ethereal realm of the Internet. A collection of his writings are freely available at the Velikovsky Archive (varchive.org), and a parallel to *Wikipedia* (www.Velikovsky.info) offers information about the man, his theories, and his critics, but it is not open for editing by the general public, orthodoxy being maintained by (undisclosed) editors. Numerous other websites promote variants of cosmic catastrophism of a less orthodox sort, as a simple Google search will reveal. Interestingly, YouTube displays the Canadian Broadcasting Corporation's 1972 documentary *Velikovsky: The Bonds of the Past* cut up into six viewable segments, which rank tens of thousands of views and quite a few "likes." (All of these snippets are prefaced by the title "2012," a point to which I will return shortly.) For brief moments, Velikovsky emerges again in plain view, like a ghost of Christmas Past, to remind us of what came before. In March 2005, on the popular *Coast to Coast AM* radio show, there was a debate on the validity of Velikovsky's theories featuring James McCanney (a prolific writer and blogger advocating various fringe scientific theories) and David Morrison (a vigorous debunker of Velikovsky since the 1970s).[39] And out of nowhere on January 15, 2011, the "Week in Review" section of the *New York Times* printed an article on continental drift and geologic catastrophes that concluded with two paragraphs on Immanuel Velikovsky.[40] He is gone, but he won't stay forgotten.

"Velikovsky's name is all over the literature of the history, philosophy, and sociology of science," wrote chronicler and debunker of Velikovskianism Henry Bauer, "but his name never appears in the science literature."[41] If one looks for Velikovsky, he materializes in the oddest of venues. For example, in 1980 historian of science Charles C. Gillispie, in his monumental history of science in Old Regime France, included a reference to Velikovsky at the beginning of his chapter on Anton Mesmer and other "charlatans" of Louis XVI's Paris, and the mention is so casual that Gillispie clearly assumed the historians reading the book would have no trouble placing the name.[42] Yet Bauer's notion of the centrality of Velikovsky to the history and sociology of science is overstated. Velikovsky does indeed pop up now and again, but he no longer serves as a central touchstone. Instead, he became a weapon for some *scientists* in the 1990s who wished to discredit the humanistic study of science. Velikovsky's place in this episode, called by some the "Science Wars," allows me to make good on an early promise: to use the cases of demarcation in this book to explore the nature of science itself.

The Science Wars were not pretty. This is not the place to survey the accusations and recriminations that rocked the science studies community during the mid-1990s, and the debates have receded far enough into the past that accounts have started to emerge purged of much of the vitriol of the day.[43] At issue in these largely academic but rather public debates— especially the "Sokal Hoax," a parody of science studies penned by physicist Alan Sokal and then published in the journal *Social Text*—was whether the community of historians and sociologists of science had been corrupted by a slew of academic vices, especially "postmodernism" and "relativism," into a stance hostile to science. The notion of "pseudoscience" received a notable boost in the publications opposed to science studies, as postmodernist theories were associated with "anti-science," accused of being pseudosciences themselves, or depicted as giving aid and comfort to pseudoscience under a veneer of academic respectability.[44]

At this moment, about a decade after he began his descent into obscurity in the popular mind, Immanuel Velikovsky returned, his name revived by the very scientists who had cheered the eclipse of his theories. Invocations of Velikovsky appear across the Science Wars literature—always on the side of the warriors opposed to contemporary trends in the sociology and history of science—as an important trope in a cautionary tale. Velikovsky's heyday had been the mid-1960s through the 1970s, and it was precisely at that moment that the germs of relativism began to burrow into the academic community, according to Paul Gross and Norman Levitt, whose book *Higher Superstition* was one of the loudest salvos in the Science Wars. "Implicitly, what had taken shape [in the 1960s] was an informal *conspiracy of the heretical*," they claimed, "a community of defiance toward the conventional, a range of wacky doctrines linked mainly by the glee they all took in spitting in the face of received opinion."[45] In their footnote to this passage, Gross and Levitt specifically cited Immanuel Velikovsky as "an example of the hostility of some social scientists to the authority of the 'exact' sciences," and contended that it "was an important harbinger of postmodern relativism and antiscientism."[46] This interpretation of the Velikovsky battles as being between the humanities and social sciences on one side and the sciences on the other, although not historically accurate, was quite common among science warriors.[47] Bauer, who produced the most extended science-studies analysis of Velikovsky to date, allied decisively with the science warriors, and he surprisingly thought Velikovsky was *better* than the relativist academics: "Velikovsky thought that much of accepted science happened to be wrong, but he did not believe the enterprise of science a wrongheaded activity vitiated from the outset by the impossibility of knowing anything, as all too many contemporary pundits do."[48]

Why Velikovsky? Was there any substance to seeing sociologists of science as more or less open to Velikovskians hoping to bring down the temple of science? Here, the evidence is sparse. Interestingly, the ur-text of academic postmodernism, Jean-François Lyotard's *Postmodern Condition*, is preceded by a preface authored by Duke University critical theorist Fredric Jameson that in fact does invoke Velikovsky. "'Doing science,' for instance, involves its own kind of legitimation (why is it that our students do not do laboratory work in alchemy? why is Immanuel Velikovsky considered to be an eccentric?) and may therefore be investigated as a subset of the vaster political problem of the legitimation of a whole social order," Jameson wrote.[49] This general point—that science is a form of authority within our culture—seems unobjectionable, and it was also completely ignored by the science warriors, who apparently did not notice the passing reference. The solitary piece cited to prove the Velikovskian perfidy of the science-studies community was a 1969 article by Michael Mulkay, which used the Velikovsky affair to make the rather uncontroversial point that scientists did not typically behave in accordance with sociologist Robert K. Merton's then decades-old notion of "norms" of proper conduct.[50] As we have seen in these pages, historians and sociologists of science typically sided quite firmly on the anti-Velikovskian side of debates, rushing in where mainstream historians feared to tread. For all the danger that Velikovsky seemed to pose to the science warriors, it turns out there was very little behind it. The threat did not lie here.

It was elsewhere, in places not obviously linked with Velikovsky or, crucially, with any notion of "pseudoscience." These days, when mainstream science is attacked in politically credible ways, the assault does not come from sociologists of science (who have never held such power) or from popular and populist purveyors of doctrines labeled "pseudoscientific," "crackpot," or "crank" by that same establishment. Rather, it hails from scientists with all the credentials of the establishment who argue against consensus points of view—that tobacco smoking is not addictive or harmful, that acid rain is not environmentally destructive, that humans are not responsible for the expanding ozone hole, and that anthropogenic climate change is not occurring.[51] These claims are no less at variance with the dominant scientific consensus, but the rhetoric works rather differently, in ways that these anti-consensus scientists—sometimes called "denialists," which I will use for lack of a better term—connect with the more far-flung fringe.

"Pseudoscience" is an empty category, a term of abuse, and there is nothing that necessarily links those dubbed pseudoscientists besides their separate alienation from science at the hands of the establishment. Denialists, on the other hand, do have something in common: a unified approach to defend industry against scientific claims that their products are causing harm to citizens. The similarity is part of a conscious strategy to generate controversy where there had been scientific consensus, to destabilize the authority of mainstream science. The first instance of this denialist strategy, and the one that spawned the rest, was centered at the Tobacco

Industry Research Committee (TIRC), an organization established by Big Tobacco to "pledg[e] aid and assistance to the research effort into all phases of tobacco use and health"—as articulated in the "Frank Statement," a text authored by the public relations firm Hill & Knowlton and published as an advertisement in 448 newspapers in 258 cities on January 4, 1954. The TIRC was clearly mostly an effort to spin the devastating anti-smoking findings that had reached popular attention in the early 1950s; in the early years, for every dollar spent on research by TIRC, $200 was spent on public relations, and the research funded focused on the basic science of carcinogenesis, not on the major issue—whether smoking caused cancer. The purpose of TIRC and equivalent groups, according to historian Allan Brandt, was "to produce and sustain scientific skepticism and controversy in order to disrupt the emerging consensus on the harms of cigarette smoking."[52]

It worked brilliantly. Instead of a debate about whether to ban cigarettes due to their negative impact on public health—a fact considered well established by 1954—America in the 1960s witnessed a manufactured controversy over whether cigarette smoking was in fact injurious to health. By calling for more research, the industry cast doubt about whether the extant research was correct, reliable, and objective. As recent studies have extensively documented, denialist scientists who advocate for Big Tobacco and polluters argue that "junk science" has demonized certain industries (such as the fossil fuel industry, in the case of global warming), and that more research is needed to reach scientific consensus. They have set themselves up as the arbiters of "sound science," a term that seems unobjectionable but emerged only in the past few decades specifically in the context of calling for deregulation of these industries and a relaxation of environmental safeguards. Denialists have a common discourse, are funded by a specific set of industries, and are affiliated with particular think tanks with a common (strongly conservative) political ideology.[53] There is a unity here that "pseudoscientists" lack.

There is, however, a relationship between denialists and the further fringes. As long as there are Velikovskians or Reichians or UFOlogists out there who can be cast out as lunatic, unhinged, or conspiracy-theorist— and such accusations, as we have seen, were and remain common—then the denialists can assume a neutral posture and claim a level of respectability for their no less radical repudiation of mainstream science. "At least we are not Velikovsky" is the implied slogan here. For if the highly conspiratorial-minded doctrines that characterize the fringe in the wake of the pseudoscience wars are taken to be emblematic of *unreasonable* dissent from consensus, then the endless calls for "sound science" and "more research" sound eminently *reasonable*.

The connection can be borne out in personal relationships as well. In a recent book, historians of science Naomi Oreskes and Erik Conway have traced the connections between a small group of establishment scientists who transformed their careers into crusaders against regulation and for right-wing political causes, accepting industry funding and parlaying their scientific credentials into cover for a nakedly political argument, one claiming to be *opposed* to the politicization of science.[54] Oreskes and Conway concentrate on a coherent assortment of four scientists who all possessed personal and institutional connections across a broad array of debates on acid rain, tobacco, ozone depletion, and climate change—but some of them also had a connection to anti-Velikovskianism. Two in particular, Robert Jastrow and S. Fred Singer, entered the fray against Velikovsky on numerous occasions, their interventions characterized by a calm but firm insistence on demarcation.[55] Demarcation is vital for denialists: it is only through having a group that they can exile to the fringe that they make their stand for legitimacy. Albeit implicitly, the particularly strong demonization of the fringe modeled on a depiction of Velikovsky and his followers, born of the pseudoscience wars, remains a feature of the American science-policy landscape to this day.

At this point, you might very well be asking: This is all well and good, but what should be done about pseudoscience? That is a good question, but like most good questions, it does not possess a satisfying answer. We see various doctrines, propounded at different times, each of which has a different degree of offensiveness to separate individuals or groups within the scientific community. There is no uniform solution for the same reason that the central dilemma is intrinsic to the process of rational inquiry—we can either set the bar high to save ourselves from crackpots but exclude new ideas, or we can set the bar low to promote innovation but allow in some deeply misguided doctrines. If we want to have a science, we need to accept that there will be doctrines on the fringe fighting for acceptance (and some of them may very well be correct). The mimetic features of these fringe doctrines make it impossible to come up with bright-line demarcation criteria that would make differentiating among them straightforward.

If not a bright line, some have argued we should turn to social institutions to do our demarcation. Just about all mainstream science has for decades been published using the institution of peer review: scientists in a particular field evaluate manuscripts written by their peers, and then filter out the valuable from the nonsense. Given that much of the denialist writing comes through think-tank publications that do not pass through this process, some have pushed for yet broader application of peer review to weed it out.[56] This will not wash, for reasons that the Velikovsky example makes clear. In just about every chapter of this book, we have seen peer review in action, but in ways that do not serve to demarcate as the core scientific community would wish. *Worlds in Collision* was peer-reviewed *twice* and passed both times, largely because reviewers had different standards for what counted as publishable material (the central dilemma again). The situation is worse than a single book squeaking through. Ever since peer review began to be held up as a marker for "good science," peer-reviewed journals have proliferated on the fringe. Creation-science journals are peer-reviewed; so were *Pensée* and *Kronos*. And in several instances, Velikovsky's articles were frozen out of establishment journals because of editorial manipulation of the peer-review process. Peer review is a social convention, a way of arranging to enforce a broad consensus while allowing for some innovation, and it only works if there is prior agreement about what the consensus is in a particular community—as there was among the Velikovskians and other separate groups. (It is also notoriously bad at weeding out fraud.) Peer review is not a magic solution.

What is? Nothing. As long as we have science, we will have a process of demarcation that happens every day in the laboratories, field sites, and classrooms of the world. Scientists will decide that some claims are relevant for their research and that some doctrines are not—sometimes so much so that they will be dubbed "pseudoscientific." This is inevitable, and it is ineradicable. Scientists will always demarcate, because part of what science *is* is an exclusion of some domains as irrelevant, rejected, outdated, or incorrect. And the more successful science becomes, the more outsiders will want to participate in the process. Some of these will be hailed as brilliant; some others will be run out of town on a rail; most will simply sink without a trace. "Pseudoscience" is not some invasive pathogen that has contaminated contemporary science but that can be fully expunged from the organism with more scientific literacy or better peer review. Pseudoscience is the shadow of science; it is cast by science itself through the very fact that demarcation happens. If pseudoscience is inevitable, then combating it becomes problematic. Either the combatants resemble Sisyphus, pushing the rock up the hill only to have it tumble back down again, or the Æsir, battling the forces of darkness that besiege Valhalla at Ragnarök (and eventually losing). One common lesson drawn by veterans of the Velikovsky affair—and one of the main contexts in which the affair is still invoked—is that sometimes attempts at opposition backfire and only make the demonized doctrine more popular.[57] Fringe theories proliferate because the status of science is high and is something worthy of imitating. They are a sign of health, not disease.

And just as they come, they also go, to be adopted, adapted, and replaced by later generations. The Velikovsky episode may seem to be over, except for a few stray gladiators who clash swords in comment threads on webpages or on the airwaves of late-night talk radio. Yet sometimes it resurfaces, in rather different contexts, as a key to the future. On December 21 or 23, 2012, depending on how you calibrate it, the thirteen-baktun cycle of the Mayan traditional calendar (also called the "Long Count") will conclude. The Long Count is a series of astronomical cycles observed by Mayan astronomers, and lasts 1,872,000 days, or roughly 5,125 years. Later this year, it will end.

A great deal has been written about the end of the Mayan calendar and what it might or might not mean. Much of that material has been labeled "pseudoscientific" by debunkers of this appropriation of ancient astronomy.[58] Even advocates for something special happening have urged restraint. For example, John Major Jenkins, one of the most prolific writers on this topic, notes that "in academia as well as in the skeptical popular press, 2012 is rendered meaningless to the extent that it is misunderstood. . . . [G]ood information on 2012 has been either seriously limited or buried under the endless bric-a-brac of the spiritual marketplace."[59] The end of the Long Count means the end of what the Mayas called a World Age, and it correlates—so he argues—with the alignment of the solar system, which travels in a sinusoidal path above and below the galactic plane, with the center of that plane. For him, the "year 2012 is not about apocalypse, it's about *apocatastasis*, the restoration of the true and original conditions," an opportunity for spiritual rebirth, not disaster.[60] His book on the subject contains many astronomical calculations, many citations of ancient lore (mostly Mayan, some not), and represents what he calls an "interdisciplinary synthesis"—which is, ironically, also what Velikovsky named his system.[61] Velikovsky is not invoked and has nothing to do with Jenkins's version of what lies in our immediate future.

Others present a more sensational vision of what awaits us in December of this year. Erich von Däniken returned to print in 2010 to declare that the gods *may* return in 2012, although he left plenty of wiggle room in his prediction.[62] He does not, however, cite Velikovsky, a stark contrast to his heyday in the 1970s. Velikovsky does appear, however, in Geoff Stray's widely read catalog of the 2012 prophecies, which cites the Venus scenario of *Worlds in Collision* as if it were a completely unproblematic account of the history of the solar system.[63] This is our first indication that Velikovsky has found a new home in another fringe discourse, a point already signaled by the flagging of his YouTube appearances with a "2012" prefix.

Velikovsky was most forcefully resurrected by Randolph Weldon in his vividly written 2003 book *Doomsday 2012*, which predicts that the world will be destroyed that year—whether by a direct astronomical impact or by a near miss that destabilizes the world's nuclear reactors and poisons the population with radioactive fallout. His scenario is explicitly astronomical, and he devotes chapter 10 ("Catastrophism") to a dramatized retelling of the terrors of Venus that plagued the world around 1500 B.C.[64] Weldon devotes additional attention to *Earth in Upheaval* but ignores the mass of Velikovsky's writings on historical reconstruction (some of which even address the Mayas) in favor of more cinematic fare. *Worlds in Collision* comes highly praised, couched in a language of conspiracy and contempt for mainstream science that we have observed blossom over the years of the pseudoscience wars: "Laboratory honchos and their legions of yes-men as good as stamped on the thing, tore it to ribbons, burned it at the stake—they may well have wished to inflict similar punishment on its author were it not against the law, vengeful as such snobs are and loath to admit they are wrong."[65] A silent legacy of the McCarthy and Lysenko debates, to be sure, but no less a legacy for that. In his most direct contradiction of *Worlds in Collision*, Weldon supplements Velikovsky's account with a mechanism that disturbed the solar system in antiquity, issuing the comet that became Venus (whose chemical composition excludes an origin in Jupiter).[66] *This* hidden force was what the Mayas had calculated would return at the end of the Long Count, and a new force will soon terrorize and destroy Earth.

Weldon's book comes supplemented with descriptions of fossils, planetary trajectories, and radioactive tracers—all the trappings of a scientific discussion of our impending predicament. Velikovsky, the unlikely diviner of the ancient past, has been transformed into a prophet of our future. If

December 2012 passes and we all remain unharmed, Weldon's book will likely end up moldering on the shelves of used bookstores, but that does not mean the appropriation of Velikovsky and *Worlds in Collision*, or debates over pseudoscience, will cease. For it is not the fate of ideas, good or bad, to simply come to an end, although such is the destiny of books.

ABBREVIATIONS
AND ARCHIVES

The notes make reference to several archival collections in abbreviated form. Their full citations are presented here, with their accompanying abbreviation, when necessary.

American Philosophical Society (Philadelphia, Pennsylvania)

[AESR] American Eugenics Society Records. 575.06 Am3.
[FOP] Frederick Osborn Papers. MS. Coll. 24.
[FBP] Franz Boas Papers. B:B61.
[LCDP] Leslie Clarence Dunn Papers. B:D917.
[Menzel UFO Papers] Donald H. Menzel Papers Concerning UFOs, 1952–1976. 629.4:M52.
[TDP] Theodosius Dobzhansky Papers. B:D65.

Harvard University (Cambridge, Massachusetts)

[GSP] George Sarton Papers. MS Am 1803. Houghton Library.
Harlow Shapley Papers. HUG 4773.10. Harvard University Archives.

The New York Public Library (New York City, New York)

[JPP] Arthur James Putnam Papers. Miscellaneous Personal Name File, Box 165. Manuscripts and Archives Division. Astor, Lenox and Tilden Foundations.
[MR] Macmillan Company Records, Manuscripts and Archives Division. Astor, Lenox and Tilden Foundations.

Princeton University Library (Princeton, New Jersey)

[HFC] Hanna Fantova Collection of Albert Einstein, Letters and Manuscripts. Manuscripts Division, Department of Rare Books and Special Collections C0703.
[IVP] Immanuel Velikovsky Papers, 1920–1996. Manuscripts Division, Department of Rare Books and Special Collections C0968.

NOTES

Introduction

1. Some philosophers maintain that there is a continuum between "bad science" and "pseudoscience." For example, Philip Kitcher, *Abusing Science: The Case Against Creationism* (Cambridge, MA: MIT Press, 1982), 48, has argued: "Given these marks of successful science [independent testability, unification, fecundity], it is easy to see how sciences can fall short, and how some doctrines can do so badly that they fail to count as science at all. . . . Where bad science becomes egregious enough, pseudoscience begins."

2. R. J. Cooter, "The Conservatism of 'Pseudoscience,'" in *Philosophy of Science and the Occult*, 2nd ed., ed. Patrick Grim (Albany: State University of New York Press, 1990), 156–69.

3. The word "science," with meanings much like its current one, predates the early nineteenth century, but the term "scientist" for one who conducts science professionally was coined in 1833. For a discussion of this issue, see Sydney Ross, "Scientist: The Story of a Word," *Annals of Science* 18 (1962): 65–85.

4. "Pseudoscience," *Oxford English Dictionary Online*, http://dictionary.oed.com, accessed March 17, 2011. For more on the history of the term and its usage, see Daniel P. Thurs and Ronald L. Numbers, "Science, Pseudoscience, and Science Falsely So-Called," in *Wrestling with Nature: From Omens to Science*, ed. Peter Harrison, Ronald L. Numbers, and Michael H. Shank (Chicago: University of Chicago Press, 2011), 281–306.

5. An early use of "pseudoscience" to demonize science that was considered overly speculative can be found in 1901 in "Pseudoscience," *American Naturalist* 35 (January 1901): 53. As a more recent case in point, in 1974 the editor of *Science*, the most prestigious American scientific journal, penned an opening editorial on the problem of pseudoscience, which he perceived as a particularly dangerous threat in the context of the 1970s: Philip H. Abelson, "Pseudoscience," *Science* 184, no. 4143 (June 21, 1974): 1233. Some of his respondents, such as University of Washington physicist Seth Neddermeyer, considered that direct confrontations with perceived pseudosciences would only pour oil on the flames. Neddermeyer in J. Eric Holmes et al., "Pseudoscience,"

Science 186, no. 4163 (November 8, 1974): 480. Such disagreements over tactics are endemic to these discussions.

6. Astrology was, for a long time, a perfectly acceptable mode of inquiry into the heavens, dating from antiquity through the Renaissance (although never wholly without its critics). Its contemporary demonization is a casualty of the transformations in European natural philosophy in the seventeenth century. The revival of astrology's popularity in the twentieth century—especially in syndicated newspaper columns—has triggered repeated (and unsuccessful) attempts at debunking. See, for example, "Astrology Lacks Every Scientific Foundation," *Science News-Letter* 39, no. 5 (February 1, 1941): 78–79; and Bart J. Bok and Lawrence E. Jerome, *Objections to Astrology* (Buffalo, NY: Prometheus Books, 1975).

7. Mark B. Adams, "Toward a Comparative History of Eugenics," in *The Wellborn Science: Eugenics in Germany, France, Brazil, and Russia*, ed. Mark B. Adams (New York: Oxford University Press, 1990), 220. For an excellent discussion of "pseudoscience" as a "battle concept (*Kampfbegriff*)," and thus its utility for exploring mainstream science, see the essay by Michael Hagner, "Bye-Bye Science, Welcome Pseudoscience?: Reflexionen über einen beschädigten Status," in *Pseudowissenschaft: Konzeptionen von Nichtwissenschaftlichkeit in der Wissenschaftsgeschichte*, ed. Dirk Rupnow et al. (Frankfurt am Main: Suhrkamp, 2008), 21–50. See also the call for historicization in Mitchell G. Ash, "Pseudowissenschaft als historische Größe: Ein Abschlusskommentar," in ibid., 452.

8. For useful bibliographies, see Marcello Truzzi, "Crank, Crackpot, or Genius? Pseudoscience or Science Revolution?: A Bibliographic Guide to the Debate," *Zetetic Scholar* 1 (1978): 20–22; and Andrew Fraknoi, "Scientific Responses to Pseudoscience: An Annotated Bibliography," *Mercury* (July–August 1984): 121–24.

9. The examples are too numerous to mention, but as one instance, consider the alleged cancer drug, Krebiozen. This debate raised many of the same themes of elitism and anti-establishment thought, as discussed in Keith Wailoo, *How Cancer Crossed the Color Line* (New York: Oxford University Press, 2011), 122–29.

10. The concept of "paranoid style" is borrowed, of course, from Richard Hofstadter, *The Paranoid Style in American Politics and Other Essays* (New York: Vintage, 2008 [1965]), and I mean it in essentially the same sense. Paranoia and conspiracy theory have engendered a vast literature of their own; especially useful are Daniel Hellinger, "Paranoia, Conspiracy, and Hegemony in American Politics," in *Transparency and Conspiracy: Ethnographies of Suspicion in the New World Order*, ed. Harry G. West and Todd Sanders (Durham, NC: Duke University Press, 2003), 204–32; and Mark Fenster, *Conspiracy Theories: Secrecy and Power in American Culture* (Minneapolis: University of Minnesota Press, 1999). One history of twentieth-century conspiracy theories outlines a seven-point checklist for what "counts" as such a theory, and the parallels to equivalent lists for "pseudoscience" are striking. See David Aaronovitch, *Voodoo Histories: The Role of the Conspiracy Theory in Shaping Modern History* (New York: Riverhead Books, 2010), 10–14. Martin Gardner, in his classic debunking text, considered "paranoia" as a common characteristic of those fields he labeled pseudosciences: Gardner, *Fads and Fallacies in the Name of Science* (New York: Dover, 1957 [1952]), 11.

11. See, for example, Henry Bauer, "Velikovsky and Social Studies of Science," *4S Review* 2 (1984): 2–3; Bauer, "Passions and Purposes: A Perspective," *Skeptical Inquirer* 5 (Fall 1980): 28–31; Bauer, *Beyond Velikovsky: The History of a Public Controversy* (Urbana: University of Illinois Press, 1984), 133; Fred Warshofsky, *Doomsday: The Science of Catastrophe* (New York: Reader's Digest Press, 1977), 63; Gregory J. Feist, *The Psychology of Science and the Origins of the Scientific Mind* (New Haven, CT: Yale University Press, 2006), 220; and Gregory N. Derry, *What Science Is and How It Works* (Princeton, NJ: Princeton University Press, 1999), 163–67. This list is far from exhaustive.

12. Paul R. Thagard, "Resemblance, Correlation, and Pseudoscience," in *Science, Pseudo-Science and Society*, ed. Marsha P. Hanen, Margaret J. Osler, and Robert C. Weyant (Waterloo, Ontario: Wilfrid Laurier University Press, 1980), 21.

13. Aaronovitch, *Voodoo Histories*, 218.

14. Gardner, *Fads and Fallacies*, 31–32.

15. Eric H. Christianson, "Pseudoscience: A Bibliographic Guide," *Choice* 16 (January 1980), 1412. L. Ron Hubbard was, of course, the founder of Scientology, which has spawned a staggeringly large literature, especially as the doctrines of his *Dianetics*—published the same year as *Worlds in Collision*—have been transformed into a religion. See Christopher Evans, *Cults of Unreason* (London: Harrap, 1973); and Hugh B. Urban, *The Church of Scientology: A History of a New Religion* (Princeton, NJ: Princeton University Press, 2011). Charles Fort was a fascinating individual who proposed to explain mysterious appearances and disappearances through the proliferation of wormhole-like vortices around the globe. On his heyday in Depression-era America, see Damon Knight, *Charles Fort: Prophet of the Unexplained* (Garden City, NJ: Doubleday, 1970). Knight mentions Velikovsky repeatedly.

16. Martin Gardner, *Science: Good, Bad and Bogus* (Amherst, NY: Prometheus Books, 1989), xiv.

17. Frederic B. Jueneman, *Velikovsky: A Personal View* (Glassboro, NJ: Kronos Press, 1980 [1975]), 59.

18. Immanuel Velikovsky, *Worlds in Collision* (New York: Macmillan, 1950).

19. For biographies, see Ruth Velikovsky Sharon, *Aba: The Glory and the Torment: The Life of Dr. Immanuel Velikovsky* (Dubuque, IA: Times Mirror Higher Education Group, 1995); Ion Degen, *Immanuil Velikovskii: Rasskaz o zamechatel'nom cheloveke* (Rostov-on-Don: Feniks, 1997); and Duane Leroy Vorhees, "The 'Jewish Science' of Immanuel Velikovsky: Culture and Biography as Ideational Determinants" (PhD diss., Bowling Green State University, 1990). The first is more a memoir, written by Velikovsky's younger daughter. The second is a Russian-language biography, penned in the mid-1980s with the cooperation of his elder daughter, Shulamit Kogan, but not published until after the collapse of the Soviet Union. Both are based very heavily on Velikovsky's own interpretation of his past. Vorhees's lengthy dissertation displays significant original research, especially on Velikovsky's earlier life. For bibliographies of the debates over Velikovsky, which contain both debunkings and responses, see Marcello Truzzi, "Velikovsky and His Critics: A Basic Bibliography," *Zetetic Scholar* 1 (1978): 100–101; as well as the many works cited below. For analyses of the history, most of which date from the time of the controversy itself, see especially Bauer, *Beyond Velikovsky*; Bauer, "Im-

manuel Velikovsky," in *The Encyclopedia of the Paranormal*, ed. Gordon Stein (Amherst, NY: Prometheus Books, 1996), 781–88; R. G. A. Dolby, "What Can We Usefully Learn from the Velikovsky Affair?" *Social Studies of Science* 5 (1975): 160–75; Robert McAulay, "Velikovsky and the Infrastructure of Science: The Metaphysics of a Close Encounter," *Theory and Society* 6 (1978): 313–42; McAulay, "Substantive and Ideological Aspects of Science: An Analysis of the Velikovsky Controversy" (MA thesis, University of New Mexico, Albuquerque, 1975); Ronald H. Fritze, *Invented Knowledge: False History, Fake Science, and Pseudo-Religions* (London: Reaktion Books, 2009); and David Stove, "Velikovsky in Collision," *Quadrant* (October–November 1964): 35–44.

20. Karl Popper, *Conjectures and Refutations: The Growth of Scientific Knowledge* (New York: Routledge, 2002 [1963]), 48; emphasis in original.

21. Kitcher, *Abusing Science*, 42. See also A. F. Chalmers, *What Is This Thing Called Science?*, 3rd ed. (Berkshire, UK: Open University Press, 1999 [1978]); A. A. Derksen, "The Seven Sins of Pseudo-Science," *Journal for General Philosophy of Science* 24 (1993): 26; and the highly developed alternative framework (which has its own troubles) in Imre Lakatos, "Falsification and the Methodology of Scientific Research Programs," in *Criticism and the Growth of Knowledge*, ed. Imre Lakatos and Alan Musgrave (Cambridge: Cambridge University Press, 1970), 91–196.

22. Imre Lakatos, *The Methodology of Scientific Research Programmes*, ed. John Worrall and Gregory Currie (Cambridge: Cambridge University Press, 1978), 125.

23. Thomas S. Kuhn, *The Structure of Scientific Revolutions*, 3rd ed. (Chicago: University of Chicago Press, 1996 [1962]).

24. Thomas S. Kuhn, "Logic of Discovery or Psychology of Research?" in *Criticism and the Growth of Knowledge*, ed. Lakatos and Musgrave, 1–23.

25. Velikovsky to Thomas S. Kuhn, February 10, 1967, IVP 85:36.

26. William Mullen to Velikovsky, July 15, 1977, IVP 90:2.

27. Lynn Rose, in Velikovsky and Lynn E. Rose, "The Sins of the Sons: A Critique of Velikovsky's A.A.A.S. Critics," [late 1970s], IVP 53:7, p. 96.

28. André M. Bennett, "Science: The Antithesis of Creativity," *Perspectives in Biology and Medicine* (Winter 1968): 237.

29. Harry Collins and Robert Evans, *Rethinking Expertise* (Chicago: University of Chicago Press, 2007), 132.

30. Fred J. Gruenberger, "A Measure for Crackpots," *Science* 145, no. 3639 (September 25, 1964): 1413–15; Michael Shermer, *Why People Believe Weird Things: Pseudoscience, Superstition, and Other Confusions of Our Time* (New York: W. H. Freeman, 1997), 48–54; Mario Bunge, "What Is Pseudoscience?" *Skeptical Inquirer* 9 (Fall 1984): 36–46; Derksen, "The Seven Sins of Pseudo-Science"; James S. Trefil, "A Consumer's Guide to Pseudo-science," *Saturday Review*, April 29, 1978, 16–21; and J. W. Grove, *In Defence of Science: Science, Technology, and Politics in Modern Society* (Toronto: University of Toronto Press, 1989), 147–50. A more subtle version of this approach is developed in Paul R. Thagard, "Why Astrology Is a Pseudoscience," *PSA: Proceedings* 1 (1978): 228. The point about isolation is particularly popular—as in Gardner, *Fads and Fallacies*, 8; Derry, *What Science Is*, 162—but to some extent it blames those excluded from the channels of main-

stream science for the fact of their exclusion; establishment science won't listen to them and marginalizes them, and their marginalization produces their isolation. See Ingo Grabner and Wolfgang Reiter, "Guardians at the Frontiers of Science," in *Counter-Movements in the Sciences: The Sociology of Alternatives to Big Science*, ed. Helga Nowotny and Hilary Rose (Dordrecht: D. Reidel, 1979), 75.

31. Larry Laudan, "Views of Progress: Separating the Pilgrims from the Rakes," *Philosophy of the Social Sciences* 10 (1980): 275.

32. Larry Laudan, "The Demise of the Demarcation Problem" (1983), in *But Is It Science?: The Philosophical Question in the Creation/Evolution Controversy*, updated ed., ed. Robert T. Pennock and Michael Ruse (Amherst, NY: Prometheus Books, 2009), 320. Laudan applied his views on demarcation directly to the legal issues surrounding creation science in his "Science at the Bar: Causes for Concern," *Science, Technology, & Human Values* 7 (Fall 1982): 16–19. See also R. G. A. Dolby, "Science and Pseudo-Science: The Case of Creationism," *Zygon* 22 (June 1987): 202.

33. Robert T. Pennock, "Can't Philosophers Tell the Difference between Science and Religion?: Demarcation Revisited," in *But Is It Science?*, ed. Pennock and Ruse, 566.

34. Laudan, "The Demise of the Demarcation Problem," 328.

35. Dirk Rupnow et al., "Einleitung," in *Pseudowissenschaft*, ed. Rupnow et al., 8; Cooter, "The Conservatism of 'Pseudoscience,'" 164; Henry H. Bauer, *Fatal Attractions: The Troubles with Science* (New York: Paraview, 2001), 82–83; Bauer, *Science or Pseudoscience: Magnetic Healing, Psychic Phenomena, and Other Heterodoxies* (Urbana: University of Illinois Press, 2001); Seymour H. Mauskopf, introduction to *The Reception of Unconventional Science*, ed. Seymour H. Mauskopf (Boulder, CO: Westview Press, 1979), 1–9; and Marcello Truzzi, "Discussion: On the Reception of Unconventional Scientific Claims," in *The Reception of Unconventional Science*, ed. Mauskopf, 127.

36. On fraud, see Trevor J. Pinch, "Normal Explanations of the Paranormal: The Demarcation Problem and Fraud in Parapsychology," *Social Studies of Science* 9 (1979): 329–48; C. L. Hardin, "Table-Turning, Parapsychology and Fraud," *Social Studies of Science* 11 (1981): 249–55; Harriet Zuckerman, "Norms and Deviant Behavior in Science," *Science, Technology, & Human Values* 9 (1984): 7–13; Zuckerman, "Deviant Behavior and Social Control in Science," in *Deviance and Social Change*, ed. Edward Sagarin (Beverly Hills: Sage, 1977), 87–138; and William Broad and Nicholas Wade, *Betrayers of the Truth* (New York: Simon & Schuster, 1982). On hoaxes, see Jim Schnabel, "Puck in the Laboratory: The Construction and Deconstruction of Hoaxlike Deception in Science," *Science, Technology, & Human Values* 19 (1994): 459–92; and Robert Silverberg, *Scientists and Scoundrels: A Book of Hoaxes* (New York: Thomas Y. Crowell, 1965).

37. Irving Langmuir, "Pathological Science: Scientific Studies Based on Non-Existent Phenomena," ed. R. N. Hall, *Speculations in Science and Technology* 8 (1985): 77–94. For criticisms of this notion, see Bauer, *Science or Pseudoscience*, 116–18; and Bauer, *Fatal Attractions*, 91.

38. Helga Nowotny, "Science and Its Critics: Reflections on Anti-Science," in *Counter-Movements in the Sciences*, ed. Nowotny and Rose, 15.

39. Amateur science is a large topic, with an especially rich tradition in astronomy.

For a recent historical analysis of the impact of amateur science on professional astronomy, see W. Patrick McCray, *Keep Watching the Skies!: The Story of Operation Moonwatch and the Dawn of the Space Age* (Princeton, NJ: Princeton University Press, 2008).

40. Philip Kitcher, *Living with Darwin: Evolution, Design, and the Future of Faith* (New York: Oxford University Press, 2007), 115. See also Kitcher, "Darwins Herausforderer: Über *Intelligent Design* oder: Woran man Pseudowissenschaft erkennt," in *Pseudowissenschaft*, ed. Rupnow et al., 430–31; and Derksen, "The Seven Sins of Pseudo-Science," 21. Kitcher first advanced aspects of this psychological line in *The Advancement of Science: Science without Legend, Objectivity without Illusions* (New York: Oxford University Press, 1993), 195–96.

41. Martin Gardner, *Did Adam and Eve Have Navels?: Debunking Pseudoscience* (New York: Norton, 2000), 1. See also Gardner, "The Hermit Scientist," *Antioch Review* 10, no. 4 (December 1950): 456n4; and Gardner, *Science*, xiii.

42. Massimo Pigliucci, *Nonsense on Stilts: How to Tell Science from Bunk* (Chicago: University of Chicago Press, 2010), 304.

43. Thomas F. Gieryn, "Boundary-Work and the Demarcation of Science from Non-Science: Strains and Interests in Professional Ideologies of Scientists," *American Sociological Review* 48 (1983): 792. See also R. G. A. Dolby, "Reflections on Deviant Science," in *On the Margins of Science: The Social Construction of Rejected Knowledge*, ed. Roy Wallis, Sociological Review Monograph 27 (Keele: University of Keele, 1979), 10.

44. Michael Martin, "The Use of Pseudo-Science in Science Education," *Science Education* 55 (1971): 53–56; also David B. Resnick, "A Pragmatic Approach to the Demarcation Problem," *Studies in History and Philosophy of Science* 31 (2000): 264.

45. Charles Alan Taylor, *Defining Science: A Rhetoric of Demarcation* (Madison: University of Wisconsin Press, 1996), 53.

46. L. Sprague de Camp, "Orthodoxy in Science," *Astounding Science-Fiction* 53, no. 3 (May 1954): 123. Or, in the words of fellow science-fiction author Isaac Asimov: "What, then, would you have the orthodox do? Is it better to reject everything and be wrong once in fifty times—or accept everything and be wrong forty-nine out of fifty times and, in the meantime, send science down endless blind alleys?" Asimov, "Foreword: The Role of the Heretic," in *Scientists Confront Velikovsky*, ed. Donald Goldsmith (New York: Norton, 1977), 9. Philosopher of science Michael Polanyi articulated a variant of this central dilemma on several occasions: Polanyi, *The Tacit Dimension* (Chicago: University of Chicago Press, 2009 [1966]), 65; Polanyi, "The Republic of Science: Its Political and Economic Theory," *Minerva* 1 (1962): 61.

47. Kendrick Frazier, "Science and the Parascience Cults," *Science News* 109 (May 29, 1976): 346. See also Gardner, *Fads and Fallacies*, 6; and Kitcher, *Abusing Science*, 122–23.

48. Grabner and Reiter, "Guardians at the Frontiers of Science," 73.

49. Thagard, "Resemblance, Correlation, and Pseudoscience," 24.

50. Gieryn, "Boundary-Work," 781. Gieryn expands on these issues in *Cultural Boundaries of Science: Credibility on the Line* (Chicago: University of Chicago Press, 1999).

51. Ronald L. Numbers, *The Creationists: From Scientific Creationism to Intelligent De-*

sign, exp. ed. (Cambridge, MA: Harvard University Press, 2006 [1992]), 12; emphasis in original.

52. Frank Cioffi, "Freud and the Idea of a Pseudo-Science," in *Explanation in the Behavioural Sciences*, ed. Robert Borger and Frank Cioffi (Cambridge: Cambridge University Press, 1970), 471.

53. For excellent historical studies of these topics, see Roger Cooter, *The Cultural Meaning of Popular Science: Phrenology and the Organization of Consent in Nineteenth-Century Britain* (Cambridge: Cambridge University Press, 1984); Alison Winter, *Mesmerized: Powers of Mind in Victorian Britain* (Chicago: University of Chicago Press, 1998); and Seymour H. Mauskopf and Michael R. McVaugh, *The Elusive Science: Origins of Experimental Psychical Research* (Baltimore: Johns Hopkins University Press, 1980).

54. This was noted in passing specifically about Velikovsky by the controversial philosopher of science Paul Feyerabend in his *Against Method: Outline of an Anarchistic Theory of Knowledge* (London: Verso, 1978 [1975]), 40n5, 298n11.

55. Frederic B. Jueneman, "The Search for Truth," *Analog* 94, no. 2 (October 1974): 25; emphasis in original.

56. Much of the correspondence is intemperate and inflammatory, and in quoting from it I am following a practice advocated by Velikovsky's prime defender, Lynn E. Rose, in a letter to Stephen Talbott, September 8, 1973, IVP 93:1: "The documents in question are indicative of the state of twentieth century science; as such, they are fair game for historians. I cannot think of any reason why they should not be published. No issues of national security mandate that they be suppressed! No one will be damaged by publication. Some may be embarrassed by the indications that they used poor judgement, but even they profess to be concerned with the discovery of the truth, and if they are truly seekers after truth, they should welcome anything that enables them to correct any false assumptions that have been distorting their investigations." Velikovsky had a similar point of view when discussing the 1950 controversy over *Worlds in Collision* (addressed in chapter 1 below): "These letters from [Harlow] Shapley to [Ted] Thackrey were not intended for publication when written; the first was marked 'Not for publication,' and the second 'Confidential.' But whether sooner or later, they belong to history. They do not contain any personal or intimate matter that the writer could possibly regard as of a private nature, which ought to remain behind a veil. Their writer considered himself to be performing a public service by writing them. Since they were ostensibly a public service, they are, therefore, a public affair." Velikovsky, *Stargazers and Gravediggers: Memoirs to "Worlds in Collision"* (New York: Quill, 1984), 112. The same point obviously applies to letters written by Velikovsky's supporters.

57. Velikovsky to George Sarton, December 21, 1950, IVP 94:16. He corresponded along similar lines to a distinguished historian of science at Yale University: Velikovsky to Derek J. de Solla Price, September 18, 1962, and Price to Velikovsky, September 27, 1962, IVP 91:23.

58. Velikovsky to Coleman Morton, December 11, 1964, IVP 126:3. In sending the third-party letters, Morton (who supported Velikovsky generally), noted: "Inciden-

tally, in this respect, I have violated the normal rules of confidence in sending you a copy of Dr. [Louis] Slichter's comments of January 14, 1964 only because I know that you would not use such information to embarrass Dr. Slichter, Dr. [Franklin] Murphy [Chancellor of UCLA], or for that matter, myself." Morton to Velikovsky, December 18, 1964, IVP 126:3.

59. Velikovsky to President and Board Members of FOSMOS, December 1, 1968, IVP 83:1.

60. Velikovsky to Howard B. Gottlieb, September 29, 1966, IVP 79:7.

61. There are many excellent examples of such histories read "against the grain," but to pick an example from each case: Carlo Ginzburg, *The Cheese and the Worms: The Cosmos of a Sixteenth-Century Miller*, trans. John Tedeschi and Anne Tedeschi (Baltimore: Johns Hopkins University Press, 1980); and Igal Halfin, *Terror in My Soul: Communist Autobiographies on Trial* (Cambridge, MA: Harvard University Press, 2003).

62. The richness of the archive shaped another choice I have made in constructing this book: the absence of personal interviews. I have spoken and corresponded with several Velikovskians, former Velikovskians, anti-Velikovskians, and people who knew him more or less intimately. Those communications have been valuable in coming to understand both the person and the controversy, but I have chosen to rely on the written record. As I write these words, it has been thirty-one years since Velikovsky's death—longer than the twenty-nine years of controversy elicited by *Worlds in Collision* during his lifetime. Velikovskianism elicited strong emotions on all sides, those emotions have in many ways hardened since his death, and memory is notoriously fickle. I am interested in capturing the first incarnation of these debates, and I invite another sociologist or historian to follow the legacy after 1979.

63. In science studies, the principle of not taking a position on the correctness or incorrectness of various scientific claims is called *methodological symmetry*. It is particularly difficult to maintain when examining debates on the fringe, as noted by one of its principal advocates: H. M. Collins, "The Sociology of Scientific Knowledge: Studies of Contemporary Science," *Annual Reviews of Sociology* 9 (1983): 279. Many historians of similar demarcation disputes consider asking for their own views on the subject matter "certainly a legitimate question." Mauskopf and McVaugh, *The Elusive Science*, xiv. Following the general position of cultural historians of science for the last few decades, however, I do not consider such a *profession de foi* necessary to provide an accurate and illuminating account.

Chapter One

1. This should not be confused with the 1932 science-fiction novel by Edwin Balmer and Philip Wylie, *When Worlds Collide* (New York: A. L. Burt, 1932), although the sensationalism of Velikovsky's title and its similarity to science-fiction tropes was no doubt an aggravating factor.

2. Henry H. Bauer, *Beyond Velikovsky: The History of a Public Controversy* (Urbana: University of Illinois Press, 1984), 12.

3. Velikovsky, *Stargazers and Gravediggers: Memoirs to "Worlds in Collision"* (New York: Quill, 1984), 41.

4. Velikovsky, *Worlds in Collision* (New York: Macmillan, 1950), vii.

5. Ibid., 226.

6. Ibid., 286.

7. Ibid., ix.

8. Ibid., 173.

9. Ibid., 306.

10. This identification of Venus with Athena (not Aphrodite) remained a sticking point with numerous critics, such as Bob Forrest, *Velikovsky's Sources*, 6 vols. (Manchester: By the author, 1981–83), 1:46–47; and Carl Sagan, "An Analysis of *Worlds in Collision*," in *Scientists Confront Velikovsky*, ed. Donald Goldsmith (New York: Norton, 1977), 54.

11. Velikovsky, *Worlds in Collision*, 358.

12. Ibid., viii.

13. Since 1950, the Velikovskian controversies have drawn a certain amount of attention from scholars. For a detailed discussion of the religious context of the reception of *Worlds in Collision*, see James Gilbert, *Redeeming Culture: American Religion in an Age of Science* (Chicago: University of Chicago Press, 1997), chap. 8. A very thorough, although somewhat partisan, pro-Velikovskian account of the reviews of Velikovsky's first catastrophist work can be found in Duane Vorhees, "'Worlds in Collision': Reviews and Reviewers," *Aeon* 3 (December 1994): 15–34. Almost all Velikovskian histories of the events in this chapter, with the exception of Vorhees, are derived from Velikovsky's own version in *Stargazers and Gravediggers*. The same narrative is repeated almost exactly, and with much of the same quotation of correspondence, in Ruth Velikovsky Sharon, *Immanuel Velikovsky: The Truth Behind the Torment* (Princeton, NJ: By the author, 2003). Likewise, a lengthy 1996 essay—Irving Wolfe, "The Original Velikovsky Affair: An Idea That Just Would Not Go Away," in *Stephen J. Gould and Immanuel Velikovsky: Essays in the Continuing Velikovsky Affair*, ed. Dale Ann Pearlman (Forest Hills, NY: Ivy Press, 1996), 1–50—consists almost exclusively of footnotes to *Stargazers*, a text mostly composed in 1956.

14. Eric Larrabee, "The Day the Sun Stood Still," *Harper's Magazine*, January 1950, 20, 23.

15. Larrabee's foreword to Velikovsky, *Stargazers and Gravediggers*, 14–15.

16. See the details of the advanced press reviews in J. R. Williams to George Brett, February 7, 1950, MR 96:1.

17. Velikovsky to George Sarton, December 21, 1950, IVP 94:16.

18. Velikovsky, "The Heavens Burst," *Collier's*, February 25, 1950, 24. The second article is Velikovsky, "World on Fire," *Collier's*, March 25, 1950, 24, 82–85.

19. Gilbert, *Redeeming Culture*, 195. Martin Gardner, at the beginning of a half-century career of debunking what he considered pseudosciences, concurred, writing in late 1950 that the whole purpose of *Worlds in Collision* was to justify the Old Testament. Gardner, "The Hermit Scientist," *Antioch Review* 10, no. 4 (December 1950): 450.

20. As Fulton Oursler put it: "To science, *Worlds in Collision* opens up a vast new debate; to millions of true believers in the Old Testament, it will come as an unintended and reassuring answer to the rationalist criticism of the last 75 years." Oursler, "Why the Sun Stood Still," *Reader's Digest* 56 (March 1950): 148.

21. Velikovsky, "The Heavens Burst," 43.

22. Velikovsky to Marshall McClintock, January 20, 1950, IVP 88:21. For the record, Eric Larrabee hated the *Collier's* version as well, writing to Velikovsky that "it is my personal conviction that the publication of the articles in the state in which I saw them would have a severely damaging effect on the initial reception of your theory." Eric Larrabee to Velikovsky, January 23, 1950, IVP 124:13.

23. Marshall McClintock to John Lear, May 13, 1966, IVP 124:13; emphasis in original.

24. Harlow Shapley to the editorial department of the Macmillan Company, January 18, 1950, IVP 124:2. Shapley specifically noted that this letter was not for publication.

25. Velikovsky to Ted O. Thackrey, April 5, 1950, IVP 124:2.

26. Velikovsky to Harlow Shapley, April 15 and April 17, 1946, IVP 124:2.

27. A. D. Walker to Velikovsky, May 15, 1946, IVP 124:2. No one agreed to perform these tests for Velikovsky in the late 1940s. All of these measurements were, however, taken as a matter of course during the explorations of the space age, as discussed in chapter 4 below.

28. On Kallen's involvement with Velikovsky, see Gilbert, *Redeeming Culture*, 181. For Kallen's own account of the events that follow, see Horace M. Kallen, "Shapley, Velikovsky, and the Scientific Spirit," in *Velikovsky Reconsidered*, ed. Editors of *Pensée* (New York: Warner Books, 1976), 52–66.

29. Horace Kallen to Velikovsky, October 14, 1941, IVP 84:4.

30. Kallen to Shapley, May 23, 1946, IVP 124:2.

31. Shapley to Kallen, May 27, 1946, IVP 124:2.

32. Velikovsky to Otto Struve, June 17, 1946, and Struve to Velikovsky, June 21, 1946, IVP 125:16; Velikovsky to Rupert Wildt, August 27, 1946, IVP 125:16; Walter S. Adams to Velikovsky, September 9, 1946, and Wildt to Velikovsky, September 13, 1946, IVP 125:16.

33. Kallen to Velikovsky, May 31, 1946, IVP 124:2.

34. James Putnam to Shapley, January 24, 1950, IVP 124:2.

35. Shapley to Putnam, January 25, 1950, IVP 124:2.

36. Shapley to Ted Thackrey, February 20, 1950, IVP 124:2.

37. Thackrey to Shapley, March 7, 1950, IVP 124:2.

38. Shapley to H. S. Latham, July 7, 1950, IVP 124:2. Actually, the American Astronomical Society did issue a protest to Macmillan: Alfred H. Joy (president of the AAS, from Caltech) to Macmillan Company, July 25, 1950, MR 96:2. By this point, as will be explained below, Macmillan was no longer the publisher of *Worlds in Collision*.

39. Shapley to Thackrey, March 8, 1950, IVP 124:2.

40. "Theories Denounced," *Science News-Letter* 57 (February 25, 1950): 119. Shapley's memoir, *Through Rugged Ways to the Stars* (New York: Charles Scribner's Sons, 1969), makes no mention of Velikovsky or the Macmillan dispute, an omission lamented by

at least one reviewer: Earl Ubell, review of *Through Rugged Ways to the Stars*, by Harlow Shapley, *New York Times*, July 13, 1969, BR7.

41. Velikovsky, "How I Arrived at My Concepts," in "Thoughts on Various Subjects," undated bound typescript, IVP 40:4, p. 10.

42. Horace Kallen to Alfred A. Knopf, June 26, 1941, IVP 84:4.

43. Alfred A. Knopf to Velikovsky, July 2, 1941, IVP 84:4.

44. Velikovsky to Morgan Shuster (president of Appleton-Century Publishing Co.), June 4, 1946, IVP 97:5.

45. Velikovsky hired a copy editor, Marion Kuhn, a graduate of Smith College, to clean up his English but to leave the punchy style intact. Velikovsky, *Stargazers and Gravediggers*, 66. With respect to Fadiman's investigation of Velikovsky, his associate reported from Paul Federn, a Viennese psychologist and friend of Velikovsky's who had emigrated to New York: "Knows him well. A genius—a great man. Excellent psychoanalyst. An M.D., member of the Palestine group. Some revolutionary scientific ideas that some people think are crazy, but he is a genius. Would not consider him for a teacher; but as an analyst have sent him some of my most difficult cases." Lawrence S. Kubie to Clifton Fadiman, October 23, 1947, IVP 53:1.

46. Clifton Fadiman to Velikovsky, March 10, 1947, IVP 76:11.

47. Clifton Fadiman to Velikovsky, April 19, 1948, IVP 76:11.

48. John J. O'Neill, "Atomic Energy Charging Globe Held Able to Erupt at Any Time," *New York Herald Tribune*, August 11, 1946, secs. II–IV, p. 10.

49. Gordon Atwater to Velikovsky, October 10, 1946, IVP 69:21.

50. Gamow to James Putnam, March 24, [1950], JPP 165:2; and Roy K. Marshall to the Macmillan Company, February 21, 1950, IVP 126:16.

51. Macmillan also sent out *Ages in Chaos* for at least one referee report. See the undated reader's report on "Ages in Chaos" by Robert H. Pfeiffer of Harvard University, MR 96:1.

52. John J. O'Neill, undated referee report [but before March 1947], MR 96:1.

53. John J. O'Neill to James Putnam, March 1, 1947, MR 96:1. See also John J. O'Neill to Ferris J. Stephens, November 24, 1950, IVP 128:2.

54. Gordon Atwater to James Putnam, May 18, 1948, MR 96:1.

55. Velikovsky, *Stargazers and Gravediggers*, 66–67.

56. See George Brett to Harlow Shapley, February 1, 1950, IVP 124:2.

57. Henry B. McCurdy to C. Leroy Ellenberger, June 1, 1983, courtesy of C. Leroy Ellenberger. On the Einstein suggestion, see J. R. Williams to George Brett, February 7, 1950, MR 96:1.

58. Ed Thorndike, "Notes on *Worlds in Collision*," February 13, 1950, IVP 123:10.

59. C. W. van der Merwe to Henry B. McCurdy, February 14, 1950, IVP 123:10.

60. Clarence S. Sherman to Henry B. McCurdy, February 14, 1950, IVP 123:10. Two additional but brief quasi-positive reports—focusing on its commercial potential—also arrived at Macmillan's offices. See C. L. Skelly to George Brett, February 17, 1950, IVP 124:2; and Theodore Dunham Jr. to W. Holt Seale, March 23, 1950, JPP 165:2.

61. See David Kaiser, *American Physics and the Cold War Bubble* (Chicago: University of Chicago Press, forthcoming), chap. 3.

62. Cecilia Payne-Gaposchkin, *The Dyer's Hand: An Autobiography* (Privately printed, 1979), 115; and Harlow Shapley to James Putnam, January 25, 1950, IVP 124:2.

63. Cecilia Payne-Gaposchkin, "Nonsense, Dr. Velikovsky!," *Reporter* 2, no. 6 (March 14, 1950): 40.

64. Editor's foreword in Cecilia Payne-Gaposchkin, "'Worlds in Collision,'" *Popular Astronomy* 58, no. 6 (June 1950): 278.

65. Payne-Gaposchkin, "'Worlds in Collision,'" (1950), 283, 285.

66. See the discussion in Edwin G. Boring, "The Validation of Scientific Belief: A Conspectus of the Symposium," *Proceedings of the American Philosophical Society* 96, no. 5 (October 15, 1952): 535–39.

67. Velikovsky, *Stargazers and Gravediggers*, 250.

68. Cecilia Payne-Gaposchkin, "'Worlds in Collision,'" *Proceedings of the American Philosophical Society* 96 (October 15, 1952): 519. The published version of the critique was followed by a brief calculation by Donald Menzel refuting Velikovsky's hypothesis that electromagnetic interactions among the planets could explain the processes astronomers deemed impossible according to the laws of celestial mechanics. Donald H. Menzel, "The Celestial Mechanics of Electrically Charged Planets," *Proceedings of the American Philosophical Society* 96 (October 15, 1952): 524–25. Menzel had also reviewed *Worlds in Collision* in his "In the Daze of the Comet," *Physics Today* 3 (July 1950): 26–27.

69. Velikovsky to his children, April 25, 1952, IVP 123:2; emphasis in original.

70. Velikovsky to George Sarton, July 22, 1953, GSP, folder: "Velikovsky, Immanuel, 1895–."

71. For example, Orville Prescott, "Books of the Times," *New York Times*, April 3, 1950, 32; Herbert B. Nichols, "The Velikovsky Excursion," *Christian Science Monitor*, March 29, 1950, 18; and Wilton Krogman, "Evidence That the Sun Did Stand Still," *Chicago Daily Tribune*, April 9, 1950, E3. In a lone scientific quasi-endorsement, a biographer of future Nobel Laureate geneticist Barbara McClintock has noted that she was "said to have told friends not to dismiss Velikovsky so quickly, that there might be a grain of truth in his work." Nathaniel C. Comfort, *The Tangled Field: Barbara McClintock's Search for the Patterns of Genetic Control* (Cambridge, MA: Harvard University Press, 2001), 152.

72. Edward U. Condon, "Velikovsky's Catastrophe," *New Republic* 122, no. 17 (April 24, 1950): 24. British geneticist J. B. S. Haldane, writing upon the publication of the British edition, suscribed to the hoax theory. Haldane, "St. Quetzalcoatl and St. Fenris," *New Statesman and Nation*, November 11, 1950, 433. Velikovsky responded in "Worlds in Collision: Letter to the Editor," *New Statesman and Nation* (February 3, 1951): 131. For a similar British mockery of *Worlds in Collision* as an "American" peculiarity, see A. W. Haslett, "Stars Off Their Courses," *Observer*, September 10, 1950, 7.

73. Paul Herget, "It Is Just as Barnum Said," *Cincinnati Enquirer*, April 1, 1950, 7.

74. Otto Struve, "Copernicus? Who Was He?" *New York Herald Tribune Book Review*, April 2, 1950, 4. Science-fiction writer L. Sprague de Camp's review in *Astounding Science-Fiction* placed the work in a pseudoscientific trajectory going back to earlier cosmic catastrophists, including the notorious Hanns Hörbiger. Review of *Worlds in*

Collision, by Velikovsky, *Astounding Science-Fiction*, September 1950, 138–41. The link to science fiction was also made by Harvard president James Bryant Conant, who called *Worlds in Collision* a "grotesque account" distorting evidence to produce "a fantasia which is neither history nor science." Conant, *Science and Common Sense* (New Haven, CT: Yale University Press, 1951), 278.

75. Harrison Brown, "Venus and the Scriptures," *Saturday Review of Literature*, April 22, 1950, 19.

76. Waldemar Kaempffert, "The Tale of Velikovsky's Comet," *New York Times Book Review*, April 2, 1950, 16. Velikovsky responded to this claim, as well as Kaempffert's statement that an unbroken chain of recorded eclipses proved that there had been no recent changes in the solar system. Velikovsky and Waldemar Kaempffert, "Dr. Velikovsky vs. Mr. Kaempffert: A Collision of Author and Reviewer," *New York Times*, May 7, 1950, 224. The two debated the issue impromptu when Kaempffert challenged Velikovsky after the latter spoke before students at Columbia in May 1950, described briefly in Hal Levine, "Dr. Velikovsky Reiterates New Theories Here," *Columbia Daily Spectator*, May 10, 1950, 1.

77. Alfred Kazin, "On the Brink," *New Yorker*, April 29, 1950, 104–5. Haldane, who had recently resigned from the British Communist Party, extended this point of view, considering excitement about Velikovsky to presage American military adventurism: Haldane, "St. Quetzalcoatl and St. Fenris," 433.

78. Harold L. Ickes, "A Distinct Choice," *New Republic*, March 6, 1950, 16.

79. Harlow Shapley to Ted Thackrey, March 8, 1950, IVP 124:2.

80. Harold S. Latham, *My Life in Publishing* (London: Sidgwick & Jackson, 1965), 74.

81. Quoted in Frederic Babcock, "Among the Authors," *Chicago Daily Tribune*, July 2, 1950, H5.

82. Dean B. McLaughlin to George Brett, May 20, 1950, IVP 124:2; emphasis in original. McLaughlin also wrote an intemperate letter to Fulton Oursler, the author of the *Reader's Digest* extract, on June 16, 1950, IVP 124:18. Echoing McLaughlin's suggested remedies, George Gamow wrote to *Harper's* to cancel his subscription in response to Larrabee's piece, declaring he had also written to Macmillan, where he was an author, "with the suggestion that they withdraw that book from the circulation." George Gamow to *Harper's*, March 18, 1950, IVP 123:14.

83. Roy K. Marshall to the Macmillan Company, February 21, 1950, IVP 126:16.

84. Marshall to Joseph M. McGarry, January 25, 1950, IVP 87:9.

85. Frank K. Edmondson to Brett, April 5, 1950, IVP 126:16. A similar lament against the aggressive publicity campaign was articulated by an associate professor of chemistry at Amherst College, David C. Grahame, in his letter to Macmillan, April 20, 1950, MR 96:1. The point about catalogs was also made in Charles L. Skelley to C. Leroy Ellenberger, April 18, 1983, courtesy of C. Leroy Ellenberger. Harold Latham later claimed that the publisher "went to great lengths in all announcements and advertisements to make it perfectly clear that this was a book for the general reader, not a text for the astronomer or scientist." Latham, *My Life in Publishing*, 74. An examination of Mac-

millan's science catalogs from 1950, however, demonstrates that this assertion is false: Velikovsky *was* in fact advertised as "science" by Macmillan. See C. Leroy Ellenberger, "*Worlds in Collision* in Macmillan's Catalogues," *Kronos* 9, no. 2 (1983): 46–57.

86. Raymond S. Haupert to George Brett, April 6, 1950; Charles W. Ramsey to Macmillan Publishing Group, February 23, 1950; and George R. Stibitz to Macmillan, June 10, 1950; all in MR 96:1.

87. Paul Herget to Boyd T. Harris, May 29, 1950, IVP 126:16.

88. Written atop ibid.

89. Harris to Herget, May 31, 1950, IVP 126:16.

90. Herget to Harris, June 6, 1950, IVP 126:16. Historian of chemistry Henry Guerlac of Cornell University, who had been recommended to Macmillan by Harvard president James Bryant Conant to write a general history of science, also threatened to withdraw his promised manuscript from any press that would also publish *Worlds in Collision*, on the grounds that he had lost confidence in their probity, further deflating Boyd Harris's spirits. Guerlac to Harris, May 23, 1950, MR 96:1.

91. Wasley S. Krogdahl to Boyd Harris, May 15, 1950, IVP 126:16.

92. K. Aa. Strand to Macmillan, February 10, 1950, MR 96:1.

93. P. Kusch to Macmillan, May 22, 1950, MR 96:1.

94. Boyd T. Harris, "Memorandum to the Travellers," March 16, 1950, MR 96:1.

95. Harold S. Latham to Roy K. Marshall, February 27, 1950, IVP 126:16. See also Brett to Latham, February 27, 1950, IVP 126:16.

96. Generic letter over signature of H. S. Latham to "Professor," undated, MR 96:1.

97. F. Sims McGrath to Brett, May 29, 1950, MR 96:1.

98. Velikovsky, *Stargazers and Gravediggers*, 135.

99. George Brett, "Velikovsky: WORLDS IN COLLISION," [early June 1950], MR 96:1.

100. Quoted in ibid.

101. These were remembered by J. Randall Williams, the director of sales for the general trade department at Macmillan, as mostly demands not to give in to the boycott. J. Randall Williams to C. Leroy Ellenberger, November 30, 1983, courtesy of C. Leroy Ellenberger. Such letters retained in Macmillan's files arrived after June 1950, when the decision had already been made, which accords with the fact that the threat to boycott was not public knowledge.

102. Wasley S. Krogdahl to H. S. Latham, July 21, 1950, MR 96:1.

103. John Pfeiffer, "Illiteracy Triumphant," *Science* 114, no. 2950 (July 13, 1951): 47.

104. Louise Thomas to David C. Grahame, July 10, 1950, IVP 123:20. For protest letters, see David C. Grahame to Doubleday & Company, June 23, 1950, and George B. Cressey to Doubleday & Company, June 26, 1950, in IVP, 123:20. For the Whipple protest and response, see Fred L. Whipple to Eunice Stevens, June 30, 1950, and Ken McCormick to Fred Whipple, July 12, 1950, IVP 124:2.

105. Ken McCormick to George B. Cressey, July 18, 1950, IVP 123:20.

106. Geoffrey Parsons to Velikovsky, March 16, 1951, IVP 123:8. There were, indeed, some echoes of the Velikovsky debates in 1951, but they were sporadic and did not generate repeated follow-ups. See, for example, Laurence J. Lafleur, "Cranks and Sci-

entists," *Scientific Monthly* 73 (November 1951): 284–90; and Velikovsky, "Answer to My Critics," *Harper's Magazine* 202 (June 1951): 51–57. Velikovsky beefed up his account into a book manuscript, later entitled *Stargazers and Gravediggers*, but was unable to publish it in his lifetime, in part because of its one-sided accounting of events and occasionally acerbic tone. Ronald F. Probstein wrote to Velikovsky when turning down the manuscript: "I have concluded that this difficulty stems mainly from the fact that it is you who is forced to write your own 'exposé' of the 'injustice.' It becomes very difficult, therefore, not to insert things which coming from a third party would be acceptable—but from the author appears sometimes bitter, sometimes small, and at times even out of place." Ronald F. Probstein to Velikovsky, February 7, 1954, IVP 91:25.

107. Boring, "The Validation of Scientific Belief," 537.

108. "Professors as Suppressors," *Newsweek*, July 3, 1950, 13, 16, 19. The *Newsweek* account was the most widely cited, but a version of the suppression story had already appeared over a week earlier in Joseph Henry Jackson, "Bookman's Notebook," *Los Angeles Times*, June 26, 1950, A5.

109. "The 1950 Silly Season Looks Unusually Silly," *Saturday Evening Post*, November 18, 1950, 10. Yale geologist Chester Longwell wrote a rebuttal to this piece in *Science*, which marked the first time that journal had mentioned the Velikovsky affair. Chester R. Longwell, "The 1950 Silly Season," *Science* 113, no. 2937 (April 13, 1951): 418.

110. Fulton Oursler to Dean McLaughlin, June 27, 1950, IVP 90:22.

111. Velikovsky to René Gallant, December 9, 1960, IVP 78:4.

112. J. W. Grove, *In Defence of Science: Science, Technology, and Politics in Modern Society* (Toronto: University of Toronto Press, 1989), 128. The same point was made even more recently in a 2009 history of alternative histories: "What the scientists really wanted was for Velikovsky's book to be pulled completely so that it was no longer available to purchase and to read." Ronald H. Fritze, *Invented Knowledge: False History, Fake Science, and Pseudo-Religions* (London: Reaktion Books, 2009), 179. The record does not support this claim.

113. Gordon Atwater, "Explosion in Science," *This Week*, April 2, 1950, 10–11, 20, 40. This piece was illustrated by artist and sci-fi set designer Chesley Bonestell, an astronomy buff himself. On Bonestell's work in popularizing astronomy, see W. Patrick McCray, *Keep Watching the Skies!: The Story of Operation Moonwatch and the Dawn of the Space Age* (Princeton, NJ: Princeton University Press, 2008), 24.

114. Gordon Atwater to Ralph Juergens, May 25, 1968, IVP 69:24; Atwater to John Fischer (editor in chief), January 2, 1964, Menzel UFO Papers, folder: "VC. Atwater, Gordon A."; Atwater to Warner Sizemore, December 19, 1963, IVP 97:8; and Atwater in "Comments from Our Readers on 'The Politics of Science and Dr. Velikovsky,'" *American Behavioral Scientist* 7, Supplement (November 1963): 22. For secondary accounts of this episode, see Clark Whelton, "The Gordon Atwater Affair," *SIS Review* 4, no. 4 (1980): 75–76; and Gilbert, *Redeeming Culture*, 188.

115. Frank K. Edmondson to George Brett, April 5, 1950, IVP 126:16.

116. George Brett, "Velikovsky: WORLDS IN COLLISION," [early June 1950], MR 96:1.

117. Velikovsky to James Putnam, August 18, 1963, IVP 91:27.

118. J. Randall Williams to C. Leroy Ellenberger, December 12, 1983, courtesy of C. Leroy Ellenberger.

119. Harold S. Latham to Velikovsky, June 18, 1963, IVP 86:9.

120. Roy K. Marshall to Joseph M. McGarry (Curtis Publishing Company), January 25, 1950, IVP 87:9.

121. Krogdahl to Harris, May 15, 1950, IVP 126:16. Edmondson called the reasons behind Atwater's appointment a "mystery." Edmondson to Brett, April 5, 1950, IVP 126:16.

122. Menzel to Andreas Grotewold, January 27, 1964, Menzel UFO Papers, Folder: "V. C. Grotewold Andreas."

123. Attempts to obtain retroactive compensation for Atwater and/or Putnam in the 1960s did not progress very far: W. T. Crouch in "Comments from Our Readers," 23; Warner Sizemore to Alexander M. White (president of AMNH), August 30, 1963, IVP 97:8.

124. Velikovsky to Robert H. Pfeiffer, December 6, 1950, IVP 91:7. (Suffice it to say that Haldane's letter was completely characteristic of his own distinctive style, widely recognizable through his extensive popular science writing.) Similar accusations are almost too numerous to mention. For a sample, see Thomas M. Howell to Ken Mc-Kormick [sic], November 29, 1950, IVP 128:2; John J. O'Neill to Ferris J. Stephens, November 24, 1950, IVP 128:2; and Fulton Oursler to Dean McLaughlin, June 27, 1950, IVP 90:22. The standard version of the Harvard/Shapley conspiracy theory is recounted in Velikovsky, *Stargazers and Gravediggers*, 207.

125. Harlow Shapley to Ted Thackrey, June 6, 1950, IVP 99:6; Judith Fox, "Immanuel Velikovsky and the Scientific Method," *Synthesis* 5 (1980): 53. Shapley was defending himself against accusations that he and Payne-Gaposchkin had "made extensive and successful efforts to suppress the book." Ted Thackrey to Harlow Shapley, April 10, 1950, IVP 124:2.

126. Horace M. Kallen, *Creativity, Imagination, Logic: Meditations for the Eleventh Hour* (New York: Gordon and Breach, 1973), 115. Kallen had made a similar point as early as July 1950 in "Democracy's True Religion," *Saturday Review of Literature*, July 28, 1951, 7.

127. Wasley S. Krogdahl to C. Leroy Ellenberger, August 20, 1983, and Frank K. Edmondson to C. Leroy Ellenberger, June 23, 1983, both courtesy of C. Leroy Ellenberger.

128. Donald Menzel to Andreas Grotewald, January 27, 1964, Menzel UFO Papers, folder: "V. C. Grotewold Andreas."

129. James Putnam to Bassett Jones, August 11, 1950, JPP 165:2.

130. Eric Larrabee, memorandum of phone conversation with Harold Lavine, June 24, 1950, IVP 86:6.

131. Shapley to Frederick Lewis Allen, January 6, 1951, reproduced in Duane Leroy Vorhees, "The 'Jewish Science' of Immanuel Velikovsky: Culture and Biography as Ideational Determinants" (PhD diss., Bowling Green State University, 1990), 716, 718. Vorhees does not provide a citation to the archive where he obtained this letter.

132. Harlow Shapley to Ted Thackrey, June 6, 1950, IVP 99:6.

133. Chester R. Longwell, review of *Worlds in Collision*, by Velikovsky, *American Journal of Science* 248, no. 8 (August 1950): 589.

Chapter Two

1. Velikovsky to Horace Kallen, June 16, 1946, IVP 124:2.

2. Velikovsky, afterword to *Recollections of a Fallen Sky: Velikovsky and Cultural Amnesia. Papers Presented at the University of Lethbridge, May 9 and 10, 1974*, ed. E. R. Milton (Lethbridge, Alberta: Unileth Press, 1978), 150.

3. Velikovsky, *Ramses II and His Time: A Volume in the Ages in Chaos Series* (Garden City, NY: Doubleday, 1978), xii.

4. See Velikovsky, "Answer to Professor Stewart," *Harper's Magazine* 202 (June 1951): 66.

5. Much of the information in what follows is drawn from Ruth Velikovsky Sharon, *Aba: The Glory and the Torment: The Life of Dr. Immanuel Velikovsky* (Dubuque, IA: Times Mirror Higher Education Group, 1995); Ion Degen, *Immanuil Velikovskii: Rasskaz o zamechatel'nom cheloveke* (Rostov-on-Don: Feniks, 1997); and Duane Leroy Vorhees, "The 'Jewish Science' of Immanuel Velikovsky: Culture and Biography as Ideational Determinants" (PhD diss., Bowling Green State University, 1990). Velikovsky also wrote an autobiography, entitled "Days and Years," which covered his life until 1939, when he was aged forty-four, and can be found at IVP 55:8. A second volume, tentatively entitled "Off the Mooring," was never written.

6. See, respectively, Gabriella Safran, *Wandering Soul: The Dybbuk's Creator, S. An-sky* (Cambridge, MA: Belknap Press, 2010); and Benjamin Harshav, *Marc Chagall and His Times: A Documentary Narrative* (Stanford, CA: Stanford University Press, 2004).

7. Velikovsky, "Days and Years," undated typescript, IVP 55:8, p. 6.

8. See Benjamin Nathans, *Beyond the Pale: The Jewish Encounter with Late Imperial Russia* (Berkeley: University of California Press, 2002); and Kenneth Moss, *Jewish Renaissance in the Russian Revolution* (Cambridge, MA: Harvard University Press, 2009).

9. Velikovsky, "Days and Years," undated typescript, IVP 55:8, p. 23.

10. Immanuel' Ramio [Velikovsky], *Tretii iskhod* (Moscow, 1917), 17 (venereal disease), 24 (the war). For further discussion, see Vorhees, "The 'Jewish Science' of Immanuel Velikovsky," 263–64. Part II of this dissertation comprises the most complete account of Velikovsky's Jewish roots.

11. Sharon, *Aba*, 12.

12. Velikovsky to S. K. Vsekhsviatsky, February 19, 1963, IVP 100:12.

13. Emanuil Ram [Velikovsky], *Tridtsat' dnei i nochei Diego Piresa na mostu Sviatogo angela* (Paris: Parabola, 1935).

14. Velikovsky, "Three Fires," IVP 60:22. It was not uncommon for Russian Jews of this period to imaginatively explore the world of Italian Jewry of an earlier epoch. See, for example, Mikhail Krutikov, *From Kabbalah to Class Struggle: Expressionism, Marxism, and Yiddish Literature in the Life and Work of Meir Wiener* (Stanford, CA: Stanford University Press, 2011). I thank Yaacob Dweck for bringing this point to my attention.

15. See Michael Brenner, *Renaissance of Jewish Culture in Weimar Germany* (New Haven, CT: Yale University Press, 1996).

16. Velikovsky, "Days and Years," undated typescript, IVP 55:8, p. 122.

17. Velikovsky and Heinrich Loewe, eds., *Scripta universitatis atque bibliothecae hierosolymitanarum*, vol. 1: *Orientalia et judaica* (Jerusalem, 1923). See also Velikovsky, "Genesis of the first Jerusalem 'Scripta,'" *Jewish Quarterly* 26 (Spring 1978): 15–19.

18. Sigmund Freud to Velikovsky, June 18, 1922, IVP 77:9.

19. Velikovsky, "Days and Years," undated typescript, IVP 55:8, p. 128; Gerard H. Wilk, "The Meteoric Velikovsky: The Man Who Gave the World Pause," *Commentary* 13 (1952): 382; "Solar System 'Born Recently,' Writer Asserts," *Chicago Daily Tribune*, February 17, 1950, 19. On Velikovsky's reported interactions with poet Hayim Bialik, see Degen, *Immanuil Velikovskii*, 61.

20. Velikovsky to Chaim Weizmann, August 19, 1935, IVP 100:12.

21. Franz Boas to Velikovsky, February 10, 1940, FBP, folder: "Velikovsky, Immanuel"; and Velikovsky to Paul Epstein, March 2, 1945, IVP 76:6.

22. IVP 62:17.

23. Velikovsky to William Safire, May 22, 1977, IVP 94:10, and other letters in this file. See also William Safire, "The Catastrophists," *New York Times*, April 25, 1974, 39.

24. Velikovsky to Horace Kallen, October 30, 1962, IVP 84:6.

25. Sharon, *Aba*, 170.

26. Velikovsky, "Before the Day Breaks," unbound typescript, [1976], IVP 40:7, quotation on p. 21.

27. Sharon, *Aba*, 69.

28. Velikovsky, "Days and Years," undated typescript, IVP 55:8, p. 142.

29. Velikovsky, "Über die Energetik der Psyche und die physikalische Existenz der Gedankenwelt: Ein Beitrag zur Psychologie des gesunden und somnambulen Zustandes," *Zeitschrift für die gesammte Neurologie und Psychiatrie* 133 (1931): 422–37.

30. Sigmund Freud to Velikovsky, June 24, 1931, reproduced in Velikovsky, *Mankind in Amnesia* (Garden City, NY: Doubleday, 1982), 17. See also Velikovsky, "'Very Similar, Almost Identical,'" in *Psychoanalysis and the Future: A Centenary Commemoration of the Birth of Sigmund Freud* (New York: National Psychoanalytic Association, 1957), 14–17.

31. Velikovsky, "Psychoanalytische Ahnungen in der Traumdeutungskunst der alten Hebräer nach dem Traktat Brachoth," *Psychoanalytische Bewegung* 5 (1933): 66–69.

32. Velikovsky, "Can a Newly Acquired Language Become the Speech of the Unconscious?: Word-Plays in the Dreams of Hebrew-Thinking Persons," *Psychoanalytic Review* 21 (1934): 329–35.

33. Velikovsky, "Psychic Anaphylaxis," *Psychoanalytic Review* 23B (1936): 187–94; and Velikovsky, "Psychic Anaphylaxis and Somatic Determination of the Affects," trans. J. V. Coleman, *British Journal of Medical Psychology* 17, no. 1 (1937): 98–104.

34. Velikovsky, "Introduction," July 3, 1977, IVP 53:1.

35. Velikovsky, "Days and Years," undated typescript, IVP 55:8, p. 154.

36. Sigmund Freud, *Moses and Monotheism*, trans. Katherine Jones (New York: Vintage Books, 1967 [1939]). For an illuminating essay on Freud's connection to the Jewish religion that takes *Moses and Monotheism* as its starting point, see Yosef Hayim Ye-

rushalmi, *Freud's Moses: Judaism Terminable and Interminable* (New Haven, CT: Yale University Press, 1991). (Yerushalmi cites Velikovsky's paper on Freud's dreams on 115n25, without further comment.) On the long tradition of "Egyptianizing" Moses, see Jan Assmann, *Moses the Egyptian: The Memory of Egypt in Western Monotheism* (Cambridge, MA: Harvard University Press, 1997).

37. Velikovsky, "Days and Years," undated typescript, IVP 55:8, p. 151.

38. Velikovsky, *Mankind in Amnesia*, 140–41.

39. Velikovsky, "Zu Tolstois Kreutzersonate," *Imago* 13, no. 3 (1937): 363–70.

40. For example, Velikovsky, *Oedipus and Akhnaton: Myth and History* (Garden City, NY: Doubleday, 1960), 48–49.

41. Jan Sammer, "The Velikovsky Archive," in *Fifty Years after "Worlds in Collision" by Velikovsky: Classical and New Scenarios on the Evolution of the Solar System*, ed. Emilio Spedicato and Antonio Agriesti (Bergamo: Università degli Studi di Bargamo, 2002), 44. Sammer wrote that in 1976, when he began his work, Velikovsky showed him a binder labeled "All Books" that listed forty titles, five of which were published either before Velikovsky died or posthumously. The remaining thirty-five are summarized in this article.

42. Alfred De Grazia, *Cosmic Heretics: A Personal History of Attempts to Establish and Resist Theories of Quantavolution and Catastrophe in the Natural and Human Sciences, 1963 to 1983* (Princeton, NJ: Metron, 1984), 107. Another later acolyte would recall that Velikovsky refused to be interviewed by *Penthouse* magazine, unlike his perceived nemesis Carl Sagan (who had been interviewed in *Oui*), on the grounds that it would negatively impact his "pride." Frederic B. Jueneman, *Velikovsky: A Personal View* (Glassboro, NJ: Kronos Press, 1980 [1975]), 71.

43. Velikovsky to Editors of the Macmillan Company, March 10, 1940, IVP 53:1.

44. Velikovsky, "Freud and His Heroes," typescript with handwritten corrections, IVP 46:4–6.

45. Velikovsky, "Summary of Freud and His Heroes," [1940], IVP 47:1.

46. Franz Boas to Velikovsky, April 29, 1940, FBP, folder: "Velikovsky, Immanuel." On Allicon, see Velikovsky to J. M. Kaplan, March 19, 1940, IVP 47:4.

47. Velikovsky, "The Dreams Freud Dreamed," *Psychoanalytic Review* 28 (1941): 490.

48. Velikovsky, *Oedipus and Akhnaton*, 73n1, 201–2.

49. H. J. Rose, a meticulous classical scholar, read the page proofs at Velikovsky's request, offering many detailed comments but concluding that he "remain[ed] totally unconvinced" and that Velikovsky offered "not nearly enough to prove that they are the same person." Rose to Velikovsky, January 6, 1960, IVP 92:20. (The one acceptance seems to be P. G. Maxwell-Stuart, "Interpretations of the Name Oedipus," *Maia* 27 [1975]: 37–43. I thank Joshua Katz for this reference.) A 1964 book on the pharaoh included Velikovsky's work in the bibliography but did not use it anywhere, and a more recent account omits any reference to Velikovsky altogether: Robert Silverberg, *Akhnaten: The Rebel Pharaoh* (Philadelphia: Chilton Books, 1964); and Donald B. Redford, *Akhenaten: The Heretic King* (Princeton, NJ: Princeton University Press, 1984). Famed art critic Meyer Schapiro likewise noted a circularity in reasoning presented in the text: "I feel at times in reading your text that you use the Oedipus legend as much

to complete the gaps in Egyptian history as you use Egyptian history to account for Greek myth." His verdict? "I must say that I'm not convinced." Schapiro to Velikovsky, March 5, 1960, IVP 95:5. (Velikovsky had attempted to enroll the critic into endorsing the embryonic *Worlds in Collision* in late 1946, but Schapiro demurred, claiming skepticism toward the science. Schapiro to Velikovsky, December 20, 1946, IVP 95:5.) Those few reviews that appeared ranged from the noncommittal (Gertrude Smith, "History, Myth Linked in an Ingenious Web," *Chicago Daily Tribune*, April 3, 1960, B10; Elizabeth C. Winship, "Another Controversy?" *Boston Globe*, April 3, 1960, A19) to the savage and sarcastic ("Myth and Moonshine," *Times Literary Supplement*, January 20, 1961, 43; and Jacquetta Hawkes, "Oedipus in Egypt?" *Observer*, June 12, 1960, 26).

50. Velikovsky, *Oedipus and Akhnaton*, 198. Anne-Marie De Grazia, the wife of one of Velikovsky's earliest disciples, argues that Velikovsky exhibited a strong psychological identification with Moses, in "The Last Days of Velikovsky," in *Fifty Years after "Worlds in Collision" by Velikovsky*, ed. Spedicato and Agriesti, 21–22.

51. Velikovsky to Claude Schaeffer, August 22/September 7, 1958, IVP 95:1.

52. Alan H. Gardiner, *The Admonitions of an Egyptian Sage from a Hieratic Papyrus in Leiden (Pap. Leiden 344 Recto)* (Leipzig: J. C. Hinrichs, 1909).

53. Velikovsky, "How I Arrived at My Concepts," in "Thoughts on Various Subjects," undated bound typescript, IVP 40:4, quotation on pp. 5–6.

54. Velikovsky, *Worlds in Collision* (New York: Macmillan, 1950), 300. For the correction, see *Worlds in Collision* (New York: Pocket Books, 1977 [1950]), 304.

55. Velikovsky, "Days and Years," undated typescript, IVP 55:8, p. 151. This method resembles the analysis of "clues" in historical research as described in Carlo Ginzburg, *Clues, Myths, and the Historical Method*, trans. John Tedeschi and Anne C. Tedeschi (Baltimore: Johns Hopkins University Press, 1992). A later critic noted that Velikovsky "sees his task as putting all mankind on the analytic couch, and digging out its buried traumatic experience (world-collision) by studying its dreams (myths)." John Sladek, *The New Apocrypha: A Guide to Strange Sciences and Occult Beliefs* (London: Panther Books, 1978), 25. For a different interpretation of Velikovsky's fragmentary presentation as a methodological choice, see R. G. A. Dolby, "What Can We Usefully Learn from the Velikovsky Affair?" *Social Studies of Science* 5 (1975): 169. On the difficulties of treating myth as either "literal" or "figurative," see the excellent discussion in Paul Veyne, *Did the Greeks Believe in Their Myths?: An Essay on the Constitutive Imagination*, trans. Paula Wissing (Chicago: University of Chicago Press, 1988 [1983]), 62. See also Elizabeth Wayland Barber and Paul T. Barber, *When They Severed Earth from Sky: How the Human Mind Shapes Myth* (Princeton, NJ: Princeton University Press, 2005).

56. Velikovsky, "How I Arrived at My Concepts," in "Thoughts on Various Subjects," undated bound typescript, IVP 40:4, p. 7.

57. William Whiston, *A New Theory of the Earth, from Its Original, to the Consummation of All Things* (London: R. Roberts, 1696).

58. Velikovsky, "In the Beginning," undated, IVP 40:1, p. 43; and Velikovsky, *Stargazers and Gravediggers: Memoirs to "Worlds in Collision"* (New York: Quill, 1984), 42 (which also mentions Ignatius Donnelly).

59. See the excellent account in Christina Wessely, "Welteis: Die 'Astronomie des Unsichtbaren' um 1900," in *Pseudowissenschaft: Konzeptionen von Nichtwissenschaftlichkeit in der Wissenschaftsgeschichte*, ed. Dirk Rupnow et al. (Frankfrut a/M: Suhrkamp, 2008), 163–93; and also Robert Bowen, *Universal Ice: Science and Ideology in the Nazi State* (London: Belhaven Press, 1993). Nonetheless, the theory remained popular in the United States through the 1950s, at that time eclipsing Velikovskianism. Martin Gardner, *Fads and Fallacies in the Name of Science* (New York: Dover, 1957 [1952]), 41.

60. Velikovsky, *Earth in Upheaval* (Garden City, NY: Doubleday, 1955), 84.

61. Donnelly's book on Shakespeare is *The Great Cryptogram: Francis Bacon's Cipher in the So-Called Shakespeare Plays* (Chicago: R. S. Peale, 1888). He was not the first to propose this solution to the problem of authorship. On the history of these disputes, see Warren Hope and Kim Holston, *The Shakespeare Controversy: An Analysis of the Authorship Theories*, 2nd ed. (Jefferson, NC: McFarland, 2009 [1992]).

62. Ignatius Donnelly, *Ragnarok: The Age of Fire and Gravel* (New York: D. Appleton, 1885), 252.

63. Ibid., 106.

64. Velikovsky, *Worlds in Collision*, 42n.

65. See Henry H. Bauer, *Beyond Velikovsky: The History of a Public Controversy* (Urbana: University of Illinois Press, 1984), 221.

66. John J. O'Neill to James Putnam, March 1, 1947, MR 96:1.

67. De Grazia, *Cosmic Heretics*, 337.

68. On the patients, see Velikovsky, "Introduction," July 3, 1977, IVP 53:1; on finances in general, see Vorhees, "The 'Jewish Science' of Immanuel Velikovsky," 461.

69. De Grazia, *Cosmic Heretics*, 12.

70. Velikovsky, *Worlds in Collision*, viii.

71. Velikovsky, *Ages in Chaos: Volume 1: From the Exodus to King Akhnaton* (Garden City, NY: Doubleday, 1952), v–vi.

72. Ibid., xxii, 72, 100–101, and many other places in the text. Bizarrely, one journalist got this backward and claimed that the book attempted to refute the biblical chronology: Howard W. Blakeslee, "Book Disputes History in Old Testament," *Washington Post*, April 17, 1952, 22.

73. Velikovsky, *Ages in Chaos*, xxi.

74. Ibid., 25, 29.

75. Gardiner, *The Admonitions of an Egyptian Sage*, 18.

76. Velikovsky, *Ages in Chaos*, 97; Velikovsky, *Mankind in Amnesia*, 151.

77. Velikovsky to Robert H. Pfeiffer, June 29, 1945, IVP 91:7. Velikovsky discovered through an erroneously forwarded envelope that the negative Oxford referee was E. A. Speiser, author of the one extant Hurrian grammar, whose work was specifically rejected in *Ages in Chaos*. Velikovsky, signed note dated March 30, 1977, IVP 91:7.

78. Robert H. Pfeiffer to Harry A. Wolfson, July 22, 1942, IVP 91:7. Again, because of an envelope from the second (negative) referee that was mistakenly included with the returned manuscript, Velikovsky learned that his second negative reviewer in 1945 was historian William Scott Ferguson of Harvard, and he wrote to the professor to

complain. Ferguson was highly displeased at the breach of anonymity. Velikovsky to William Scott Ferguson, November 25, 1945, and Ferguson to Velikovsky, December 6, 1945, IVP 91:7.

79. Pfeiffer to Velikovsky, August 24, 1942, IVP 91:7.

80. Pfeiffer to H. T. Hatcher (trade editor at Oxford University Press), July 12, 1944, IVP 91:7.

81. [Pfeiffer], reader's report for Velikovsky, AGES IN CHAOS, [spring 1949], IVP 91:7.

82. Velikovsky to Claude Schaeffer, August 22/September 7, 1958, IVP 95:1.

83. In a September 1, 1955, letter to Pfeiffer, Velikovsky wrote that he "will explain the delay of the book as the result of my wish to see the [radiocarbon] test performed." Reproduced in Velikovsky, "Ash," *Pensée*, 4, no. 1 (Winter 1973–74): 9.

84. Velikovsky to Schaeffer, February 10, 1960, IVP 95:1.

85. This text can be accessed on http://www.varchive.org/ce/theses.htm.

86. Velikovsky, *Peoples of the Sea: The Concluding Volume of the Ages in Chaos Series* (Garden City, NY: Doubleday, 1977), xiv. For contemporary reviews, see "All Wrong?" *Irish Times*, May 20, 1977, 8; and Dietrick E. Thomsen, "Velikovsky Lives Again," *New York Times*, April 17, 1977, BR3. This latter hostile review sparked a series of angry letters and a caustic author response: Shane Mage et al., "Letters: Velikovsky," *New York Times*, July 3, 1977. For a later engagement with the arguments in *Peoples of the Sea*, see Horst Friedrich, *Velikovsky, Spanuth und die Seevölker-Diskussion: Argumente für eine Abwanderung atlanto-europäischer spät-bronzezeitlicher Megalith-Völker gegen 700 v. Chr. in den Mittelmeerraum*, 2nd ed. (Wörthsee: By the author, 1990).

87. Velikovsky to Sune Hjorth, September 7, 1967, IVP 81:5.

88. Velikovsky, "The Assyrian Conquest," undated typescript, IVP 44:1; and Velikovsky, "New Light on the Dark Ages of Greece," undated typescript, IVP 42:10. Eddie Schorr, a graduate student in ancient history who was a close companion of Velikovsky's in the late 1960s and early 1970s, wrote to the recently widowed Elisheva that the "Dark Ages" book had so many problems that it was not ready to be published without extensive revision. Eddie Schorr to Elisheva Velikovsky, April 30, 1980, IVP 96:4. Both manuscripts are quite short, and Velikovsky toyed with the idea of unifying the two into one volume, to be entitled "In the Time of Isaiah and Homer." Velikovsky, *Peoples of the Sea*, 193.

89. Velikovsky, "In the Beginning," undated, IVP 40:1, p. v.

90. Ibid., pp. 19, 78 (quotation), 119–21, and part 4.

91. Velikovsky, *Cosmos without Gravitation: Attraction, Repulsion, and Electromagnetic Circumduction in the Solar System* (New York/Jerusalem: Scripta Academica Hierosolymitana, 1946), 3. For a detailed response to the physical arguments, see Bauer, *Beyond Velikovsky*, chap. 7.

92. Charles Ginenthal, "Henry H. Bauer and Immanuel Velikovsky," in *Stephen J. Gould and Immanuel Velikovsky: Essays in the Continuing Velikovsky Affair*, ed. Dale Ann Pearlman (Forest Hills, NY: Ivy Press, 1996), 217.

93. Velikovsky to Harlow Shapley, March 31, 1947, IVP 124:2; and Donald H. Menzel to Velikovsky, April 3, 1947, IVP 124:2. On the missives to Britain, see Velikovsky to

Bertrand Russell, December 8, 1946, IVP 94:7; and Velikovsky to Arthur S. Eddington, December 8, 1946, IVP 74:23.

94. Shapley to George Brett, February 9, 1950, IVP 124:2; and Wildt in Chester R. Longwell, review of *Worlds in Collision*, by Velikovsky, *American Journal of Science* 248, no. 8 (August 1950): 586–87.

95. Robert J. Schadewald, "If Continents Can Wander, Why Not Planets?" *Isaac Asimov's Science Fiction Magazine* 5 (September 28, 1981): 92. Irving Michelson, who would appear in 1974 on a panel at the American Association for the Advancement of Science as a partial advocate for Velikovsky (specifically to argue that *Worlds in Collision* did not necessitate a rejection of celestial mechanics), demanded in 1973 to see the pamphlet. "The title is a bit spine-chilling," he wrote, "and I don't doubt that it has alienated a great many astronomers and other scientists." Michelson to Velikovsky, December 19, 1973, IVP 89:7.

96. Velikovsky, *Cosmos without Gravitation*, 21–22.

97. Velikovsky, "King Solomon's Radium," *This Month*, December 1946, 58–59; and V. Immanuel [Velikovsky], "Why NOT Do Something about the Weather?" *Daily Compass* 1, no. 38 (June 28, 1949): 1. The speed of light is especially interesting. In 1944 he came up with his plan for the experiment (which involved rotating mirrors to measure reflections off moving bodies), and copyrighted it as a lecture on February 23, 1945, as well as registering it at the National Academy of Sciences that December. He renewed his copyright in 1961 and communicated it to Dr. H. K. Zeigler, chief scientist of the U.S. Army Signal Research and Development Laboratory in Fort Monmouth, New Jersey. Velikovsky, "Velocity of Light in Relation to Moving Bodies," [1961], IVP 50:6. The Advanced Research Projects Agency (ARPA) expressed some interest in his ideas in 1960, but no trace remains of any follow-up. Lt. Gen. Arthur G. Trudeau to Velikovsky, July 21, 1960, IVP 99:16.

98. Velikovsky to Claude Schaeffer, April 11, 1963, IVP 95:2.

99. Velikovsky, *Worlds in Collision*, 300.

100. Freud, *Moses and Monotheism*, 127.

101. This is elaborated upon in Velikovsky, *Mankind in Amnesia*, 104–5.

102. George Grinnell to Velikovsky, May 4, 1972, IVP 80:6. See also the early criticism in Maximilian Berners, "Catastrophic Comet Tale Draws Fire," *Los Angeles Times*, May 7, 1950, D6.

103. For example, Velikovsky, *Mankind in Amnesia*, 124 (Byron), 129 (Poe).

104. Bob Forrest, *Velikovsky's Sources*, vols. 1–6 (Manchester: By the author, 1981–83), I:[iii]. See also Bob Forrest, *A Guide to Velikovsky's Sources* (Santa Barbara, CA: Stonehenge Viewpoint, 1987).

105. Velikovsky, *Mankind in Amnesia*, 38–39; see also 85 and Velikovsky, "Cultural Amnesia: The Submergence of Terrifying Events in the Racial Memory and Their Later Emergence," in *Recollections of a Fallen Sky*, ed. Milton, 28. Interestingly, archaeologist Claude Schaeffer offered a similar interpretation to Velikovsky in 1956: "Perhaps it is good, at present, to establish only the reality of those crises and tremendous upheavals during the last millennia before our time, or B.C. and leave the study of the causes to later research. For the historians and the general public are not yet ready to accept

the thought that the earth is a much less safe place than they were accustomed to be-lieve. With the removal of the troublesome warlords in some of the modern nations, with Hitler, Mussolini and the Communists finally removed, they think eternal peace and security will automatically be attained on earth everywhere." Schaeffer to Veli-kovsky, July 23, 1956, IVP 95:1.

106. For example: "Therefore, we arrive at the inevitable conclusion that the rejec-tion of Velikovsky is a result of deep seated and no doubt subconscious anti-Semitism. Velikovsky brings the Jews and ancient Hebrew culture into its correct focus as part of ancient history and it is this, rather than his scientific theories, that is subconsciously rejected." S. Ted Isaacs to John Fischer (editor of *Harper's*), December 25, 1963, Menzel UFO Papers, folder: "VC. Isaacs, S Ted."

107. Lynn E. Rose, "The A.A.A.S. Affair: From Twenty Years After," in *Stephen J. Gould and Immanuel Velikovsky*, ed. Pearlman, 147; emphasis in original. See also Livio C. Stec-chini, "The Inconstant Heavens: Velikovsky in Relation to Some Past Cosmic Perplexi-ties," *American Behavioral Scientist* 7 (September 1963): 28; John M. MacGregor, "Psy-chological Aspects of the Work of Immanuel Velikovsky," in *Recollections of a Fallen Sky*, ed. Milton, 51; and Harold Graff, "Scientific Prejudice: The Velikovsky Incident," *Bulletin of the Philadelphia Association for Psychoanalysis* 23 (December 1973): 301. For an interpretation of the affair based on the structure of academia, see the recent pro-Velikovsky book by David Marriott, *The Velikovsky Inheritance: An Essay in the History of Ideas* (Cambridge: Vanguard Press, 2006), 240.

108. Velikovsky interview by WHRB radio, taped February 18, 1972, broadcast Feb-ruary 27, 1972, IVP 64:21.

109. Velikovsky, *Ages in Chaos*, vi.

110. William A. Irwin, review of *Worlds in Collision*, by Velikovsky, *Journal of Near Eastern Studies* 11 (April 1952): 146–47. For a (significantly) later critique of Velikovsky's history from the point of view of archaeology, see William H. Stiebing Jr., "Velikov-sky's Historical Revisions," in *The Universe and Its Origins: From Ancient Myth to Present Reality and Fantasy*, ed. S. Fred Singer (New York: Paragon House, 1990), 57–61.

111. Kenneth A. Kitchen (University of Liverpool) to Mr. du Sautoy, February 19, 1977, IVP 85:17.

112. "Theories Denounced," *Science News-Letter* 57 (February 25, 1950): 119.

113. Robert Gorham Davis, "Velikovsky's World," *New Leader*, January 31, 1977, 18. Indeed, the most prominent reviews of the work were by an anthropologist and a sci-ence editor: Wilton Krogman, "Author Finds a Time-Dislocation and Takes Ancient History for a Ride," *Chicago Daily Tribune*, April 20, 1952, B4; and Waldemar Kaemp-ffert, "Solomon, the Queen of Sheba, and the Egypt of Exodus," *New York Times* April 20, 1952, BR23.

114. See N. M. Swerdlow, "Otto Neugebauer: May 26, 1899–February 19, 1990," available at http://www.mat.ufrgs.br/~portosil/neugebau.html. Neugebauer was not opposed to investigations of fringe topics, as he noted in his brief but famous riposte against George Sarton's denigration of the history of astrology: "The Study of Wretched Subjects," *Isis* 42 (1951): 111.

115. Wildt in Longwell, review of *Worlds in Collision*, by Velikovsky, 587.

116. Otto Neugebauer, review of *Worlds in Collision*, by Velikovsky, *Isis* 41 (July 1950): 245.

117. Franz Xaver Kugler, *Die babylonische Mondrechnung: Zwei Systeme der Chaldäer über den Lauf des Mondes und der Sonne* (Freiburg: Herder, 1900), original reference to 90.

118. The German text reads: "Um dies zeigen zu können, müssen wir, der spätern Erörterung des Verhältnisses der chaldäischen Ekliptik von Nr. 272 zur beweglichen Ekliptik vorgreifend, schon jetzt erwähnen, dass die Neumondlängen auf der erstern gezählt durchschnittlich um 3° 14′ grösser ausfallen als nach der Zählung auf der letztern."

119. Neugebauer, review of *Worlds in Collision*, by Velikovsky, 245.

120. Harlow Shapley loved this review: "I do not see how you can go on with your plan in the light of this analysis by America's highly competent historian of science. This reviewer suggests definite dishonesty, as well as incompetence." Shapley to Frederick Lewis Allen (editor of *Harper's*), January 6, 1951, reproduced in Vorhees, "The 'Jewish Science' of Immanuel Velikovsky," 717.

121. Velikovsky to George Sarton, November 28, 1950, GSP, folder: "Velikovsky, Immanuel, 1895-."

122. Sarton to Velikovsky, [December 18, 1950], IVP 128:7.

123. Velikovsky, "Answer to Professor Stewart," 65.

124. See, for example, Ralph E. Juergens, "Minds in Chaos: A Recital of the Velikovsky Story," *American Behavioral Scientist* 7 (September 1963): 10; and Velikovsky to Lewis Kaplan, July 9, 1964, IVP 85:2.

125. De Grazia to Velikovsky, January 8, 1968, IVP 128:7.

126. Neugebauer to De Grazia, [January 1965], IVP 128:7.

127. Neugebauer to Mr. Dilliplane, October 6, 1967, IVP 128:7.

128. Velikovsky, *Stargazers and Gravediggers*, 171. See also Velikovsky to Robert H. Pfeiffer, December 6, 1950, IVP 91:7; and Velikovsky to Julius S. Miller, April 29, 1951, IVP 89:11.

129. Velikovsky to Atwater, October 6, 1950, IVP 69:21.

130. Velikovsky, *Mankind in Amnesia*, 86–87 (see also 93). Velikovsky repeated this point in his address to the American Association for the Advancement of Science meeting in February 1974, as reprinted in Velikovsky, "My Challenge to Conventional Views in Science," *Kronos* 3, no. 2 (1977): 7. Several commentators later argued that Velikovsky was popular precisely because of the fear of nuclear catastrophes that imbued his American context. See, for example, David Morris, *The Masks of Lucifer: Technology and the Occult in Twentieth-Century Popular Literature* (London: B. T. Batsford, 1992), 59.

Chapter Three

1. In addition to the literature cited below, see Nikolai Krementsov, *Stalinist Science* (Princeton, NJ: Princeton University Press, 1997); Valery Soyfer, *Lysenko and the Tragedy of Soviet Science*, trans. Leo Gruliow and Rebecca Gruliow (New Brunswick, NJ: Rutgers University Press, 1994); E. I. Kolchinskii, *Biologiia Germanii i Rossii-SSSR v usloviiakh*

sotsial'no-politicheskikh krizisov pervoi poloviny XX veka (mezhdu liberalizmom, kommunizmom i natsional-sotsializmom) (St. Petersburg: Nestor-Istoriia, 2007); and E. S. Levina, *Vavilov, Lysenko, Timofeev-Resovskii* . . . : *Biologiia v SSSR: Istoriia i istoriografiia* (Moscow: AIRO-XX, 1995).

2. On Soviet genetics in the 1920s, see Mark B. Adams, "Sergei Chetverikov, the Kol'tsov Institute, and the Evolutionary Synthesis," in *The Evolutionary Synthesis: Perspectives on the Unification of Biology*, ed. Ernst Mayr and William B. Provine (Cambridge, MA: Harvard University Press, 1980), 242–78.

3. In the West, this doctrine is often called "Lysenkoism," but it was never called that in the Soviet Union. I use the terms interchangeably here, since the story below is primarily about the American reaction.

4. T. D. Lysenko, *Heredity and Its Variability*, trans. Theodosius Dobzhansky (New York: King's Crown Press, 1946), 46. For a technical discussion and partial debunking of Lysenko's arguments, see Johann-Peter Regelmann, *Die Geschichte des Lyssenkoismus* (Frankfurt a/M: Rita G. Fischer Verlag, 1980).

5. For example: "Thus Darwin had recourse to the reactionary, pseudoscientific Malthusian doctrine of intraspecific struggle to gloss over the obvious incongruity between evolutionism and the real development of the plant and animal world." T. D. Lysenko, *New Developments in the Science of Biological Species* (Moscow: Foreign Languages Publishing House, 1951), 9.

6. On the tragic Vavilov case, see Mark Popovsky, *The Vavilov Affair* (Hamden, CT: Archon Books, 1984); and Peter Pringle, *The Murder of Nikolai Vavilov: The Story of Stalin's Persecution of One of the Great Scientists of the Twentieth Century* (New York: Simon & Schuster, 2008).

7. For some early reactions, see J. W. Pincus, "The Genetic Front in the U.S.S.R.," *Journal of Heredity* 31 (1940): 165–68; Paul G. 'Espinasse, "Genetics in the U.S.S.R.," *Nature* 148 (December 20, 1941): 739–43; and Kirtley Mather, "Genetics and the Russian Controversy," *Nature* 149 (April 18, 1942): 427–30.

8. Loren R. Graham, *Science, Philosophy, and Human Behavior in the Soviet Union* (New York: Columbia University Press, 1987), 102. This particular statement by Graham was vigorously contested in Nils Roll-Hansen, "A New Perspective on Lysenko?" *Annals of Science* 42 (1985): 262 (but see partial retraction on 276); and Roll-Hansen, "Genetics under Stalin," *Science* 227 (1985): 1329. But even Roll-Hansen, in later writings, does not hesitate to call Lysenko's notions in genetics "pseudoscience," although he insists on the relative intellectual merit of the work in plant physiology. See Roll-Hansen, "The Practice Criterion and the Rise of Lysenkoism," *Science Studies* 1 (1989): 3–4; and Nils Roll-Hansen, *The Lysenko Effect: The Politics of Science* (Amherst, NY: Humanity Books, 2005), 12.

9. The term is used extensively by both David Joravsky and Zhores Medvedev in their classic histories of the Lysenko affair. See, for example, David Joravsky, *The Lysenko Affair* (Cambridge, MA: Harvard University Press, 1970), 227; Zhores A. Medvedev, *The Rise and Fall of T. D. Lysenko*, trans. I. Michael Lerner (New York: Columbia University Press, 1969), 243; Roy A. Medvedev and Zhores A. Medvedev, *Khrushchev: The Years in Power*, trans. Andrew R. Durkin (New York: Columbia University Press,

1976), 130; and Zhores A. Medvedev, *Soviet Agriculture* (New York: Norton, 1987), ix. Retrospective memoirs by Russian participants have also invoked *"Izhenauka"* in connection with Lysenko, as in V. Ia. Aleksandrov, *Trudnye gody sovetskoi biologii: Zapiski sovremennika* (St. Petersburg: Nauka, 1992), 29; S. M. Gershenson, "Difficult Years in Soviet Genetics," *Quarterly Review of Biology* 65 (1990): 447–56; and Academician V. A. Strunnikov's preface to *Zarozhdenie i razvitie genetiki*, by A. E. Gaisinovich (Moscow: Nauka, 1988), 4. See also the Poland-centric account in Waclaw Gajewski, "Lysenkoism in Poland," *Quarterly Review of Biology* 65 (1990): 423–34.

10. In that book, perhaps with an eye to Lysenko's criticisms of the "idealistic" character of the gene as a concept, Dobzhansky wrote that "the existence of genes is as well established as that of molecules and atoms in chemistry." Theodosius Dobzhansky, *Genetics and the Origin of Species*, 3rd ed. (New York: Columbia University Press, 1951 [1937]), 38.

11. Dunn's case presents its own interest. For Dunn's views on Soviet-American friendship, genetics, and other topics, see Melinda Gormley, "Geneticist L. C. Dunn: Politics, Activism, and Community" (PhD diss., Oregon State University, 2006). For additional background on Dobzhansky and his relation to Lysenkoism, see M. B. Konashev, "Lysenkoizm pod okhranoi spetskhrana," in *Repressirovannaia nauka*, vol. 2, ed. M. G. Iaroshevskii (St. Petersburg: Nauka, 1994), 97–112.

12. Dobzhansky to Dunn, December 21, 1936, LCDP, box 6, folder: "Dobzhansky, Theodosius—Dunn Correspondence #1, 1936–37."

13. L. C. Dunn and Th. Dobzhansky, *Heredity, Race and Society* (New York: Penguin Books, 1946); and *Heredity, Race and Society*, rev. ed. (New York: Mentor, 1952).

14. Dobzhansky to Dunn, January 4, 1937, LCDP, box 6, folder: "Dobzhansky, Theodosius—Dunn Correspondence #1, 1936–37."

15. Eric Ashby, *Scientist in Russia* (New York: Penguin Books, 1947), 83; Dunn to Henry Wallace, January 30, 1946, LCDP, box 21, folder: "Lysenko Controversy—Correspondence 1945–46"; Michael Lerner to Dunn, June 27, 1950, LCDP, box 25, folder: "U.S.S.R.—Correspondence with Geneticists 1944–45." See also Nikolai L. Krementsov, "A 'Second Front' in Soviet Genetics: The International Dimension of the Lysenko Controversy," *Journal of the History of Biology* 29 (1996): 229–50; and Oren Solomon Harman, "C. D. Darlington and the British and American Reaction to Lysenko and the Soviet Conception of Science," *Journal of the History of Biology* 35 (2003): 309–52.

16. This debate happened in microcosm in an exchange in *Science* between Karl Sax and Anton Zhebrak. In 1944 Sax wrote an article that fiercely criticized Lysenkoism in terms that left the Soviet Union open to condemnation. Karl Sax, "Soviet Biology," *Science* 99, no. 2572 (April 14, 1944): 298–99. Zhebrak responded ("Soviet Biology," *Science* 102, no. 2649 [October 5, 1945]: 357–58), after a delay due to the war, by defending the good geneticists still working within the Soviet Union, and thus exonerating the state from allegations of repression—precisely in order not to antagonize Party elites. The same approach was taken by Nikolai Dubinin in "Work of Soviet Biologists: Theoretical Genetics," *Science* 105, no. 2718 (January 31, 1947): 109–12, where he chronicled advances in Soviet genetics without mentioning Lysenko. This did not work. Zhebrak was labeled "antipatriotic" in *Pravda* in 1947: I. Laptev, "Antipatrioticheskie postupki

pod flagom 'nauchnoi' kritiki," *Pravda*, no. 230 (10621) (September 2, 1947): 2. Laptev indicted Zhebrak and his Western allies for disguising blatant anti-Soviet attitudes with a veneer of "scientific" critique, and he singled out Dobzhansky as part of this company.

17. Dobzhansky to Dunn, July 4, 1945, LCDP, box 6, folder: "Dobzhansky, Theodosius—Dunn Correspondence #4, 1943–45"; emphasis in original.

18. Dunn to Waldemar Kaempffert, May 22, 1946, LCDP, box 2, folder: "American-Soviet Science Society Correspondence, 1945–46, n.d."

19. Dunn to editor of the *Saturday Review of Literature*, March 11, 1946, LCDP, box 21, folder: "Lysenko Controversy—Correspondence 1945–46."

20. Dunn to Walter Bara (of McGraw-Hill), May 26, 1945, and Bara to Dunn, June 1, 1945, LCDP, box 2, folder: "American-Soviet Science Society Correspondence, 1945–46, n.d."; and Dunn to Henry Silver (of King's Crown Press), August 17, 1945, LCDP, box 21, folder: "Lysenko Controversy—Correspondence 1945–46."

21. Dobzhansky to Dunn, July 24, 1945, LCDP, box 6, folder: "Dobzhansky, Theodosius—Dunn Correspondence #4, 1943–45."

22. Dobzhansky to Dunn, July 31, 1945, in ibid.

23. Dobzhansky to Dunn, August 20, 1945, in ibid.

24. Dobzhansky, *Reminiscences* (1962, 1965), Oral History Research Office, Butler Library, Columbia University, New York City, New York, p. 322.

25. Dobzhansky's preface to Lysenko, *Heredity and Its Variability*. Dobzhansky's translation is much more fluid to read and faithful to the original Russian than the Soviet-sponsored "official" translation of a later edition of the same work: T. D. Lysenko, *Heredity and Its Variability* (Moscow: Foreign Languages Publishing House, 1951).

26. Krementsov, *Stalinist Science*, 121–22; Krementsov, "A 'Second Front' in Soviet Genetics"; Gormley, "Geneticist L. C. Dunn," 337–38. See also Dunn to Lerner, June 29, 1945, LCDP, box 25, folder: "U.S.S.R.—Correspondence with Geneticists 1944–45."

27. Dunn to L. J. Stadler, December 22, 1945, LCDP, box 21, folder: "Lysenko Controversy—Correspondence 1945–46."

28. Stadler to Dunn, December 28, 1945, in ibid.

29. Dunn to editor of *Science*, January 15, 1946, in ibid.

30. For example, L. C. Dunn, *Science* 103, no. 2667 (February 8, 1946): 180–81; Richard B. Goldschmidt, *Physiological Zoology* 19 (July 1946): 332–34; Bentley Glass, *Quarterly Review of Biology* 21 (September 1946): 279; Curt Stern, *American Naturalist* 80 (March 1, 1946): 241–43; and Conway Zirkle, *Isis* 37 (May 1947): 108. One review that took the opportunity to attack the Soviet Union (angering Dunn), was Robert Simpson, "Science, Totalitarian Model," *Saturday Review of Literature*, March 9, 1946, 28–32.

31. Dobzhansky, "Lysenko's 'Genetics,'" *Journal of Heredity* 37 (1946): 5, 9.

32. P. S. Hudson and R. H. Richens, *The New Genetics in the Soviet Union* (Cambridge: Imperial Bureau of Plant Breeding and Genetics, 1946). Along the same lines as this horticulturally informed survey, a British scholar published a technical analysis of vernalization in early 1948 that likewise criticized Lysenko mildly within the context of other plant physiologists: R. O. Whyte, "History of Research in Vernalization," in

Vernalization and Photoperiodism: A Symposium, ed. A. E. Murneed and R. O. Whyte (Waltham, MA: Chronica Botanica Company, 1948), 1–38.

33. Dobzhansky, review of *The New Genetics in the Soviet Union*, by P. S. Hudson and R. H. Richens, *American Naturalist* 80 (November–December 1946): 651. In Britain, one writer in 1947 used Hudson and Richens as a foundational text to criticize Lysenko's views in terms of Marxist philosophy in one of the central Communist journals, likely intending to have the same effect back in Moscow: R. G. Davies, "Genetics in the U.S.S.R.," *Modern Quarterly* 2 (1947): 336–46.

34. Dobzhansky, "N. I. Vavilov, a Martyr of Genetics, 1887–1942," *Journal of Heredity* 38 (1947): 232.

35. Dunn to Waldemar Kaempffert, January 31, 1946, LCDP, box 21, folder: "Lysenko Controversy—Correspondence 1945–46."

36. Ethan Pollock, *Stalin and the Soviet Science Wars* (Princeton, NJ: Princeton University Press, 2006), chap. 3; Alexei B. Kojevnikov, *Stalin's Great Science: The Times and Adventures of Soviet Physicists* (London: Imperial College Press, 2004), 188 and passim. For the full proceedings of this session, see *The Situation in Biological Science: Proceedings of the Lenin Academy of Agricultural Sciences of the U.S.S.R., Session: July 31–August 7, 1948* (Moscow: Foreign Languages Publishing House, 1949).

37. T. D. Lysenko, *The Science of Biology Today* (New York: International Publishers, 1948), 33. On Stalin's editing of the speech, see Kirill O. Rossianov, "Stalin as Lysenko's Editor: Reshaping Political Discourse in Soviet Science," *Configurations* 3 (1993): 439–56; and Rossianov, "Editing Nature: Joseph Stalin and the 'New' Soviet Biology," *Isis* 84 (1993): 728–45.

38. Lysenko, *The Science of Biology Today*, 49.

39. Dunn to Richard Morford, January 14, 1949, LCDP, box 21, folder: "Lysenko Controversy—Correspondence 1948–53."

40. Milislav Demerec to *Advances in Genetics* Editorial Board, February 10, 1949, LCDP, box 6, folder: "Demerec, Milislav, 1949–51."

41. Dobzhansky, "The Suppression of a Science," *Bulletin of the Atomic Scientists* 5 (May 1949): 144. See also his further publications after 1948: "Marxist Biology, French Style," *Journal of Heredity* 40 (1949): 78–79; "Lysenko's 'Michurinist' Genetics," *Bulletin of the Atomic Scientists* 8 (February 1952): 40–44; "Lysenko Progresses Backwards," *Journal of Heredity* 44 (1953): 20–22; "Animal Breeding under Lysenko," *American Naturalist* 88 (May–June 1954): 165–67; "The Crisis of Soviet Biology," in *Continuity and Change in Russian and Soviet Thought*, ed. Ernest J. Simmons (Cambridge, MA: Harvard University Press, 1955), 329–346; "Lysenko at Bay," *Journal of Heredity* 49 (1958): 15–17; and *Mankind Evolving: The Evolution of the Human Species* (New Haven, CT: Yale University Press, 1962), 17.

42. Zirkle's most important intervention was a "sourcebook" on the rise of Lysenko, compiled from translations of *Pravda*, the official report of the VASKhNIL meeting, and several other sources, entitled *Death of a Science in Russia: The Fate of Genetics as Described in "Pravda" and Elsewhere* (Philadelphia: University of Pennsylvania Press, 1949). This book channeled early interpretations of the Lysenko affair to the orthodox view that state intervention generated pseudoscience, and it is clear from

the text that Zirkle consulted with Dobzhansky extensively. He also reviewed *Heredity and Its Variability* (see n. 30 above). Later, Zirkle argued that Marxism was inherently Lamarckian, bringing him to endorse a variant of the "ideology" thesis as well: Conway Zirkle, *Evolution, Marxian Biology, and the Social Scene* (Philadelphia: University of Pennsylvania Press, 1959).

43. Dobzhansky to Dunn, January 3, 1949, LCDP, box 6, folder: "Dobzhansky, Theodosius—Dunn Correspondence #6, 1948–49."

44. Dobzhansky to Dunn, December 12, 1948, in ibid. See also Dobzhansky to Dunn, September 9, 1948, in ibid.

45. The number of such articles is vast. For a representative selection, see E. G. Butler et al., "A Statement of the Governing Board of the A.I.B.S.," *Science* 110, no. 2848 (July 29, 1949): 124–25; Robert C. Cook, "Walpurgis Week in the Soviet Union," *Scientific Monthly* 68 (June 1949): 367–72; Cook, "Lysenko's Marxist Genetics: Science or Religion?" *Journal of Heredity* 40 (July 1949): 169–202; Joseph P. Lash, "Russian Science and Heresy," *New Republic* (January 3, 1949): 11–13; M. B. Crane, "The Moscow Conference on Genetics," *Heredity* 3 (August 1949): 252–61; and Bertram D. Wolfe, "Science Joins the Party," *Antioch Review* 10 (1950): 47–60. The term "pseudoscience" was used widely in these.

46. H. J. Muller, "It Still Isn't a Science: A Reply to George Bernard Shaw," *Saturday Review of Literature*, April 16, 1949, 61; and Muller, "The Destruction of Science in the USSR," *Saturday Review of Literature*, December 4, 1948, 13. See also Muller, "The Crushing of Genetics in the USSR," *Bulletin of the Atomic Scientists* 4 (December 1948): 369–71; and Muller, "Back to Barbarism—Scientifically," *Saturday Review of Literature*, December 11, 1948, 8–10.

47. Dunn to Muller, February 17, 1949, LCDP, box 21, folder: "Lysenko Controversy—Correspondence 1948–53." This was in response to Muller to Dunn, January 14, 1949, in ibid. See also Elof Axel Carlson, *Genes, Radiation, and Society: The Life and Work of H. J. Muller* (Ithaca, NY: Cornell University Press, 1981), 330.

48. L. C. Dunn, "Motives for the Purge," *Bulletin of the Atomic Scientists* 5 (May 1949): 142–43. The entire issue was devoted to the Lysenko affair, including an article by Dobzhansky, and also presented detailed refutations of the scientific claims (M. B. Crane, "Lysenko's Experiments," 147–49, 156), along with politically inflected articles (Sewall Wright, "Dogma or Opportunism?" 141–42; Eugene Rabinowitch, "The Purge of Genetics in the Soviet Union," 130; and Rabinowitch, "History of the Genetics Conflict," 131–40, 156).

49. C. D. Darlington, "The Retreat from Science in Soviet Russia," *Nineteenth Century and After* 142 (1947): 157–68; and Darlington, "A Revolution in Soviet Science," *Journal of Heredity* 38 (May 1947): 143–48. See the analysis in Harman, "C. D. Darlington and the British and American Reaction to Lysenko." For other criticisms after August 1948, see Eric Ashby, "Genetics in the Soviet Union," *Nature* 162 (December 11, 1948): 912–13; and Sir Henry Dale's foreword to John Langdon-Davies, *Russia Puts the Clock Back: A Study of Soviet Science and Some British Scientists* (London: Victor Gollancz, 1949), 7.

50. Julian Huxley, *Heredity East and West: Lysenko and World Science* (New York: Henry Schuman, 1949), viii, 63.

51. Haldane, "A Note on Genetics in the U.S.S.R.," *Modern Quarterly* 1 (1938): 393 (quotation); and Haldane, "Lysenko and Genetics," *Science and Society* 4 (1940): 433–37. That same year, Haldane had published a puff piece on Lysenko that painted him as a great scientist. Haldane, *Science and Everyday Life* (New York: Macmillan, 1940), 134–38.

52. Haldane, "In Defence of Genetics," *Modern Quarterly* 4 (1949): 194–202; Diane B. Paul, "A War on Two Fronts: J. B. S. Haldane and the Response to Lysenkoism in Britain," *Journal of the History of Biology* 16 (1983): 1–37; Ronald W. Clark, *JBS: The Life and Work of J.B.S. Haldane* (New York: Coward-McCann, 1968), 130.

53. For example, J. D. Bernal, "The Biological Controversy in the Soviet Union and Its Implications," *Modern Quarterly* 4 (1949): 203–17.

54. J. L. Fyfe, *Lysenko Is Right* (London: Lawrence & Wishart, 1950), 4. Contrast with his earlier critique: "The Soviet Genetics Controversy," *Modern Quarterly* 2 (1947): 347–51.

55. Support for Lysenko actually began to wane in 1952, as Stalin considered reorienting his science policy. See Ethan Pollock, "From *Partiinost'* to *Nauchnost'* and Not Quite Back Again: Revisiting the Lessons of the Lysenko Affair," *Slavic Review* 68 (2009): 95–115.

56. Reproduced in Dobzhansky, *The Roving Naturalist: Travel Letters of Theodosius Dobzhansky*, ed. Bentley Glass (Philadelphia: American Philosophical Society, 1980), 195 (quotation), 309. On blacklisting see ibid., 16; and Mark B. Adams, "Introduction: Theodosius Dobzhansky in Russia and America," in *The Evolution of Theodosius Dobzhansky: Essays on His Life and Thought in Russia and America*, ed. Adams (Princeton, NJ: Princeton University Press, 1994), 4.

57. Dobzhansky, "Lysenko's 'Michurinist' Genetics," 44.

58. Dunn and Dobzhansky, *Heredity, Race and Society*, 104.

59. Richard B. Goldschmidt, "Research and Politics," *Bulletin of the Atomic Scientists* 5 (May 1949): 155.

60. Dunn to Goldschmidt, January 11, 1949, LCDP, box 14, folder: "Goldschmidt, Richard B. 1939–50."

61. Dobzhansky, "Lysenko's 'Michurinist' Genetics," 44.

62. This observation, which is generally noted in histories of this period, formed a touchstone of later histories of the Velikovsky affair. See, for example, Donald Goldsmith, introduction to *Scientists Confront Velikovsky*, ed. Goldsmith (New York: Norton, 1977), 19; Norman W. Storer, "The Sociological Context of the Velikovsky Controversy," in ibid., 36; and Judith Fox, "Immanuel Velikovsky and the Scientific Method," *Synthesis* 5 (1980): 54.

63. Gormley, "Geneticist L. C. Dunn," 386. On Dobzhansky, see Dobzhansky to Dunn, September 26, 1946, and March 10, 1947, both in LCDP, box 6, folder: "Dobzhansky, Theodosius—Dunn Correspondence #5, 1946–47."

64. Jessica Wang, *American Science in an Age of Anxiety: Scientists, Anticommunism, and the Cold War* (Chapel Hill: University of North Carolina Press, 1999), 125. For Shapley's views on international cooperation, see his "Astronomy and International Cooperation," *Proceedings of the American Philosophical Society* 91 (February 25, 1947):

73–74; and Peter J. Kuznick, *Beyond the Laboratory: Scientists as Political Activists in 1930s America* (Chicago: University of Chicago Press, 1987).

65. Wang, *American Science in an Age of Anxiety*, 118–19, 128–29. See also Shapley, *Through Rugged Ways to the Stars* (New York: Charles Scribner's Sons, 1969), 156.

66. Shapley to Ted Thackrey, June 6, 1950, IVP 124:2.

67. For his views on anti-Communism, see L. C. Dunn, Harlow Shapley, Alice Hamilton, Leo Loeb, and Duncan A. MacInnes, "The American-Soviet Science Society," *Science* 108, no. 2802 (September 10, 1948): 279; and Don K. Price, "The Scientist as Politician," *Bulletin of the American Academy of Arts and Sciences* 26, no. 7 (April 1973): 27.

68. "Professors as Suppressors," *Newsweek*, July 3, 1950, 16. See also Duane Leroy Vorhees, "The 'Jewish Science' of Immanuel Velikovsky: Culture and Biography as Ideational Determinants" (PhD diss., Bowling Green State University, 1990), who analyzes the reaction as "left-wing McCarthyism" (526); and "Velikovsky: Odd Man Out of Science," *Irish Times*, November 4, 1978, 9.

69. Cecilia Payne-Gaposchkin, "'Worlds in Collision,'" *Proceedings of the American Philosophical Society* 96 (October 15, 1952): 523.

70. Velikovsky to Horace Kallen, October 30, 1962, IVP 84:6. See also Robert E. McAulay, "Substantive and Ideological Aspects of Science: An Analysis of the Velikovsky Controversy" (MA thesis, University of New Mexico, Albuquerque, 1975), 90–91.

71. Muller, "Back to Barbarism," 10. A very similar point was echoed by Otto Struve, as quoted in C. M. Huffer, "Otto Struve on the Freedom of Science," *Science* 111, no. 2896 (June 30, 1950): 726.

72. Dean B. McLaughlin to George P. Brett, May 20, 1950, IVP 124:2.

73. McLaughlin, "Science and Freedom," [January–February 1950], IVP 124:2; emphasis in original.

74. Struve to Editor at *Harper's*, March 3, 1950, IVP 98:11. The tone is similar in a review of *Worlds in Collision* in the *Journal of Near Eastern Studies*: "The great peril of our age is not imperialist communism—that is an acute but transient aberration; our real danger is medievalism. Its assault is peculiarly pernicious because its roots are deep in every one of us; man is a superstitious animal. When organized and triumphant, it seeks to deny our glorious gains of recent centuries and enthral us once more beneath a despotism worse than that of the Kremlin." William A. Irwin, *Journal of Near Eastern Studies* 11 (April 1952): 147.

75. Quoted in Velikovsky, "Answer to My Critics," *Harper's Magazine* 202 (June 1951): 51.

76. Martin Gardner, "The Hermit Scientist," *Antioch Review* 10, no. 4 (December 1950): 456. See also the analogy between Lysenko, Velikovsky, and other heretics in Mark B. Adams, "Genetics and the Soviet Scientific Community, 1948–1965" (PhD diss., Harvard University, 1972), 363.

77. For example: "The wild tenor of the attack and the accusations of base profiteering [in the pro-Velikovsky articles of Frederick Jueneman] smack of nothing less than Lysenkoism." John D. G. Rather to Robert R. Jones, August 16, 1976, IVP 82:18.

78. Velikovsky, *Earth in Upheaval* (Garden City, NY: Doubleday, 1955), 247–48; Velikovsky, *Mankind in Amnesia* (Garden City, NY: Doubleday, 1982), 28.

79. Michael Polanyi, "The Growth of Science in Society," *Minerva* 5 (Summer 1967): 534.

80. Velikovsky to Thackrey, June 2, 1953, IVP 99:7.

81. M. Abramovich to editor of the *Scientific Monthly*, November 20, 1951, IVP 89:4. The piece referred to is Laurence J. Lafleur, "Cranks and Scientists," *Scientific Monthly* 73 (November 1951): 284–90.

82. Leroy Ellenberger, "A Point of View: The Cold War, McCarthy and Velikovsky," *SIS Workshop* 1, no. 5 (April 1979): 6. This point is worded almost identically in Ellenberger to Paul de Forest, January 16, 1979, IVP 76:3.

83. Eric Larrabee, "Science, Poetry, and Politics," *Science* 117, no. 3042 (April 17, 1953): 395; J. P. Schaefer and Larrabee, "Two Spheres Collide," *Science* 118, no. 3058 (August 7, 1953): 167.

84. J. W. Grove, *In Defence of Science: Science, Technology, and Politics in Modern Society* (Toronto: University of Toronto Press, 1989), 125; Theodore J. Gordon, *Ideas in Conflict* (New York: St. Martin's Press, 1966), 234.

85. TDP, box 3, folders: "Crackpot File #1, #2, and #3."

86. I. Michael Lerner to William Shockley, April 22 and May 1, 1970, TDP, box 7, folder: "Lerner, I. Michael #1."

87. On the resilience of genetics under Lysenko's reign, see Adams, "Genetics and the Soviet Scientific Community." As an indication of the rehabilitation, the Soviets published some of Mendel's writings in 1965, complete with a 1935 essay by Nikolai Vavilov: Gregor Mendel, *Opyty nad rastitel'nymi gibridami*, ed. A. E. Gaisinovich (Moscow: Nauka, 1965).

88. Dunn, *A Short History of Genetics: The Development of Some of the Main Lines of Thought, 1864–1939* (Ames: Iowa State University Press, 1991 [1965]), xi; Dobzhansky, "Russian Genetics: Emerging from Lysenko," *Science* 158, no. 3801 (November 3, 1967): 577; Dobzhansky, "Revival of Genetics in the U.S.S.R.," *Quarterly Review of Biology* 43 (1968): 56–59.

Chapter Four

1. Harrison Brown, review of *Earth in Upheaval* (1955), by Velikovsky, *Scientific American* 194 (March 1956): 127.

2. George Gaylord Simpson to David Arons, December 27, 1950, IVP 123:2.

3. Ira S. Bowen to Arons, December 1, 1950, IVP 123:2. Bowen made a point of mentioning that he was glad to have the opportunity to read the book. See also Joel Stebbins to Arons, February 8, 1951, Eliot Blackwelder to Arons, November 7, 1950, and Frederick Seitz to David Arons, April 20, 1951, IVP 123:2. Yale primatologist Robert Yerkes begged off commenting, claiming he had no competence in the topics discussed in the book. Yerkes to Arons, January 15, 1951, IVP 123:2.

4. B. F. Skinner to Arons, April 30, 1951, IVP 123:2.

5. Walter S. Adams to Velikovsky, July 28, 1950, IVP 69:4.

6. Velikovsky to Adams, August 10, 1950, IVP 69:4.

7. Velikovsky, "The Climatic Influence upon the Intellect of Races with Special

Consideration of the Eugenic Question of the Cerebral Development of the Jews in Palestine," [early to mid-1940s], IVP 61:9, p. 5.

8. On the German-American relationship, see Stefan Kühl, *The Nazi Connection: Eugenics, American Racism, and German National Socialism* (New York: Oxford University Press, 1994); and Diane B. Paul, *Controlling Human Heredity: 1865 to the Present* (Amherst, NY: Humanity Books, 1995), 84, 91.

9. Kenneth M. Ludmerer, *Genetics and American Society: A Historical Appraisal* (Baltimore: Johns Hopkins University Press, 1972), 45; Edwin Black, *War against the Weak: Eugenics and America's Campaign to Create a Master Race* (New York: Four Walls Eight Windows, 2003), 380, 385; Thomas M. Shapiro, *Population Control Politics: Women, Sterilization, and Reproductive Choice* (Philadelphia: Temple University Press, 1985), 40; Allan Chase, *The Legacy of Malthus: The Social Costs of the New Scientific Racism* (New York: Knopf, 1977), xv; Richard A. Soloway, *Demography and Degeneration: Eugenics and the Declining Birthrate in Twentieth-Century Britain* (Chapel Hill: University of North Carolina Press, 1995 [1990]), xi; and, in a qualified form, Nils Roll-Hansen, "Norwegian Eugenics: Sterilization as Social Reform," in *Eugenics and the Welfare State: Sterilization Policy in Denmark, Sweden, Norway, and Finland*, ed. Gunnar Broberg and Nils Roll-Hansen (East Lansing: Michigan State University Press, 1996), 185; and Roll-Hansen, "Conclusion: Scandinavian Eugenics in the International Context," in ibid., 261. Mark Adams, for one, objected to this designation: Adams, "Toward a Comparative History of Eugenics," in *The Wellborn Science: Eugenics in Germany, France, Brazil, and Russia*, ed. Adams (New York: Oxford University Press, 1990), 219.

10. George Gaylord Simpson, *The Meaning of Evolution: A Study of the History of Life and of Its Significance for Man* (New Haven, CT: Yale University Press, 1949), 333–34. See also James V. Neel and William J. Schull, *Human Heredity* (Chicago: University of Chicago Press, 1954), 337.

11. Dobzhansky, *The Biological Basis of Human Freedom* (New York: Columbia University Press, 1956), 22, 45. Many of Dobzhansky's publications raise this theme of antiracism: "The Race Concept in Biology," *Scientific Monthly* 52 (February 1941): 161–65; "An Outline of Politico-Genetics," *Science* 102, no. 2644 (August 31, 1945): 234–36; "What Is Heredity?" *Science* 100, no. 2601 (November 3, 1944): 406; "Natural Selection and the Mental Capacities of Mankind," *Science* 105, no. 2736 (June 6, 1947): 587–90 (with Ashley Montagu); *Mankind Evolving: The Evolution of the Human Species* (New Haven, CT: Yale University Press, 1962), 13; *Heredity and the Nature of Man* (London: George Allen & Unwin, 1964), 103, 161–62; and "A Geneticist's View of Human Equality," *The Pharos of Alpha Omega Alpha* 29 (1966): 12–16. Historians have generally interpreted Dobzhansky as absolutely anti-eugenics (which is, we shall see, not quite correct), rooted in his "balance" interpretation of heterozygosity: Costas B. Krimbas, "The Evolutionary Worldview of Theodosius Dobzhansky," in *The Evolution of Theodosius Dobzhansky: Essays on His Life and Thought in Russia and America*, ed. Mark B. Adams (Princeton, NJ: Princeton University Press, 1994), 185; and John Beatty, "Weighing the Risks: Stalemate in the Classical/Balance Controversy," *Journal of the History of Biology* 20 (1987): 302.

12. Walter Landauer to Dobzhansky, February 29, 1936, TDP, box 7, folder: "Landauer, Walter."

13. Daniel J. Kevles, *In the Name of Eugenics: Genetics and the Uses of Human Heredity* (Cambridge, MA: Harvard University Press, 1995 [1985]), 251.

14. For example, C. P. Blacker, *Eugenics: Galton and After* (London: Gerald Duckworth, 1952), 141, 239; Shapiro, *Population Control Politics*, 66; Nancy Leys Stepan, *The Idea of Race in Science: Great Britain 1860–1960* (Hamden, CT: Archon Books, 1982), 143; Black, *War against the Weak*, xvii; Stephen Jay Gould, *The Mismeasure of Man*, rev. ed. (New York: Norton, 1996 [1981]), 54; Pauline M. Mazumdar, *Eugenics, Human Genetics and Human Failings: The Eugenics Society, Its Sources and Its Critics in Britain* (London: Routledge, 1992); and Soloway, *Demography and Degeneration*, 350. The truncation at 1945 is especially sharp for histories of German eugenics, like Paul Weindling, *Health, Race and German Politics between National Unification and Nazism, 1870–1945* (Cambridge: Cambridge University Press, 1989). But even in Germany, as Weindling notes on p. 566, there was still active discussion of eugenics until 1948.

15. Moya Woodside, *Sterilization in North Carolina: A Sociological and Psychological Study* (Chapel Hill: University of North Carolina Press, 1950), 24, 161.

16. This time lag has been noted by some: Alexandra Minna Stern, *Eugenic Nation: Faults and Frontiers of Better Breeding in Modern America* (Berkeley: University of California Press, 2005); Roll-Hansen, "Conclusion," 267; Mark A. Largent, *Breeding Contempt: The History of Coerced Sterilization in the United States* (New Brunswick, NJ: Rutgers University Press, 2008), 130; Diane B. Paul, *The Politics of Heredity: Essays on Eugenics, Biomedicine, and the Nature-Nurture Debate* (Albany: State University of New York Press, 1998), 135, 141; and Edward J. Larson, *Sex, Race, and Science: Eugenics in the Deep South* (Baltimore: Johns Hopkins University Press, 1995), 147, 160.

17. Carl Jay Bajema, ed., *Eugenics: Then and Now* (Stroudsburg, PA: Dowden, Hutchinson & Ross, 1976), 269. From the perspective of old-school eugenicists, Osborn was a terror who consistently shot down favorite projects, as noted in Leon Whitney's oral history reproduced in Marvin D. Miller, *Terminating the "Socially Inadequate": The American Eugenicists and the German Race Hygienists, California to Cold Spring Harbor, Long Island to Germany* (Commack, NY: Malamud-Rose, 1996), 235.

18. Garland E. Allen and Barry Mehler, "Sources in the Study of Eugenics #1: Inventory of the American Eugenics Society Papers," *Mendel Newsletter* 14 (1977): 12; Kühl, *The Nazi Connection*, 75–76; Ludmerer, *Genetics and American Society*, 117.

19. Frederick Osborn, *Preface to Eugenics* (New York: Harper & Brothers, 1940), x, 261.

20. Osborn, *Preface to Eugenics*, rev. ed. (New York: Harper & Brothers, 1951 [1940]), 322. For a discussion of Osborn's "eugenic hypothesis"—that rational citizens will naturally and voluntarily improve their genetic stock if provided with accurate information—see this book and Osborn and Carl Jay Bajema, "The Eugenic Hypothesis," *Social Biology* 19 (1972): 337–45. For a contemporary critique, see Charles Galton Darwin, "Osborn's Eugenic Hypothesis," *Eugenics Review* 48 (1956): 130–32.

21. Maurice A. Bigelow, "Brief History of the American Eugenics Society," *Eugenical News* 31, no. 4 (December 1946): 49–51; Osborn, "History of the American Eugenics Society," *Social Biology* 21 (Summer 1974): 115–26. On the sanitization of American eugenics through invocations of discontinuity, see Garland E. Allen, "Old Wine in

New Bottles: From Eugenics to Population Control in the Work of Raymond Pearl," in *The Expansion of American Biology*, ed. Keith R. Benson, Jane Maienschein, and Ronald Rainger (New Brunswick, NJ: Rutgers University Press, 1991), 231–61; Allen, "Eugenics and American Social History, 1880–1950," *Genome* 31 (1989): 887; and Molly Ladd-Taylor, "Eugenics, Sterilisation and Modern Marriage in the USA: The Strange Career of Paul Popenoe," *Gender and History* 13 (2001): 298–327.

22. Osborn to F. C. Schwarz, April 28, 1966, AESR, box 5, folder: "AES—'Crack-Pot-Literature #1.'" For example, he wrote that "after 1921, Laughlin had, I think, little influence with the American Eugenics Society." Osborn to Frances Hassencahl, July 2, 1969, AESR, box 4, folder: "AES: Correspondence, September 1969."

23. Osborn to Kenneth Ludmerer, November 5, 1970, AESR, box 4, folder: "AES: Correspondence, November 1970." See also Donald K. Pickens, *Eugenics and the Progressives* (Nashville: Vanderbilt University Press, 1968), 210–11. This over-dependence on Osborn is thoroughly documented in Allen and Mehler, "Sources in the Study of Eugenics #1," 13.

24. Wickliffe P. Draper to Osborn, May 31, 1956, and Osborn to Draper, June 14, 1956, FOP, folder: "Draper, Wicliffe [sic] P."

25. Osborn to Anders Lunde, February 16, 1971, AESR, box 4, folder: "AES: Correspondence, February 1971"; Osborn, *Atomic Impasse 1948* (Washington, DC: Department of State, 1948); and Osborn, "The USSR and the Atom," *International Organization* 5 (1951): 480–98. On the collapse of atomic arms control in this period, see Michael D. Gordin, *Red Cloud at Dawn: Truman, Stalin, and the End of the Atomic Monopoly* (New York: Farrar, Straus & Giroux, 2009), chap. 1.

26. Osborn, "Heredity and Practical Eugenics Today," *Eugenical News* 33 (March–June 1948): 3; Osborn, "Heredity and Genetics Today: Is Eugenics Practical?" speech at Cooper Union, March 2, 1948, AESR, box 17, folder: "Osborn, Frederick—Papers #12." Eugenics was tied up with the Lysenko affair in multiple ways: it was invoked by Lysenko as proof of the pseudoscientific status of genetics; geneticists defended eugenics as an argument against Lysenkoism; and so on. For some of these connections in the contemporary sources, see H. J. Muller, "Our Load of Mutations," *American Journal of Human Genetics* 2 (June 1950): 111; Goodwin Watson's foreword to Nicholas Pastore, *The Nature-Nurture Controversy* (New York: Garland, 1984 [1949]), vii; Theodosius Dobzhansky, "The Suppression of a Science," *Bulletin of the Atomic Scientists* 5 (May 1949): 146; and S. Kaftanov, "In Support of Michurin's Biological Theory in Higher Institutions of Learning," *Science* 109, no. 2822 (January 28, 1949): 90.

27. Paul, *Controlling Human Heredity*, 120–21. On Muller, see also Diane B. Paul, "'Our Load of Mutations' Revisited," *Journal of the History of Biology* (1987): 328.

28. Osborn to Bernard [sic: David] R. Weir, June 20, 1966, AESR, box 13, folder: "Brush Foundation." Nonetheless, in a clear exception, Charles Lindbergh was elected to the board in 1955 to replace Paul Popenoe, a prominent eugenicist and the founding practitioner in America of marriage counseling. Osborn to Lindbergh, January 21, 1955, AESR, box 16, folder: "Lindbergh, Charles."

29. Minutes of the Meeting of the Board of Directors of the AES, January 27, 1953,

AESR, box 7, folder: "AES—Minutes, 1952–1956"; Louise Everby to Kenneth A. Bennett, February 24, 1967, AESR, box 3, folder: "AES: Correspondence, February 1967."

30. Osborn to M. H. Mothersill, October 7, 1964, AESR, box 18, folder: "Osborn, Frederick—Letters on Eugenics."

31. Osborn to Shibdas Burman, May 19, 1966, AESR, box 2, folder: "AES: Correspondence, May 1966."

32. Osborn to Paul Popenoe, March 25, 1965, AESR, box 18, folder: "Osborn, Frederick—Letters on Eugenics"; Osborn to Charles Brush, March 2 1966, AESR, box 13, folder: "Brush Foundation." It is interesting to note that heredity counseling could be "dysgenic," by providing individuals who had overestimated the risk of a child with a hereditary defect with accurate information so they might undertake a risk that was still rather substantial. See Sheldon C. Reed, *Counseling in Medical Genetics* (Philadelphia: W. B. Saunders, 1955), 15. There was a prewar history to this medicalization story, as described in Nathaniel C. Comfort, "'Polyhybrid heterogeneous bastards': Promoting Medical Genetics in America in the 1930s and 1940s," *Journal of the History of Medicine and Allied Sciences* 61 (2006): 415–55.

33. Osborn to Booth-Ferris Foundation, May 24, 1962, AESR, box 13, folder: "Booth-Ferris Foundation."

34. Osborn to Robertson, December 14, 1965, AESR, box 16, folder: "Milbank Memorial Fund Grants, 1963–1967."

35. L. C. Dunn and Th. Dobzhansky, *Heredity, Race and Society*, rev. ed. (New York: Mentor, 1952 [1946]), 100.

36. Osborn to Dunn, January 4, 1952, LCDP, box 23, folder: "Osborn, Frederick 1952."

37. Dunn to Osborn, January 7, 1952, in ibid.

38. Dobzhansky, "Two Recent Versions of Eugenics," *American Naturalist* 86 (January–February 1952): 61–62.

39. Dobzhansky, review of *Eugenics and the Progressives*, by Donald K. Pickens, *Man* 4 (1969): 460. See also Dobzhansky's foreword to *The Future of Human Heredity: An Introduction to Eugenics in Modern Society*, by Osborn (New York: Weybright and Talley, 1968), v–vi; and Diane B. Paul, "Dobzhansky in the 'Nature-Nurture' Debate," in *The Evolution of Theodosius Dobzhansky*, ed. Adams, 223.

40. Osborn to Lee R. Dice, January 4, 1965, AESR, box 1, folder: "American Eugenics Party." For Dobzhansky's objections, see Osborn to C. C. Aronsfeld, December 6, 1967, AESR, box 3, folder: "AES: Correspondence, December 1967."

41. Osborn to Shibdas Burman, September 8, 1965, and Osborn to P. S. Barrows, March 25, 1965, AESR, box 18, folder: "Osborn, Frederick—Letters on Eugenics."

42. Osborn to Amram Scheinfeld, August 25, 1965, AESR, box 1, folder: "American Eugenics Party."

43. Osborn to K. Hodson, March 30, 1966, AESR, box 3, folder: "AES: Correspondence, February–March 1966."

44. Osborn to J. P. Scott, May 27, 1969, AESR, box 3, folder: "AES: Correspondence, May 1969."

45. See Osborn to Richard H. Post, June 26, 1969, AESR, box 3, folder: "AES: Correspondence, June, 1969"; Osborn to Kirk, July 26, 1972, AESR, box 4, folder: "AES: Correspondence, July–September 1972."

46. Osborn to Bajema, September 15, 1969, AESR, box 15, folder: "Harvard Growth Study, 3rd #2."

47. L. Erlenmeyer-Kimling to Members for the Society of the Study of Social Biology (formerly the American Eugenics Society), [early 1973], AESR, box 20, folder: "Society for the Study of Social Biology—Newsletter." For many of the suggested alternatives, see Bajema to Osborn, June 14, 1972, AESR, box 7, folder: "AES—Name Change." On the legal process and the reservation of the name, see Osborn to Dudley Kirk, March 8, 1972, AESR, box 4, folder: "AES: Correspondence, April 1972"; and Osborn to Bruce Eckland, January 16, 1973, AESR, box 4, folder: "AES: Correspondence, January–February 1973."

48. Osborn to C. Lalor Burdick, March 22, 1973, AESR, box 13, folder: "Burdick, C. Lalor."

49. See Naomi Oreskes, *The Rejection of Continental Drift: Theory and Method in American Earth Science* (New York: Oxford University Press, 1999), 267; H. E. Le Grand, *Drifting Continents and Shifting Theories: The Modern Revolution in Geology and Scientific Change* (Cambridge: Cambridge University Press, 1988), 197, 267.

50. Velikovsky, "Tests and Measurements Proposed for Inclusion in the Program of the International Geophys. Year," December 5, 1956, IVP 80:33. At least one later Velikovskian supporter implied that Velikovsky's ideas about electromagnetism in the solar system, as presented in these proposals, led to the discovery of the Van Allen belts in 1958: James P. Hogan, *Kicking the Sacred Cow: Questioning the Unquestionable and Thinking the Impermissible* (Riverdale, NY: Baen, 2004), 173.

51. H. H. Hess to Velikovsky, March 15, 1963, IVP 80:33. Velikovsky and his followers repeatedly cited the second sentence of this quotation without its context. In an interview with the *New York Times* science editor in 1966, Hess insisted that he did not agree with Velikovsky's theories, but insisted on defending him from a sense of "fair play." Quoted in Walter Sullivan, "Science: 'The Velikovsky Affair,'" *New York Times*, October 2, 1966.

52. Velikovsky, "H. H. Hess and My Memoranda," in *Velikovsky Reconsidered*, ed. Editors of *Pensée* (New York: Warner Books, 1976), 71; Hess to Velikovsky, January 20, 1965, IVP 80:33.

53. Albert Shadowitz and Peter Walsh, *The Dark Side of Knowledge: Exploring the Occult* (Reading, MA: Addison-Wesley, 1976), 224.

54. Elisheva Velikovsky to Frederic Jueneman, December 29, 1980, IVP 82:19; emphasis in original.

55. M. S. Wyeth Jr. (editor in chief, Harper & Row) to Velikovsky, October 28, 1977, IVP 101:5. Portions of this text were published posthumously in Velikovsky, *Stargazers and Gravediggers: Memoirs to "Worlds in Collision"* (New York: Quill, 1984).

56. Velikovsky to Albert Einstein, January 7, 1939, IVP 75:3.

57. Quoted in Velikovsky, "Before the Day Breaks," unbound typescript, [1976], IVP 40:7, p. 6.

58. Einstein to Velikovsky, July 8, 1946, IVP 75:3. Translations from the original letters (written in German) are my own; those quoted from "Before the Day Breaks" are Velikovsky's.

59. Velikovsky to Einstein, November 29, 1946, and Einstein to Velikovsky, July 10, 1950, IVP 75:3.

60. Quoted in Velikovsky, "Before the Day Breaks," unbound typescript, [1976], IVP 40:7, p. 31.

61. Quoted in ibid., p. 5.

62. Velikovsky to Einstein, August 26, 1952, IVP 75:3.

63. Einstein to Velikovsky, August 27, 1952, IVP 75:3.

64. Quoted in Velikovsky, "Before the Day Breaks," unbound typescript, [1976], IVP 40:7, p. 15; emphasis in original.

65. Hanna Fantova, "Gespräche mit Einstein," HFC, 1:6, p. 6. All translations from this source are mine.

66. See Ze'ev Rosenkranz, *Einstein Before Israel: Zionist Icon or Iconoclast?* (Princeton, NJ: Princeton University Press, 2011).

67. Quoted in Velikovsky, "Before the Day Breaks," unbound typescript, [1976], IVP 40:7, p. 59.

68. Einstein to Velikovsky and Elisheva Velikovsky, March 27, 1955, IVP 75:4.

69. Hanna Fantova, "Gespräche mit Einstein," HFC, 1:6, p. 15.

70. Quoted in Velikovsky, "Before the Day Breaks," unbound typescript, [1976], IVP 40:7, pp. 62–63.

71. Hanna Fantova, "Gespräche mit Einstein," HFC, 1:6, p. 39.

72. Quoted in Velikovsky, "Before the Day Breaks," unbound typescript, [1976], IVP 40:7, pp. 101, 118.

73. Velikovsky, *Earth in Upheaval* (Garden City, NY: Doubleday, 1955), ix–x.

74. Livio C. Stecchini, "The Inconstant Heavens: Velikovsky in Relation to Some Past Cosmic Perplexities," *American Behavioral Scientist 7* (September 1963): 19.

75. I. Bernard Cohen, "An Interview with Einstein," *Scientific American* (July 1955): 70.

76. Velikovsky to I. B. Cohen, July 18, 1955, IVP 123:11.

77. Velikovsky to Otto Nathan, June 26, 1955, IVP 123:11.

78. Nathan to Velikovsky, June 28, 1955, IVP 123:11.

79. Dennis Flanagan to Velikovsky, September 4, 1956, IVP 125:15.

80. Otto Nathan and I. Bernard Cohen, Letters to the Editor, *Scientific American*, September 1955, 12, 14, 16.

81. See Seymour H. Mauskopf and Michael R. McVaugh, *The Elusive Science: Origins of Experimental Psychical Research* (Baltimore: Johns Hopkins University Press, 1980). For one important contemporary attack on these experiments, see George R. Price, "Science and the Supernatural," *Science* 122, no. 3165 (August 26, 1955): 359–67.

82. Edwin G. Boring, "The Validation of Scientific Belief: A Conspectus of the Symposium," *Proceedings of the American Philosophical Society* 96 (October 15, 1952): 539. Yet Martin Gardner, ever looking for pseudosciences to debunk, was kinder to Rhine, who was "clearly not a pseudoscientist to a degree even remotely comparable to that of

most of the men discussed in this book." Gardner, *Fads and Fallacies in the Name of Science* (New York: Dover, 1957 [1952]), 299.

83. J. B. Rhine to Velikovsky, November 11, 1963, IVP 92:11.

84. Velikovsky to Rhine, December 23, 1963, IVP 92:11.

85. Rhine to Velikovsky, January 27 and February 24, 1964, IVP 92:11.

86. Velikovsky to Rhine, July 14, 1965, IVP 92:11.

87. Stonehenge posed a similar problem. In 1965 Harvard astronomer Gerald Hawkins argued that the megaliths once formed a reasonably accurate observatory. Gerald S. Hawkins with John B. White, *Stonehenge Decoded* (Garden City, NY: Doubleday, 1965). All datings of the monument in Salisbury agreed on its preceding Velikovsky's eighth-century catastrophic scenario (the Mars-Venus interaction). If Hawkins's theory was correct, then Velikovsky's mechanism of the earth's axis shifting must be false, because the observatory would then be inaccurate. This was soon billed as a critical test by Velikovsky's critics. Shadowitz and Walsh, *The Dark Side of Knowledge*, 236. Velikovsky deployed numerous arguments to debunk Hawkins, as in Velikovsky, "On Decoding Hawkins' Stonehenge Decoded," in *Velikovsky Reconsidered*, ed. Editors of *Pensée*, 98–109; but Hawkins, who corresponded with Velikovsky, found none of them persuasive. See Gerald S. Hawkins to Velikovsky, June 30, 1970, IVP 100:23.

88. Velikovsky wrote to Libby personally to ask about this issue as early as October 1953. See the correspondence reproduced in Velikovsky, "Ash," *Pensée*, 4, no. 1 (Winter 1973–74): 6.

89. Other unconventional theorists distrusted carbon dating. As George McCready Price, the founder of flood geology and a correspondent of Velikovsky's discussed in the following chapter, wrote: "Your reference to radio carbon dating is interesting; but I think you would be wise not to depend upon this method too heavily. . . . In general, it seems to be fairly accurate for the first few thousands of years as we go backward, but becoming worthless (as I think) back of (let us say) two or three thousands of years before the current era. . . . These many uncertainties tend to spoil the radio carbon method completely for the very sort of work for which we need it most." Price to Velikovsky, August 14, 1956, IVP 91:24.

90. Robert H. Pfeiffer to William Christopher Hays, April 16, 1955, IVP 91:7. See also Pfeiffer to Dows Dunham, August 13, 1955, IVP 91:7.

91. Reproduced in Velikovsky, "Ash," 7. In 1961 Velikovsky acolyte Warner Sizemore conducted a correspondence with Virginia Burton, a curatorial assistant at the Metropolitan Museum, insisting that such a letter existed in their files (he enclosed a copy) and inquiring why it was never acted upon. See ibid., 11.

92. Velikovsky, "Radiocarbon," [mid-1970s], IVP 46:3, quotation on p. 24.

93. Velikovsky to Elizabeth A. Ralph, March 2, 1964, IVP 46:3.

94. Torgny Säve-Söderbergh to [Velikovsky], June 2, 1964, IVP 81:5.

95. Velikovsky, *Ramses II and His Time: A Volume in the Ages in Chaos Series* (Garden City, NY: Doubleday, 1978), 208.

96. See the October 7, 1959, letter from Froelich Rainey, director of the University Museum at the University of Pennsylvania, to Lynne O. Ramer, reproduced in Velikovsky, "Ash," 12.

97. Velikovsky, *Peoples of the Sea: The Concluding Volume of the Ages in Chaos Series* (Garden City, NY: Doubleday, 1977), xvii.

98. Velikovsky, "Radiocarbon," [mid-1970s], IVP 46:3, 36, 15–16.

99. Velikovsky, "On Radiocarbon Tests Performed and Planned to Be Performed for FOSMOS at the Laboratory of the Museum of the U. of Pa.," December 15, 1970, IVP 60:32.

100. V. Bargmann and Lloyd Motz, "On the Recent Discoveries Concerning Jupiter and Venus," *Science* 138, no. 3547 (December 21, 1962): 1352.

101. Robert McAulay, "Velikovsky and the Infrastructure of Science: The Metaphysics of a Close Encounter," *Theory and Society* 6 (1978): 320.

102. Lloyd Motz, "A Personal Reminiscence," in *The Universe and Its Origins: From Ancient Myth to Present Reality and Fantasy*, ed. S. Fred Singer (New York: Paragon House, 1990), 54. Later Motz stepped back from his support of Velikovsky with "Velikovsky—A Rebuttal," *Yale Scientific Magazine* 41, no. 7 (April 1967): 12–13. Velikovsky suggested that "in 1962 he [Motz] was disconcerted and believed in fair play. He felt the squeeze since then on." He blamed "Harvard's controlling group." Velikovsky to John Holbrook, March 28, 1967, IVP 81:9.

103. Abelson quoted in Alfred De Grazia, "The Scientific Reception System and Dr. Velikovsky," *American Behavioral Scientist* 7, no. 1 (September 1963): 55. In the wake of the famous letter, Velikovsky attempted to submit an article, "Venus: A Youthful Planet," to the *Proceedings of the American Philosophical Society*, but it was rejected "after extremely thoughtful discussion, at which every possible way of dealing with this matter was considered." It had been reviewed by "an eminent historian of science and an equally eminent sociologist, and an astronomer of very high standing completely outside the circle of Mr. Velikovsky's critics." George W. Corner to Harry Hess, January 20, 1964, IVP 123:1. This piece was eventually published in the undergraduate magazine *Yale Scientific* 41, no. 7 (April 1967): 8–11, 32, as discussed in chapter 6 below. That same issue contained a scientific refutation of Velikovsky's Venus theory penned by none other than Lloyd Motz, perhaps as atonement for the *Science* letter: "My purpose here is to show that in spite of these correct predictions, his speculations about the origin and history of Venus are wrong." Motz, "Velikovsky—A Rebuttal," 12. Velikovsky responded in the same issue with a coauthored piece: Velikovsky and Ralph E. Juergens, "A Rejoinder to Motz," 14–16, 30.

104. Poul Anderson, "Laputans and Lemurians," *Science* 139, no. 3555 (February 15, 1963): 672.

105. Velikovsky to John Holbrook, January 5, 1967, IVP 81:9. On the Einstein connection, see Judith Fox, "Immanuel Velikovsky and the Scientific Method," *Synthesis* 5 (1980): 50.

106. Velikovsky, "Some Additional Examples of Correct Prognosis," *American Behavioral Scientist* 7, no. 1 (September 1963): 50–54. On the predictions and their degree of confirmation (or lack thereof) in the ensuing decades, see the survey in James E. Oberg, "How Good Were Velikovsky's Space and Planetary Science Predictions, Really?" in *The Universe and Its Origins*, ed. Singer, 37–46.

107. Reproduced in Velikovsky, *Earth in Upheaval*, 297.

108. Bernard F. Burke and Kenneth L. Franklin to Walter I. Bradbury, April 12, 1955, IVP 71:14. This point was frequently echoed over the next decade and a half, as in Fred Whipple to Clark Whelton, July 2, 1970, IVP 124:2; V. R. Eshelman to Whelton, September 11, 1970, IVP 100:23; and M. W. Friedlander, "Some Comments on Velikovsky's Methodology," *PSA Proceedings* (1974): 481.

109. Velikovsky, "Venus—A Youthful Planet," [1963], IVP 123:1, p. 7.

110. "If it should be established that the atmosphere of Venus below the cloud-tops contains significant quantities of water vapor and no more than traces of hydrocarbon or other organic molecules, then Velikovsky's reconstruction of solar system history would be virtually disproven, at least insofar as Venus is concerned." Shane Mage, *Velikovsky and His Critics* (Grand Haven, MI: Cornelius Press, 1978), 27.

111. Eric Larrabee, "Scientists in Collision: Was Velikovsky Right?" *Harper's Magazine* 227 (August 1963): 50. This article produced a rebuttal by Donald Menzel of the Harvard College Observatory ("The Debate over Velikovsky: An Astronomer's Rejoinder," *Harper's Magazine* [December 1963]: 83–86), which threatened to escalate when Larrabee responded in "A Comment on Dr. Menzel's Rejoinder," in ibid., 87. The editors soon decided that "because of the increasing technicality of the argument, it seems necessary to leave further debate on the Velikovsky theory to the professional journals." Editor's note to Menzel, "The Debate over Velikovsky," 85. Even Gordon Atwater, the victim of the 1950 affair, joined in: Gordon Atwater to John Fischer (editor in chief), January 2, 1964, Menzel UFO Papers, folder: "VC. Atwater, Gordon A." The editor was stunned by the virulence of the dispute, writing to Menzel that "sometimes I get the impression that everyone involved in this argument displays a degree of emotion which I had not expected in a scientific discussion." Menzel claimed the "emotion comes in because our (the Velikovsky) discussion was not 'scientific.' You cannot argue with a Velikovsky or a Larrabee because they do not think or argue scientifically." Fischer to Menzel, December 17, 1963, and Menzel to Fischer, January 7, 1964, Menzel UFO Papers, folder: "VC. Harper's Magazine. Editor."

112. Carl Sagan, "An Analysis of *Worlds in Collision*," in *Scientists Confront Velikovsky*, ed. Donald Goldsmith (New York: Norton, 1977), 77.

113. Lewis Kaplan to Julian Bigelow, April 1, 1963, IVP 85:2.

114. William T. Plummer, "Venus Clouds: Test for Hydrocarbons," *Science* 163, no. 3872 (March 14, 1969): 1191–92.

115. Jim Steiner to Velikovsky, November 1, 1969, IVP 126:9; Alta Price to Velikovsky, March 29, 1971, IVP 91:1.

116. Velikovsky to William T. Plummer, September 11, 1969, IVP 91:18.

117. Velikovsky, "Venus' Atmosphere," in *Velikovsky Reconsidered*, ed. Editors of *Pensée*, 240. See also Velikovsky, "Venus and Hydrocarbons," in ibid., 202–10. On the failure to get into *Science*, see Velikovsky to John Holbrook, July 27, 1970, IVP 81:10.

118. Plummer to Whelton, June 6, 1969, IVP 91:18; emphasis in original.

119. Thomas M. Rees to Thomas O. Paine, November 24, 1969, IVP 72:15.

120. Holbrook to Velikovsky, November 10, 1965, IVP 81:8.

Chapter Five

1. John C. Whitcomb Jr. and Henry M. Morris, *The Genesis Flood: The Biblical Record and Its Scientific Implications* (Phillipsburg, NJ: Presbyterian and Reformed Publishing Company, 1961), iii.

2. Ibid., xxi; emphasis in original. This same point about the unscientific nature of historical geology was repeated by Morris in many places, such as in "Science versus Scientism in Historical Geology," in *A Symposium on Creation*, by Henry M. Morris et al. (Grand Rapids, MI: Baker Book House, 1968), 11–31.

3. Whitcomb and Morris, *The Genesis Flood*, xxvi.

4. Davis A. Young, *The Biblical Flood: A Case Study of the Church's Response to Extrabiblical Evidence* (Grand Rapids, MI: William B. Eerdmans, 1995), 257. On ID, see Ronald L. Numbers, *The Creationists: From Scientific Creationism to Intelligent Design*, 2nd exp. ed. (Cambridge, MA: Harvard University Press, 2006 [1992]), chap. 17. Advocates attempted in 2004 to introduce ID into the public schools in Dover County, Pennsylvania—in 1981 in *McClean v. Arkansas* scientific creationism had been rejected as violating the establishment clause—an effort that failed in December 2005 in federal judge John E. Jones III's decision in *Kitzmiller et al. v. Dover Area School District et al.*

5. Respectively, Christopher Toumey, *God's Own Scientists: Creationists in a Secular World* (New Brunswick, NJ: Rutgers University Press, 1994), 2; and Robert Charles Williams, "Scientific Creationism: An Exegesis for a Religious Doctrine," *American Anthropologist* 85 (1983): 93.

6. Raymond A. Eve and Francis B. Harrold, *The Creationist Movement in Modern America* (Boston: Twayne, 1991), 53. See also Toumey, *God's Own Scientists*, 53–54, 113.

7. Warren D. Dolphin, "A Brief Critical Analysis of Scientific Creationism," in *Did the Devil Make Darwin Do It?: Modern Perspectives on the Creation-Evolution Controversy*, ed. David B. Wilson (Ames: Iowa State University Press, 1983), 23.

8. The literature explicitly attacking creationism as "pseudoscience" is vast. See, for example, Philip Kitcher, *Abusing Science: The Case against Creationism* (Cambridge, MA: MIT Press, 1982), 5; Michael Ruse, "Creation Science Is Not Science," *Science, Technology, & Human Values* 7 (Summer 1982): 72–78; Ruse, *The Evolution-Creation Struggle* (Cambridge, MA: Harvard University Press, 2005); Robert J. Schadewald, "Creationist Pseudoscience: Scientific Creationism Is a Classical Pseudoscience in Every Way," *Skeptical Inquirer* 8 (Fall 1983): 22–35; and Leon H. Albert, "'Scientific' Creationism as a Pseudoscience," *Creation/Evolution* 18 (Summer 1986): 25–34. For a criticism of the "pseudoscience" designation by a historian and philosopher of science (who prefers to think of creationism as "archaic science"), see R. G. A. Dolby, "Science and Pseudo-Science: The Case of Creationism," *Zygon* 22 (June 1987): 195–212.

9. Davis A. Young, *Creation and the Flood: An Alternative to Flood Geology and Theistic Evolution* (Grand Rapids, MI: Baker Book House, 1977), 8. See also Keith Abney, "Naturalism and Nonteleological Science: A Way to Resolve the Demarcation Problem between Science and Nonscience," *Perspectives on Science and Christian Faith* 49 (September 1997): 162–69; and J. P. Moreland, "Scientific Creationism, Science, and Conceptual Problems," *Perspectives on Science and Christian Faith* 46 (March 1994): 2–13.

10. For an extremely useful annotated bibliography of creationist texts, see Tom McIver, *Anti-Evolution: An Annotated Bibliography* (Jefferson, NC: McFarland, 1988). On the battles over creationism in the public schools, see Edward J. Larson, *Trial and Error*, 3rd ed. (New York: Oxford University Press, 2003 [1985]); Marcel C. La Follette, ed., *Creationism, Science, and the Law: The Arkansas Case* (Cambridge, MA: MIT Press, 1983); and Amy J. Binder, *Contentious Curricula: Afrocentrism and Creationism in American Public Schools* (Princeton, NJ: Princeton University Press, 2002).

11. For example, Robert Price, "The Return of the Navel, the 'Omphalos' Argument in Contemporary Creationism," *Creation/Evolution* 2 (Fall 1980): 31; and Peter J. Bowler, *Monkey Trials and Gorilla Sermons: Evolution and Christianity from Darwin to Intelligent Design* (Cambridge, MA: Harvard University Press, 2007), 205–6, who compares Velikovsky to countercultural "anti-science" movements. Dorothy Nelkin, in her classic *Science Textbook Controversies and the Politics of Equal Time* (Cambridge, MA: MIT Press, 1977), likewise lumps Velikovsky in with the counterculture, thereby ignoring the direct links with creationism.

12. Velikovsky, *Earth in Upheaval* (Garden City, NY: Doubleday, 1955), vii. Despite this disclaimer, Velikovsky cited evidence from ancient records on page 151.

13. Ibid., 58–59.

14. Ibid., 108.

15. Ibid., 133–34 (on ice ages), 117–18 (on continental drift). On the slow acceptance of continental drift by establishment scientists, see H. E. LeGrand, *Drifting Continents and Shifting Theories: The Modern Revolution in Geology and Scientific Change* (Cambridge: Cambridge University Press, 1988); Naomi Oreskes, *The Rejection of Continental Drift: Theory and Method in American Earth Science* (New York: Oxford University Press, 1999); and Henry Frankel, "The Reception and Acceptance of Continental Drift Theory as a Rational Episode in the History of Science," in *The Reception of Unconventional Science*, ed. Seymour H. Mauskopf (Boulder, CO: Westview Press, 1979), 51–89.

16. Velikovsky, *Earth in Upheaval*, 28.

17. For a criticism of Velikovsky's geology, see Stephen Jay Gould, *Ever Since Darwin: Reflections in Natural History* (New York: Norton, 1979), essay 19: "Velikovsky in Collision." A follower of Velikovsky would later respond rather fiercely to this essay (Velikovsky himself had not done so when it first appeared in March 1975) in a piece that extended several times the length of the original (and relatively charitable) criticism: Charles Ginenthal, "Stephen Jay Gould and Immanuel Velikovsky," in *Stephen J. Gould and Immanuel Velikovsky: Essays in the Continuing Velikovsky Affair*, ed. Dale Ann Pearlman (Forest Hills, NY: Ivy Press, 1996), 552–90. *Earth in Upheaval* was reviewed far less widely than *Worlds in Collision*, but the tenor of the response was equally negative. See, for example, Leonard Engel, "Comets and Cataclysms," *New York Times*, November 20, 1955, BR28; and Wilton Krogman, "More Velikovsky Revisions," *Chicago Daily Tribune*, November 27, 1955, B2.

18. Velikovsky, *Earth in Upheaval*, 210.

19. Ibid., 255–59.

20. Ibid., x.

21. The debt to Price is usually unacknowledged in the voluminous commentary

on Velikovsky, both pro and con. For an exception, see Martin Gardner, *The New Age: Notes of a Fringe Watcher* (Buffalo, NY: Prometheus Books, 1988), 94.

22. See, for example, Whitcomb and Morris, *The Genesis Flood*, xxxii, 184.

23. On Price, see especially the authoritative Numbers, *The Creationists*, chap. 5; and Numbers, introduction to *Selected Works of George McCready Price*, ed. Numbers (New York: Garland, 1995), ix–xvii. Further details can be found in Young, *The Biblical Flood*, 245, 252–53.

24. Velikovsky to George McCready Price, September 7, 1951, IVP 91:24.

25. George McCready Price to Velikovsky, October 7, 1951, IVP 91:24. Price used a two-tone typewriter, and throughout I have transcribed his red letters as italics, while preserving his original underlining.

26. Velikovsky to George McCready Price, October 24, 1951, IVP 91:24.

27. Price to Velikovsky, May 20, 1952, IVP 91:24.

28. Price to Velikovsky, June 5, 1952, IVP 91:24.

29. Price to Velikovsky, November 15, 1955, IVP 91:24.

30. George McCready Price, *The Story of the Fossils* (Mountain View, CA: Pacific Press, 1954), 22. For the mammoths, see p. 9. I have found no reaction from Velikovsky to this statement.

31. Ibid., 24.

32. Ibid., 58 (and also 64). Price used exactly the same metaphor in a letter to Velikovsky: "I don't profess to know just what did happen, nor how it happened. But the plain proofs of a major cosmic disaster some time in the long ago are as plain as the finding of a body with several bullet holes through the skull. And if geologists would only approach the whole problem in the spirit of an honest coroner about to hold an inquest on the world as a whole, there could never be any question about uniformity." Price to Velikovsky, June 5, 1952, IVP 91:24.

33. Price to Velikovsky, October 11, 1956, IVP 91:24. The request can be found in Velikovsky to Price, October 7, 1956, IVP 91:24.

34. George McCready Price, review of *Earth in Upheaval*, [October 11, 1956], IVP 91:24.

35. Velikovsky to Price, October 18, 1956, IVP 91:24.

36. Edward J. Larson, *Summer for the Gods: The Scopes Trial and America's Continuing Debate over Science and Religion* (New York: Basic Books, 1997).

37. J. Laurence Kulp, "Deluge Geology," *Journal of the American Scientific Affiliation* 2 (January 1950): 15.

38. Alfred M. Rehwinkel, *The Flood in the Light of the Bible, Geology, and Archaeology* (St. Louis: Corcordia, 1951). Rehwinkel mostly wrote mainstream ethical works, and this venture into geology stands out. See, for example, Rehwinkel, *The Voice of Conscience* (St. Louis: Concordia, 1956); and *Planned Parenthood and Birth Control in the Light of Christian Ethics* (St. Louis: Concordia, 1959). On sectarianism in modern creationism, see Christopher Toumey, "Modern Creationism and Scientific Authority," *Social Studies of Science* 21 (1991): 681–99; Toumey, "Sectarian Aspects of American Creationism," *International Journal of Moral and Social Studies* 5 (1990): 116–42. On Price's Adventism as a bar to initial acceptance of Price within the creationist community,

see Henry M. Morris, *A History of Modern Creationism* (San Diego: Master Book Publishers, 1984), 79; and Gardner, *The New Age*, 94.

39. Toumey, *God's Own Scientists*, 33.

40. Numbers, *The Creationists*, 264–65; Nelkin, *Science Textbook Controversies and the Politics of Equal Time*; and John Rudolph, *Scientists in the Classroom: The Cold War Reconstruction of American Science Education* (New York: Palgrave, 2002), chap. 6.

41. George E. Webb, *The Evolution Controversy in America* (Lexington: University Press of Kentucky, 1994), 159.

42. Quoted in Numbers, *The Creationists*, 216.

43. Quoted in ibid., 223.

44. Whitcomb and Morris, *The Genesis Flood*, 98n2, 157n2. Ironically, Dutch creationist Nicolaas Rupke came to the movement by first reading *Earth in Upheaval*, which led him to Price, who in turn brought him to *The Genesis Flood*. Numbers, *The Creationists*, 307.

45. Velikovsky to F. J. Kerkhof, October 21, 1962, IVP 85:12.

46. Morris, *A History of Modern Creationism*, 195. Thomas Barnes was a creationist physicist and a Morris ally of long-standing.

47. This view of Velikovsky is widespread in creationist publications. See, for example, A. E. Wilder-Smith, *The Natural Sciences Know Nothing of Evolution* (San Diego: Master Books, 1981), 98. Interestingly, in 1951 a Christian tract attacked Velikovsky (specifically *Worlds in Collision*) as an overzealous interpretation of the Bible and thus suspect on the grounds of questioning biblical inerrancy: O. E. Sanden, *Does Science Support the Scriptures?* (Grand Rapids, MI: Zondervan, 1951).

48. Numbers, *The Creationists*, 234.

49. Dolphin, "A Brief Critical Analysis of Scientific Creationism," 21.

50. Biographical information drawn from Donald W. Patten, Ronald R. Hatch, and Loren C. Steinhauer, *The Long Day of Joshua and Six Other Catastrophes: A Unified Theory of Catastrophism* (Seattle: Pacific Meridian, 1973), xii; and pamphlet advertising *The Biblical Flood and the Ice Epoch*, IVP 91:6.

51. Patten is entirely unmentioned in Numbers, *The Creationists*; Webb, *The Evolution Controversy in America*; Morris, *A History of Modern Creationism*; Eve and Harrold, *The Creationist Movement in Modern America*; and Toumey, *God's Own Scientists*. I have found only three discussions of Patten's work, all by geologists or geographers (the first of whom is a Christian non-creationist): Young, *The Biblical Flood*, 262; James H. Shea, "Creationism, Uniformitarianism, Geology and Science," *Journal of Geological Education* 31 (1983): 105–10; and R. J. Huggett, "Cranks, Conventionalists and Geomorphology," *Area* 34 (2002): 184.

52. Patten, *The Biblical Flood and the Ice Epoch: A Study in Scientific History* (Seattle: Pacific Meridian, 1966), ix.

53. Patten et al., *The Long Day of Joshua and Six Other Catastrophes*, 189.

54. With one exception: a non-technical endorsement by a religious writer, which is the only creationist treatment of Velikovsky that leaves the role of Venus intact. See George L. Norris, *Creation—Cataclysm—Consummation* (Fort Worth: Marno, 1973).

55. Patten, *The Biblical Flood and the Ice Epoch*, 21.

56. Ibid., 22–23. Patten frequently repeated the point about images and graphs, as in ibid., 138. Patten strongly disapproved of "the sex-oriented Freud," who was as bad as, if not worse than, Karl Marx and Charles Darwin (241).

57. Patten et al., *The Long Day of Joshua and Six Other Catastrophes*, 6.

58. Patten, *The Biblical Flood and the Ice Epoch*, 99.

59. Patten et al., *The Long Day of Joshua and Six Other Catastrophes*, 123.

60. The explanation is clearest in Patten's 1988 book: *Catastrophism and the Old Testament: The Mars-Earth Conflicts* (Seattle: Pacific Meridian, 1988), 34. Patten held to the same views as late as 1998, the most recent publication I have been able to locate: Patten, "Reviewing Velikovsky's Venus and Mars Theories," *The Velikovskian* 4, no. 2 (1998): 1–25.

61. See Patten, "The Ice Age Phenomena and a Possible Explanation," *Creation Research Society Quarterly* 3 (May 1966): 63–72; Patten, "Millennial Climatology," *Symposium on Creation* 6 (1977): 29–55; Patten, "The Noachian Flood and Mountain Uplifts," in *A Symposium on Creation*, by Morris et al., 106–9; Patten, "The Ice Epoch," in ibid., 119–35; Patten, "The Pre-Flood Greenhouse Effect (The Antediluvian Canopy)," *Symposium on Creation* 2 (1970): 11–41; Patten, ed., *Symposium on Creation III* (Grand Rapids, MI: Baker Book House, 1971); and Patten, *Evangelical Flood Cosmologies* (Seattle: Pacific Meridian, 1968).

62. Charles McDowell, "Catastrophism and Puritan Thought," *Symposium on Creation* 6 (1977): 76. Likewise R. Russell Bixler, *Earth, Fire, and Sea: The Untold Drama of Creation* (Pittsburgh: Baldwin Manor Press, 1986), 144, which endorses some of Velikovsky's interpretations while differentiating him from Patten, who is described as having "a similar although creationist position" to Velikovsky. Patten wrote a foreword—the one "by a scientist," as opposed to the one "by a historian"—to this text.

63. Joseph Henson, George Mulfinger, Robert Reymond, and Emmett Williams Jr., review of *The Biblical Flood and the Ice Epoch*, by Donald W. Patten, *Creation Research Society Quarterly* 4 (March 1968): 129; emphasis in original. See also John H. Fermor, "Paleoclimatology and Infrared Radiation Traps: Earth's Antediluvian Climate," *Symposium on Creation* 6 (1977): 15–27, which compares Morris and Patten (20), to Morris's advantage.

64. By the 1970s, Morris's journal softened criticism of Velikovskian ideas on this specific point: "Acceptance of Velikovsky's theory (or a modified version), does *not* mean rejecting scripture, *provided*: (a) all parts of his theory which go contrary to scripture are modified to be consistant [*sic*] with scripture, and (b) we recognize that God governs *all* of the universe, and that so-called 'natural' events are every bit as much God's doing as the miraculous events." J. C. Keister, "A Critique and Modification of Velikovsky's Catastrophic Theory of the Solar System," *Creation Research Society Quarterly* 13 (June 1976): 8; emphasis in original. Patten is completely ignored in the text, but in the notes (12n6), Keister explicitly differentiates his own view from Patten's.

65. Henson et al., review of *The Biblical Flood and the Ice Epoch*, by Patten, 131; emphasis in original.

66. Ibid., 130.

67. Donald W. Patten to Velikovsky, October 6, 1972, IVP 91:6.

68. Velikovsky to Patten, October 13, 1972, IVP 91:6.

69. Patten to Velikovsky, December 7, 1960, IVP 91:6.

70. Velikovsky to Patten, December 14, 1960, IVP 91:6.

71. Patten to Velikovsky, July 9, 1962, IVP 91:6.

72. Velikovsky to Patten, July 29, 1962, IVP 91:6.

73. Velikovsky to Patten, August 16, 1962, IVP 91:6.

74. Patten to Velikovsky, September 13, 1962, IVP 91:6.

75. Velikovsky to Patten, September 17, 1962, IVP 91:6. This letter was marked "not sent" in Velikovsky's hand.

76. Secretary of Velikovsky to Patten, September 20, 1962 [mailed October 7], IVP 91:6.

77. For example, Velikovsky to Price, June 13, 1952, and August 10, 1956, IVP 91:24; Velikovsky, *Stargazers and Gravediggers: Memoirs to "Worlds in Collision"* (New York: Quill, 1984), 284.

78. Velikovsky to George McCready Price, May 22, 1952, IVP 91:24.

79. Larrabee foreword to *Stargazers and Gravediggers*, by Velikovsky, 16.

80. John R. Hadd, *Evolution: Reconciling the Controversy* (Glassboro, NJ: Kronos Press, 1979), 66n85; emphasis in original.

81. Patten, *Catastrophism and the Old Testament*, 63n6.

82. Patten to Velikovsky, August 4, 1966, IVP 91:6.

83. Patten to Velikovsky, December 8, 1966, IVP 91:6.

84. Patten to Velikovsky, August 19, 1967, IVP 91:6.

85. Velikovsky to Patten, October 4, 1967, IVP 91:6.

86. Velikovsky to Patten, November 8, 1968, IVP 91:6.

87. Patten to Velikovsky, June 7, 1969, and Velikovsky to Patten, June 22, 1969, IVP 91:6.

88. See, for example, T. Robert Ingram to C. J. Ransom, September 8, 1971, IVP 92:2; Ransom to Velikovsky, October 7, 1971, IVP 92:2; and Patten to Ransom, June 19, 1972, IVP 92:3.

89. Lynn E. Rose to Malcolm Lowery, October 12, 1977, IVP 93:3.

90. Lynn E. Rose and Robert W. Bass, "A Choice of Pattens," *SIS Review* 4, no. 4 (Spring 1980): 87–89.

91. Steve Talbott to editors of *Kronos*, May 2, 1976, IVP 99:2.

92. Alfred De Grazia, *Cosmic Heretics: A Personal History of Attempts to Establish and Resist Theories of Quantavolution and Catastrophe in the Natural and Human Sciences, 1963 to 1983* (Princeton, NJ: Metron, 1984), 335.

93. Patten quoted in ibid., 336.

94. Velikovsky to Mary Buckalew, January 31, 1969, IVP 71:7.

95. Juergens was one of the more independent-minded of the group, proposing embellishments far removed from the original scenario in *Worlds in Collision*. Specifically, Juergens attempted to develop a model of the solar system that was entirely powered by electromagnetism, including the thermal and radiant power of the sun: Ralph E. Juergens, "Plasma in Interplanetary Space: Reconciling Celestial Mechanics

and Velikovskian Catastrophism," in *Velikovsky Reconsidered*, ed. Editors of *Pensée* (New York: Warner Books, 1976), 176–95. Velikovsky granted Juergens far more latitude on such speculations than was typical for the inner circle.

96. De Grazia, *Cosmic Heretics*, 8.

97. Alfred de Grazia, "The Politics of Science and Dr. Velikovsky," *American Behavioral Scientist* 7 (September 1963): 3.

98. De Grazia, "The Scientific Reception System and Dr. Velikovsky," *American Behavioral Scientist* 7 (September 1963): 59; Livio C. Stecchini, "The Inconstant Heavens: Velikovsky in Relation to Some Past Cosmic Perplexities," *American Behavioral Scientist* 7 (September 1963): 28.

99. Ralph E. Juergens, "Minds in Chaos: A Recital of the Velikovsky Story," *American Behavioral Scientist* 7 (September 1963): 4–18.

100. Jacques Barzun, *Science: The Glorious Entertainment* (New York: Harper & Row, 1964), 78n.

101. Howard Margolis, "Velikovsky Rides Again," *Bulletin of the Atomic Scientists* (April 1964): 38–40.

102. Alfred de Grazia, Ralph E. Juergens, and Livio C. Stecchini, eds., *The Velikovsky Affair: Scientism vs. Science* (New Hyde Park, NY: University Books, 1966). Some disciples, such as University of Kansas chemist Albert Burgstahler, still thought the attention was insufficient: "Ironically, since its appearance in the spring of 1966, *The Velikovsky Affair* has simply been ignored by virtually all the major book reviewing media. Just what this may mean for the future of a free market place for new ideas in science is unclear, but it certainly does not augur well." Albert W. Burgstahler, review of *The Velikovsky Affair*, ed. De Grazia, Juergens, and Stecchini, *Mankind Quarterly* 7 (1966): 123. (This journal was also associated with contemporary eugenics.) Burgstahler remained an important acolyte of Velikovsky's for years, being among several who had doctorates in the sciences or engineering, but one of the very few to hold a university appointment in the sciences. He was active in several controversial causes, including vigorous opposition to the fluoridation of water: Juergens to Burgstahler, January 9, 1968, IVP 83:1; and Burgstahler to Velikovsky, December 6, 1966, IVP 71:10. He is now an active proponent of the idea that the plays of William Shakespeare were composed by Edward de Vere, the seventeenth Earl of Oxford.

103. Ruth Velikovsky Sharon, *Aba: The Glory and the Torment: The Life of Dr. Immanuel Velikovsky* (Dubuque, IA: Times Mirror Higher Education Group, 1995), 165.

104. Frederic Jueneman to Lewis Greenberg, November 11, 1975, IVP 82:17.

105. A. Bruce Mainwaring to Velikovsky, April 3, 1968, IVP 88:6.

106. On the idea for the dig, see Mainwaring to David Arons, June 28, 1968, IVP 88:6. On the debacle with the Israeli archaeologists, see John Holbrook to Velikovsky, June 10, 1968, IVP 81:9. On funding issues, see Mainwaring to Velikovsky, August 6, 1970, IVP 88:6; and Holbrook to the Trustees of FOSMOS, April 26, 1971, IVP 81:10.

107. Juergens to Velikovsky, November 25, 1968, IVP 83:1.

108. Reproduced in De Grazia, *Cosmic Heretics*, 226 (see also 221–23).

109. Velikovsky to President and Board Members of FOSMOS, December 1, 1968, IVP 83:1.

110. Later Juergens offered to store a copy of the archive in Arizona, so there would be a duplicate of these materials in case of accident or disaster. Juergens to Velikovsky, December 4, 1969, IVP 83:1. The project never got very far.

111. Juergens to Velikovsky, January 2, 1969, IVP 83:1.

112. Velikovsky to Juergens, September 11, 1969, IVP 83:1.

113. Velikovsky to Mainwaring, August 18, 1970, IVP 88:6.

114. Velikovsky to Mainwaring, April 19, 1972, IVP 88:6.

115. Christopher Evans, *Cults of Unreason* (London: Harrap, 1973), 209–12. See also the recent biography of Reich: Christopher Turner, *Adventures in the Orgasmatron: How the Sexual Revolution Came to America* (New York: Farrar, Straus & Giroux, 2011).

116. Theodore Lasar to Velikovsky, July 16, 1960, reproduced in Velikovsky, "Ash," *Pensée*, 4, no. 1 (Winter 1973–74): 9.

117. Velikovsky to Ted Lasar, September 15, 1960, IVP 86:8.

118. Lasar to Velikovsky, September 17, 1960, IVP 86:8.

119. "Earth Therapy" pamphlet, [late 1960s], IVP 86:8.

120. Velikovsky to David Stove, October 7, 1969, IVP 98:9.

121. Robert Stephanos, "Ben Franklin's Town: Where 'Lightning' Strikes Twice," *Yale Scientific Magazine* 41, no. 7 (April 1967): 26–28; and Charles H. McNamara, "The Persecution and Character Assassination of Immanuel Velikovsky as Performed by the Inmates of the Scientific Establishment," *Philadelphia*, April 1968, 96–97.

122. Velikovsky to Bruce Mainwaring and John Holbrook, September 20, 1969, IVP 88:6.

123. Warner Sizemore to Velikovsky, December 16, 1970, and Sizemore to Velikovsky, October 5, 1975, IVP 97:10. Ralph Juergens thought the root of the problem was that Stephanos was a "Marxist" and that his socialism demonstrated his inability to think clearly and to be susceptible to cultish fads: Juergens to Velikovsky, September 22, 1969, IVP 99:13.

124. Robert Stephanos to Eddie Schorr, May 23, 1969, IVP 95:11.

125. Stephanos to Mary Buckalew, September 22, 1969, IVP 98:6.

126. Velikovsky to Schorr, September 19, 1969, IVP 95:11. Ironically, Schorr was later purged as well. In 1974 his expertise in ancient history (he had gone on to graduate school in the subject) was so trusted that he was declared the responsible editor to finish *Peoples of the Sea* in case of Velikovsky's death. Velikovsky wrote: "I designate you and only you as the person to take care of and see the work completed and printed should I die or become incapacitated before the work is done and printed. I decided so because of my complete trust in your character and scholarship." Atop this letter, Velikovsky later added by hand: "Revoked January 77." Velikovsky to Schorr, October 11, 1974, IVP 96:1. The two had a falling-out over the role of Lewis Greenberg, editor of the journal *Kronos*, to be discussed in the next chapter. Lynn Rose became the executor of Velikovsky's literary estate: Velikovsky to Rose, August 1, 1976, IVP 87:8.

127. Mary Buckalew to Velikovsky, [early December 1968], IVP 71:7.

128. Buckalew to Velikovsky, September 2, 1969, IVP 71:7.

129. Buckalew to Velikovsky, September 4, 1969, IVP 71:7.

130. Stephanos did not lose his interest in cosmic catastrophism, however. In 1975

he established the Beaumont Society, to promote the ideas of Comyns Beaumont, who published an account of cosmic disaster and falsified history (the real Jerusalem was Edinburgh, the real Atlantis was Britain, which never sank) in 1925 and 1932. Stephanos came across these ideas after he was "cast free of the Velikovskians," and he claimed that Velikovsky cribbed several of his ideas from Beaumont and then concealed the influence. Robert C. Stephanos, "Catastrophists in Collision: Velikovsky vs. Beaumont," *Fate*, March 1994, 66.

131. Velikovsky to Burgstahler, September 12, 1969, IVP 71:12.

Chapter Six

1. Harold C. Urey to Alfred De Grazia, June 2, 1964, IVP 126:8. Urey's view was similar to the later analysis by noted philosopher of science Michael Polanyi on the Velikovsky affair in "The Growth of Science in Society," *Minerva* 5 (Summer 1967): 533–45. Polanyi argued that only scientists with much training and experience possessed the tacit knowledge to determine what a true "scientific problem" was.

2. Draft by Velikovsky of a response to Urey to be sent by De Grazia, undated, IVP 126:8.

3. De Grazia to Urey, July 8, 1964, IVP 126:8.

4. Donald Goldsmith, introduction to *Scientists Confront Velikovsky*, ed. Goldsmith (New York: Norton, 1977), 20.

5. Urey to Albert W. Burgstahler, June 21, 1967, IVP 126:8.

6. Urey to Eddie Schorr, September 23, 1969, IVP 95:11.

7. For example, Elwood Miller to Velikovsky, August 14, 1969, IVP 126:8, reporting one such brief conversation.

8. Stephen L. Talbott, recorded phone interview with Dr. Harold Urey, March 16, 1972, and Talbott to Urey, March 17, 1972, IVP 126:8.

9. Talbott to Urey, April 10, 1972, IVP 126:8.

10. Murray Gell-Mann, "How Scientists Can Really Help," *Physics Today* 24 (May 1971): 23. On the obvious and rapid expansion of interest in the occult in this period, see Richard G. Kyle, *The New Age Movement in American Culture* (Lanham, MD: University Press of America, 1995), esp. 47; Steven Dutch, "Four Decades of Fringe Literature," *Skeptical Inquirer* 10 (1986): 342–51; Edward A. Tiryakian, "Toward the Sociology of Esoteric Culture," *American Journal of Sociology* 78 (1972): 491–512; and Andrew W. Greeley, "Superstition, Ecstasy and Tribal Consciousness," *Social Research* 37 (Summer 1970): 203–11.

11. Theodore Roszak, *The Making of a Counter Culture: Reflections on the Technocratic Society and Its Youthful Opposition* (Garden City, NY: Anchor Books, 1969), 42.

12. Robert J. Good, "It Can't Be Our Fault, Can It?" *Chemical and Engineering News* 51 (August 20, 1973): 3. Historian of science Owen Gingerich agreed, maintaining that Velikovsky represented "a resurgence of interest associated with the disenchantment with science and the science establishment. It's part of a pattern, and has a great appeal to people looking for a literal explanation of miracles." Quoted in Robert Cooke, "Theory on Collision of Planets Sets up a Few Earthly Ripples, by Jupiter . . . ," *Boston*

Globe, February 26, 1974, 2. The "anti-pseudoscience movement," as a journalist characterized the Committee for the Scientific Investigation of Claims of the Paranormal (CSICOP), established in 1975, certainly grouped Velikovsky in with the activities it intended to investigate and debunk. See Boyce Rensberger, "The Invasion of the Pseudo-scientists," *New York Times*, November 20, 1977, E16.

13. Alvin M. Weinberg, "Science and Trans-Science," *Minerva* 10 (1972): 222. On the new sense of vulnerability of the scientific establishment beginning in the mid-1960s, see Kelly Moore, *Disrupting Science: Social Movements, American Scientists, and the Politics of the Military, 1945–1975* (Princeton, NJ: Princeton University Press, 2008), 37.

14. After many years of touch-and-go finances, 1976 was a good year for Velikovsky. Although still struggling with phone bills that reached to $200 a month, the royalties from his paperback sales in that year enabled him to enlarge his Princeton home and obtain a summer place on the Jersey Shore. Velikovsky to C. J. Ransom, August 21, 1977, IVP 92:5.

15. Peggy Constantine, "Peter Fonda Not Really a Hippie," *Los Angeles Times*, September 19, 1967, D13.

16. Velikovsky to Bruce Mainwaring and John Holbrook, September 20, 1969, IVP 88:6. And the young did come. A graduate student in Near Eastern studies at the University of Texas at Austin, armed with a magna cum laude undergraduate degree from Harvard, wrote to Velikovsky in 1970 asking whether he could transfer to Princeton and study for his doctorate with him. William Mullen to Velikovsky, March 13, 1970, IVP 126:17. Of course, Velikovsky had no association with Princeton University, but Mullen did become a member of the inner circle.

17. Velikovsky and Lynn E. Rose, "The Sins of the Sons: A Critique of Velikovsky's A.A.A.S. Critics," [late 1970s], IVP 53:7, p. 3. Similar statements appear on pages 181–82 and 185 of this manuscript, which is a detailed criticism of both the statements made against Velikovsky at the 1974 AAAS meeting and of the ensuing publications by those critics in a 1977 volume.

18. Velikovsky, "Answer to My Critics," *Harper's Magazine* 202 (June 1951): 57. See also Velikovsky to George McCready Price, October 7, 1956, IVP 91:24.

19. "Cosmos and Chronos," no. 4, December 1967, IVP 126:3.

20. One of Velikovsky's biographers, who drew much of his information from Velikovsky's elder daughter, Shulamit, presented the Harvard visit in the fashion of the conquering hero taking over the enemy's citadel: Ion Degen, *Immanuil Velikovskii: Rasskaz o zamechatel'nom cheloveke* (Rostov-on-Don: Feniks, 1997), 448–50.

21. Velikovsky to Sune Hjorth, September 7, 1967, IVP 81:5.

22. Press release from Parsons School of Design, April 14, 1970, IVP 66:3.

23. See the documents about the honorary degree reproduced in E. R. Milton, ed., *Recollections of a Fallen Sky: Velikovsky and Cultural Amnesia: Papers Presented at the University of Lethbridge, May 9 and 10, 1974* (Lethbridge, Alberta: Unileth Press, 1978).

24. Daniel Cohen, *Myths of the Space Age* (New York: Dodd, Mead & Company, 1967 [1965]), 191; Henry H. Bauer, *Beyond Velikovsky: The History of a Public Controversy* (Urbana: University of Illinois Press, 1984), 207–8. On counterculture beyond the New Left, see Peter Braunstein and Michael William Doyle, eds., *Imagine Nation: The Ameri-*

can *Counterculture in the 1960s and '70s* (New York: Routledge, 2002); Michael William Doyle, "Debating the Counterculture: Ecstasy and Anxiety over the Hip Alternative," in *The Columbia Guide to America in the 1960s*, ed. David Farber and Beth Bailey (New York: Columbia University Press, 2001), 143–56; Van Gosse, *Rethinking the New Left: An Interpretive History* (New York: Palgrave Macmillan, 2005); and Christopher Gair, *The American Counterculture* (Edinburgh: Edinburgh University Press, 2007), 8.

25. Patrick Doran, "Living with Velikovsky: Catastrophism as World View," in *Recollections of a Fallen Sky*, ed. Milton, 143.

26. Alfred De Grazia, *Cosmic Heretics: A Personal History of Attempts to Establish and Resist Theories of Quantavolution and Catastrophe in the Natural and Human Sciences, 1963 to 1983* (Princeton, NJ: Metron, 1984), 24.

27. See the October 20, 1968, extract of Albert De Grazia's diary in ibid., 219.

28. C. J. Ransom and Lynn E. Rose to "Possible Signers of the Accompanying Statement," [mid- to late 1974], IVP 93:1. See also Frederic B. Jueneman, "Velikovsky," *Industrial Research*, March 1973, 40–44.

29. Walter Kaufmann to Velikovsky, February 23, 1976, IVP 85:7.

30. I leave aside here the small group who attempted to apply Velikovsky to literary criticism. Irving Wolfe argued, citing Velikovsky, that certain artistic geniuses were able to cut through the collective amnesia and glimpse the true catastrophes, and, by reading their works correctly, we can obtain yet more evidence of the Venus and Mars events. Take, for example, Shakespeare's *Antony and Cleopatra*: "If we now take the step of transposing the action into possible astronomical and castastrophic terms . . . we can see that Antony and Cleopatra are presented as heavenly bodies, specifically Mars and Venus, who have abandoned their roles, or left their accustomed orbits, to pose a vast danger to the Roman Empire, or Earth. They are then opposed and defeated by Octavius, who may be the Sun. When they are dead, their names and memories can be safely elevated to myth, just as Dr. Velikovsky tells us that the actual planets Mars and Venus, once so prominent in the skies and so threatening, are now safely distant, in fixed orbits, presenting no living danger to the Earth, and so they too can be safely venerated." Wolfe, "Shakespeare and Velikovsky: Catastrophic Theory and the Springs of Art," in *Recollections of a Fallen Sky*, ed. Milton, 103.

31. Stanislaw Lem, "On Science, Pseudo-Science, and Some Science Fiction," trans. Franz Rottensteiner, *Science Fiction Studies* 7 (1980): 333.

32. Isaac Asimov, "Worlds in Confusion," *Magazine of Fantasy and Science Fiction*, October 1969, 53.

33. Isaac Asimov, "CP," *Analog* 94, no. 2 (October 1974): 40.

34. Isaac Asimov to Frederic Jueneman, May 22, 1972, IVP 102:10.

35. Isaac Asimov to Jerry L. Clark, January 17, 1975, IVP 82:17.

36. Chris Sherrerd to Velikovsky, July 29, 1968, IVP 97:3.

37. This sociological splitting of Velikovsky's audience draws from Robert McAulay, "Velikovsky and the Infrastructure of Science: The Metaphysics of a Close Encounter," *Theory and Society* 6 (1978): 332. As a view from industry, consider Velikovsky's close supporter Frederic Jueneman, who wrote: "Personally, I don't have a quarrel or an ax to grind with the academic community *per se*; still, as a member of the industrial

community I have a growing impatience with those members of the academic hierarchy who have taken it upon themselves to tell me what I should or shouldn't believe." Jueneman, "The Search for Truth," *Analog* 94, no. 2 (October 1974): 30. On humanists acting out resentment of the scientists, see George Grinnell (of the history department at McMaster) to Velikovsky, April 14, 1972, IVP 80:6.

38. Bruce J. Schulman, *The Seventies: The Great Shift in American Culture, Society, and Politics* (New York: Da Capo Press, 2001), 24.

39. Charles H. McNamara, "The Persecution and Character Assassination of Immanuel Velikovsky as Performed by the Inmates of the Scientific Establishment," *Philadelphia*, April 1968, 64.

40. See the announcement for the course of lectures in IVP 65:10.

41. Horace Kallen to Velikovsky, January 10, 1965, IVP 84:6.

42. W. C. Straka to editors of *Pensée*, June 22, 1972, IVP 128:13.

43. C. J. Ransom to Velikovsky, October 4, 1971, IVP 92:2.

44. Ransom to Velikovsky, October 7, 1971, IVP 92:2. Ransom later used the course to produce a popularized introduction to Velikovsky's entire thesis, as described in C. J. Ransom to Velikovsky, May 31, 1975, IVP 92:4. This book was published as C. J. Ransom, *The Age of Velikovsky* (Glassboro, NJ: Kronos Press, 1976). Other such books appeared authored by individuals outside the inner circle, such as Ralph Franklin Walworth with Geoffrey Walworth Sjostrom, *Subdue the Earth* (New York: Delacorte Press, 1977).

45. Ransom to Velikovsky, April 14, 1971, IVP 92:2.

46. Lynn E. Rose to Velikovsky, November 14, 1971, and March 16, 1973, IVP 93:1.

47. Jerry Rosenthal to Velikovsky, February 18, 1977, IVP 94:2.

48. Mary Buckalew to Velikovsky and Elisheva Velikovsky, September 8, 1970, IVP 71:8; emphasis in original.

49. Guenter Koehler to Velikovsky and Elisheva Velikovsky, July 20, 1979, IVP 85:23.

50. R. H. Good Jr. to Dr. Hilton Hinderliter, January 3, 1974, IVP 81:4. Hinderliter responded angrily and at great length on February 13. Ironically, Good had been dismissed from Berkeley in the early 1950s because his own work in abstract quantum field theory was not considered useful for graduate students, and thus pedagogically unacceptable. See David I. Kaiser, *American Physics and the Cold War Bubble* (Chicago: University of Chicago Press, forthcoming), chap. 4.

51. Mary Buckalew to Velikovsky, May 8, 1975, IVP 71:8. Buckalew did not sink without a trace, and she continued her academic career, although without Velikovsky. See, for example, Mary Buckalew, "Global Time in Lucy Boston's Green Knowe Novellas," *Children's Literature Association Quarterly* 19 (1994): 182–87.

52. Velikovsky, recommendation letter for Curtiss Hoffman, March 11, 1968, IVP 81:7.

53. Curtiss Hoffman to Velikovsky, April 20, 1968, IVP 81:7.

54. Curtiss Hoffman to Velikovsky, July 8, 1970, IVP 81:7.

55. Vine Deloria to Velikovsky, June 5, 1977, IVP 74:6. See also Vine Deloria Jr., *God Is Red: A Native View of Religion* (Golden, CO: Fulcrum, 2003 [1973]). He also published his views in the Velikovskian journal *Pensée* (discussed below): Deloria, "Myth and the

Origin of Religion," *Pensée*, no. 4 (Fall 1974): 45–50. Later still, Deloria would expand his vision to a complete rejection of the Bering land bridge and an attack on any affinity between Native tradition and Western culture and science. For a discussion and critique, see H. David Brumble, "Vine Deloria, Jr., Creationism, and Ethnic Pseudoscience," *American Literary History* 10 (1998): 335–46. In 2002 Deloria also discussed Donald Wesley Patten's model of creationist catastrophism, which further demonstrates his independence of Velikovskian orthodoxy: Deloria, *Evolution, Creationism, and Other Modern Myths* (Golden, CO: Fulcrum, 2002), 107, 165. On the relationship between the counterculture and the Native American movement (including Deloria), see Philip Deloria, "Counterculture Indians and the New Age," in *Imagine Nation*, ed. Braunstein and Doyle, 159–88; and Gosse, *Rethinking the New Left*, 135–40.

56. Vine Deloria to Dr. H. Sheinin, March 3, 1977, IVP 74:6. For fan mail to Velikovsky himself, see Deloria to Velikovsky, March 5, 1976, IVP 74:6.

57. Erich von Däniken, *Chariots of the Gods?: Unsolved Mysteries of the Past*, trans. Michael Heron (New York: Berkley Books, 1999 [1968]), 158; quotation from 75. On sales figures, see Kenneth L. Feder, "Cult Archaeology and Creationism: A Coordinated Research Project," in *Cult Archaeology and Creationism: Understanding Pseudoscientific Beliefs about the Past*, exp. ed., ed. Francis B. Harrold and Raymond A. Eve (Iowa City: University of Iowa Press, 1995), 34–48. For a critique of von Däniken's theories, see William H. Stiebing Jr., "The Nature and Dangers of Cult Archaeology," in ibid., 1–10.

58. E. C. Krupp, "Observatories of the Gods and Other Astronomical Fantasies," in *In Search of Ancient Astronomies*, ed. E. C. Krupp (Garden City, NY: Doubleday, 1978), 241–78; and David Sterritt, "We Are the Martians," *Christian Science Monitor*, April 30, 1974, 16. This lumping was contested by at least one contemporary journalist: L. J. Davis, "Velikovsky 'Pinball' Theories Put Him in Outer Space," *Chicago Tribune*, January 24, 1982, F3–F4.

59. Jim Cartwright to Velikovsky, December 5, 1969, IVP 72:14.

60. Respectively, Velikovsky, "In the Beginning," undated MS, IVP 40:1, p. 31n; and Velikovsky to George C. Wilson, November 2, 1962, IVP 100:28.

61. Velikovsky to Charles J. Jacobs, August 29, 1950, IVP 82:4.

62. Velikovsky, "In the Beginning," undated MS, IVP 40:1, p. 32.

63. Lewis Greenberg to Velikovsky, November 11, 1971, IVP 79:13.

64. Ben Bova, "The Whole Truth: Editorial," *Analog* 94, no. 2 (October 1974): 8. See also the vehement dissociation of the two in Ransom, *The Age of Velikovsky*, 238. Interestingly, the most systematic investigation of Velikovsky's sources concluded that von Däniken was *more* careful with his evidence than Velikovsky: Bob Forrest, *Velikovsky's Sources*, vols. 1–6 (Manchester: by the author, 1981–83), 1:5.

65. Jerry L. Clark to Isaac Asimov, [late 1974], IVP 82:17.

66. John R. Cole, "Cult Archaeology and Unscientific Method and Theory," in *Advances in Archaeological Method and Theory*, ed. Michael B. Schiffer (New York: Academic Press, 1980), 3:16. See also Stephen Williams, "Fantastic Archaeology: What Should We Do about It?" in *Cult Archaeology and Creationism*, ed. Harrold and Eve, 124–33; Barry Thiering and Edgar Castle, eds., *Some Trust in Chariots* (New York: Popular Library, 1972); Kenneth L. Feder, *Frauds, Myths, and Mysteries: Science and Pseudo-*

science in Archaeology, 6th ed. (Boston: McGraw-Hill Higher Education, 2008 [1990]), 248; and James Randi, *Flim-Flam!: Psychics, ESP, Unicorns, and other Delusions* (Amherst, NY: Prometheus Books, 1982), chap. 6.

67. Ben Bova to Lynn E. Rose, November 20, 1974, IVP 93:1.

68. Ben Bova, "'With Friends Like These . . . ,'" *Analog* 90, no. 5 (January 1973): 6.

69. Bova to Frederic Jueneman, March 7, 1975, IVP 82:17.

70. On the spectrum of Velikovskian journals, see Bauer, *Beyond Velikovsky*, 68–69.

71. John W. Crowley, "A Scientific Approach to Velikovsky," *Yale Scientific Magazine* 41, no. 7 (April 1967): 5.

72. Velikovsky, "Venus—A Youthful Planet," *Yale Scientific Magazine* 41, no. 7 (April 1967): 8–11, 32.

73. Albert W. Burgstahler and Ernest E. Angino, "Venus—Young or Old?" *Yale Scientific Magazine* 41, no. 7 (April 1967): 18–19; Lloyd Motz, "Velikovsky—A Rebuttal," ibid., 12–13; Velikovsky, "A Rejoinder to Burgstahler and Angino," ibid., 20–25, 32; Velikovsky and Ralph E. Juergens, "A Rejoinder to Motz," ibid., 14–16, 30; and Horace Kallen, "A Letter to the Editor," ibid., 30.

74. Press release for Velikovsky Symposium, Lewis and Clark College, [1972], IVP 65:2.

75. Stephen Talbott, "The Population Crisis Is a Put-On," *Pensée* 1, no. 4 (June 1971): 14–15, 30. For more on the debates over population and their political valence, especially among science and technology enthusiasts on college campuses, see W. Patrick McCray, *The Visioneers: How a Group of Elite Scientists Pursued Space Colonies, Nanotechnologies, and a Limitless Future* (Princeton, NJ: Princeton University Press, 2012).

76. It is unclear from material in Velikovsky's papers why he decided to cooperate. It is possible that some of the articles from the magazine—such as Andrew Kaplan's "At the Edge of War," *Pensée* 2, no. 1 (January 1972), 15–16, which offered a mainstream anti-Communist interpretation of why the United States should support Israel—suited Velikovsky's sense of the editors' probity.

77. Philip M. Boffey, "'Worlds in Collision' Runs into Phalanx of Critics," *Chronicle of Higher Education* 8, no. 22 (March 4, 1974): 7. On the enormous heterogeneity of publishing ventures spawned by countercultural enthusiasts in the 1970s, see Sam Binkley, *Getting Loose: Lifestyle Consumption in the 1970s* (Durham, NC: Duke University Press, 2007).

78. See Lynn Rose to David Morrison, November 26, 1973, and Ralph E. Juergens to Stephen L. Talbott, December 30, 1974, IVP 93:1. On the "scientization" of Velikovskianism in the *Pensée* period, see Judith Fox, "Immanuel Velikovsky and the Scientific Method," *Synthesis* 5 (1980): 49.

79. Albert Burgstahler was the intermediary. See Burgstahler to Velikovsky, December 6, 1966, IVP 71:10.

80. Carl Sagan, *Carl Sagan's Cosmic Connection: An Extraterrestrial Perspective* (Cambridge: Cambridge University Press, 2000 [1973]), 59; Albert Shadowitz and Peter Walsh, *The Dark Side of Knowledge: Exploring the Occult* (Reading, MA: Addison-Wesley, 1976), 241. Scientists (and even the odd ancient historian) did indeed begin publishing more critiques, both empirical and theoretical, of Velikovsky's ideas in the popular

scientific press, presumably out of a similar sentiment: Cyclone Covey, "Velikovsky's 'Reconstruction' of Ancient History," *Anthropological Journal of Canada* 13 (1975): 2–10; E. Öpik, "'Worlds in Collision' and the Peril of Credulity," *Irish Astronomical Journal* 10 (1972): 293–95; and James Oberg, "Ideas in Collision," *Skeptical Inquirer* 4, no. 1 (Fall 1979): 20–27.

81. Ivan King to Stephen Talbott, November 28, 1973, quoted in Velikovsky and Rose, "The Sins of the Sons: A Critique of Velikovsky's A.A.A.S. Critics," [late 1970s], IVP 53:7, pp. 37–38.

82. Velikovsky to Ivan King, December 3, 1973, quoted in ibid., p. 39.

83. For pro-establishment reportage on the AAAS meeting, see Robert Gillette, "Velikovsky: AAAS Forum for a Mild Collision," *Science* 183 (March 15, 1974): 1059–62; Boffey, "'Worlds in Collision' Runs into Phalanx of Critics"; Miranda Robertson, "Velikovsky in the Open," *Nature* 248 (March 15, 1974): 190; George Alexander, "Controversial Author, Scientists in Collision," *Los Angeles Times*, February 26, 1974, A4; Walter Sullivan, "Writer Collides with Scientists," *New York Times*, February 26, 1974, 9; David F. Salisbury, "Velikovsky Cosmology: Theories in Collision," *Christian Science Monitor*, March 12, 1974, 14; and Graham Chedd, "Velikovsky in Chaos," *New Scientist*, March 7, 1974, 624–25. For pro-Velikovskian accounts, see note 91 below.

84. Historian of science Owen Gingerich, one of the organizers, saw the problem as entirely Velikovsky's, as he wrote to arch-Velikovskian Lynn Rose in response to a critical letter: "I find you overdrawing your case like a crooked lawyer. . . . Nor do you choose to remember Velikovsky's spoiled insouciance at the AAAS, how he refused to come down to the meeting unless he was always seated on the stage and had his own microphone so that he could interrupt whenever he chose. Nor do you mention that he simply took twice as much time as had been allocated to any other speaker; thus the program got completely off the track with respect to the schedule." Gingerich to Rose, September 22, 1978, IVP 93:4.

85. Irving Michelson, "Velikovsky's Catastrophism: A Scientific View," *Analog* 95, no. 6 (June 1975): 65–76. In the letter, sent after the McMaster symposium, Michelson wrote: "You have needed a great quantity of creative conceit and it has served you well. A really big man does not let it inflate his ego to such a point that he thinks himself infallible in areas of knowledge beyond his ken—or to the point where he sees as enemies (and badly treats) any who strive to speak the truth to him. My disappointment must be very great if I must conclude that these are your failings, highly esteemed Immanuel." Michelson to Velikovsky, July 23, 1974, IVP 89:7.

86. The relevant sources are Goldsmith, ed., *Scientists Confront Velikovsky*; Velikovsky, "My Challenge to Conventional Views in Science," *Pensée* 4, no. 2 (Spring 1974): 10–14; Irving Michelson, "Mechanics Bears Witness," ibid., 15–21; and the verbatim transcripts of the discussions around each paper reproduced in Lynn E. Rose, ed., "Transcripts of the Morning and Evening Sessions of the A.A.A.S. Symposium on 'Velikovsky's Challenge to Science' Held on February 25, 1974," in *Stephen J. Gould and Immanuel Velikovsky: Essays in the Continuing Velikovsky Affair*, ed. Dale Ann Pearlman (Forest Hills, NY: Ivy Press, 1996), 727–95. Sagan's essay, long to begin with and expanded after the meeting, became the focal point of later discussions and can be found

as Carl Sagan, "An Analysis of *Worlds in Collision*," in *Scientists Confront Velikovsky*, ed. Goldsmith, 41–104. A revised and corrected version is printed in Sagan, *Broca's Brain: Reflections on the Romance of Science* (New York: Presidio Press, 1979 [1974]), chap. 7. An entire issue of *Kronos* (3, no. 2 [1977]), entitled "Velikovsky and Establishment Science," was devoted to the AAAS meeting, but almost all of it was directed against Sagan's contribution. One latter-day Velikovskian devoted an entire monograph to refuting Sagan's piece: Charles Ginenthal, *Carl Sagan & Immanuel Velikovsky*, 2nd ed. (Tempe, AZ: New Falcon Publications, 1995). The Goldsmith volume also includes an essay by David Morrison ("Planetary Astronomy and Velikovsky's Catastrophism," 145–76) which was not part of the original symposium.

87. Velikovsky, "My Challenge to Conventional Views in Science," 10.

88. Ibid., 14. Mystifyingly, Degen claims in *Immanuil Velikovskii*, 480–81, that Velikovsky never said this statement, and it was misleadingly attributed to him by the pro-Sagan press in order to discredit him.

89. Sagan, quoted in Rose, "Transcripts of the Morning and Evening Sessions of the A.A.A.S. Symposium," 757.

90. Velikovsky, quoted in ibid., 772.

91. Velikovsky and Lynn Rose wrote a whole book-length manuscript dissecting the meeting and defending Velikovsky against his critics, especially Sagan: "The Sins of the Sons: A Critique of Velikovsky's A.A.A.S. Critics," [late 1970s], IVP 53:7. According to them (p. 2, for example), the pro-Sagan press did not reflect actual events, but was part of a conspiracy to taint Velikovsky. For additional negative reviews by the Velikovskians, see "Velikovsky's Challenge to Science," *Pensée* 4, no. 2 (Spring 1974): 23–44; George Grinnell, "Trying to Find the Truth about the Controversial Theories of Velikovsky," *Science Forum* 38 (April 1974): 3–5; Frederic B. Jueneman, "A Kick in the AAAS," *Industrial Research* (August 1976): 9; Charles Ginenthal, "The AAAS Symposium on Velikovsky," in *Stephen J. Gould and Immanuel Velikovsky*, ed. Pearlman, 51–138; Lynn E. Rose, "The A.A.A.S. Affair: From Twenty Years After," in ibid., 139–85; George W. Early, "Velikovsky Confronts His Critics," *Fate* 32 (February 1979): 81–88; and Shane Mage, *Velikovsky and His Critics* (Grand Haven, MI: Cornelius Press, 1978).

92. Storer to Sidney M. Wilhelm, April 18, 1974, IVP 93:3.

93. Dennis Rawlins, "Sagan and Velikovsky," *Science News* 105, no. 17 (April 27, 1974): 267; emphasis in original.

94. See the program for McMaster in IVP 65:3 and the descriptions of the various conferences in Frederic B. Jueneman, *Velikovsky: A Personal View* (Glassboro, NJ: Kronos Press, 1980 [1975]), 42.

95. Lewis Greenberg to Velikovsky, June 8, 1974, IVP 79:14.

96. C. J. Ransom to Steve Talbott, January 26, 1975, IVP 92:4.

97. S. Talbott to Lynn E. Rose, December 30, 1974, IVP 93:1.

98. Thomas L. Ferté to S. Talbott, March 10, 1972, and S. Talbott to Ferté, March 15, 1972, IVP 77:4. For the legal proceedings, see letter from lawyer Scott McArthur to S. Talbott, March 17, 1972, IVP 77:4.

99. Quoted in Gillette, "Velikovsky," 1060.

100. Greenberg to Velikovsky, January 29, 1975, IVP 79:14.

101. Ransom to S. Talbott, January 26, 1975, IVP 92:4.

102. S. Talbott to Rose, December 30, 1974, IVP 93:1.

103. Velikovsky to Eddie Schorr, May 19, 1975, IVP 96:1.

104. Table of contents of *Chiron: Journal of Interdisciplinary Studies* 1 (Winter–Spring 1974). Winter 1974 also saw the appearance of volume 1, number 1, of *Chiron: The Velikovsky Newsletter*, which offered Velikovskians news of recent events. This folded equally quickly.

105. Greenberg to Velikovsky, December 29, 1974, IVP 79:14.

106. Robert Hewsen, preface to *Index to the Works of Immanuel Velikovsky*, vol. 1, by Alice Miller (Glassboro, NJ: Center for Velikovskian and Interdisciplinary Studies, 1977), i. See also "A Focal Point," *SIS Review* 1, no. 3 (1976): 17–18; and Jueneman, *Velikovsky*, 52–53. Here is how Hewsen characterized the venture to Velikovsky: "I would say, then, that my role in the Center, of which I am the Director, will be largely organizational. It is our goal to make your work better known, to give it publicity, and to stimulate its discussion. . . . To accomplish this aim the Center, as proper with an academic institution, cannot and should not take a *public* stand for or against your views." Robert Hewsen to Velikovsky, April 17, 1975, IVP 81:2. Kronos Press, based at Glassboro, also published several books, most of which referred to Velikovsky prominently, such as Miller's *Index* and an un-Velikovskian initial publication, H. C. Dudley, *The Morality of Nuclear Planning?* (Glassboro, NJ: Kronos Press, 1976). Velikovsky adored the index, which he believed would "serve as a guide to those who would undertake the ambitious project of preparing material for a thorough overhaul of the existing encyclopedias and textbooks in numerous fields touched upon by the heresy born in the first years of World War II and brought to the attention of the scholarly world and of the literati (the reading and thinking public) between 1950 and 1960." Velikovsky, foreword to *Index to the Works of Immanuel Velikovsky*, by Miller, v.

107. Greenberg to Velikovsky, December 29, 1974, IVP 79:14.

108. *Kronos* 1, no. 1 (Spring 1975): inside cover.

109. Ibid., 64n.

110. Velikovsky to Greenberg, June 5, [1975] and Greenberg to Velikovsky, June 9, [1975], IVP 79:14.

111. S. Talbott to Rose, December 30, 1974, IVP 93:1.

112. Editors of *Pensée*, ed., *Velikovsky Reconsidered* (New York: Warner Books, 1976). The volume opens with two pieces arguing that scientists attempted to suppress Velikovsky's arguments in the debate over the Macmillan publication: David Stove, "The Scientific Mafia," 36–44; and Lynn E. Rose, "The Censorship of Velikovsky's Interdisciplinary Synthesis," 45–52.

113. C. J. Ransom, memorandum of conversation with Stephen Talbott, March 6, 1976, IVP 92:5.

114. David Talbott to Stephen Talbott, April 27, 1976, IVP 99:2.

115. S. Talbott to editors of *Kronos*, May 2, 1976, IVP 99:2; Ransom to Velikovsky, August 8, 1977, IVP 92:5. I have been unable to locate these lists, so I cannot ascertain whether they were the full subscriber lists or some subset. It is unclear what use, if any, *Kronos* made of them.

116. Donald H. Menzel to Stephen L. Talbott, May 13, 1975, IVP 124:2.

117. Menzel to Frederic Jueneman, July 23, 1975, IVP 124:2. This was in response to Jueneman to Menzel, June 16, 1975, IVP 124:2.

118. Warner Sizemore to Velikovsky, October 5, 1975, IVP 97:10.

119. Ransom to Fred W. Dieffenbach (sales director of *Science News*), March 8, 1978, IVP 92:5.

120. Donovan A. Courville, *The Exodus Problem and Its Ramifications: A Critical Examination of the Chronological Relationships between Israel and the Contemporary Peoples of Antiquity*, 2 vols. (Loma Linda, CA: Challenge Books, 1971), 1:128.

121. Courville to Greenberg, March 24, [1977], IVP 73:3.

122. Velikovsky to Rose, May 4, 1975, IVP 93:2.

123. Jerry Rosenthal to Velikovsky, June 28, 1977, IVP 94:2.

124. Philip Glass to Velikovsky, September 9, 1979, IVP 78:19. For contemporary (and rather negative) reviews of the opera, see Donal Henahan, "City Opera: 'Akhnaten' by Glass," *New York Times*, November 5, 1984, C13; Manuela Hoelterhoff, "New Glass Opera: Death in Egypt," *Wall Street Journal*, November 9, 1984, 28. Both articles refer to the Velikovskian inspiration of the work.

125. Lewis Greenberg, ed., "Velikovsky and Establishment Science," special issue of *Kronos* 3, no. 2 (November 1977).

126. Rosenthal to Greenberg, March 10, 1978, IVP 94:2.

127. Extensive correspondence on this affair, which began when Schorr questioned Greenberg's scholarly bona fides to run a journal like *Kronos*, litters the archive: Fred Jueneman to Eddie Schorr, August 15, 1975, and Jueneman to Schorr, August 23, 1975, IVP 96:1; Greenberg to Schorr, February 22, 1977, Velikovsky to Greenberg, March 1, 1977, and Greenberg to Velikovsky, May 19, 1977, IVP 79:15.

128. De Grazia, *Cosmic Heretics*, 70.

129. Quoted in ibid., 285.

130. A sense of this is captured in the charming parody of Velikovsky in Max Apple, "The Yogurt of Vasirin Kefirovsky," in *The Oranging of America and Other Stories* (New York: Grossman, 1976), 61–78.

Conclusion

1. Quoted in Robert J. Schadewald, "Velikovsky: The Last Interview," *Fate* 33 (May 1980): 84; emphasis in original.

2. "Immanuel Velikovsky, at 84, wrote 'Worlds in Collision,'" *Boston Globe*, November 19, 1979, 22; "Author Laid 'Miracles' to Fragments from Planets," *Los Angeles Times*, November 19, 1979, B22; Martin Weil, "Immanuel Velikovsky, 84, Dies: Cosmic Evolution Theories Spurred Scientific Controversy," *Washington Post*, November 19, 1979, B7; George Goodman Jr., "Immanuel Velikovsky, Who Wrote 'Worlds in Collision,' Is Dead at 84," *New York Times*, November 19, 1979, D9; "Immanuel Velikovsky, 84, Dies; Author, Developer of Theory of Colliding Planets," *Baltimore Sun*, November 19, 1979, A12; and "Grand Old Man of the Fringe Dies," *New Scientist*, November 29, 1979, 679.

3. Kendrick Frazier, "The Distortions Continue," *Skeptical Inquirer* 5 (Fall 1980): 32–38.

4. S. F. Kogan, "Sagan versus Velikovsky," *Physics Today* 33, no. 9 (September 1980): 97–98; Robert R. Newton et al., "More on Velikovsky," *Physics Today* 34, no. 4 (April 1981): 15, 72–74; Jean C. Piquette et al., "Velikovsky Again," *Physics Today* 35, no. 6 (June 1982): 12–15, 77–81.

5. R. J. Huggett, "Cranks, Conventionalists and Geomorphology," *Area* 34 (2002): 185. For some of these earlier advocacies of Velikovsky in popular geology, see René Gallant, *Bombarded Earth: An Essay on the Geological and Biological Effects of Huge Meteorite Impacts* (London: John Baker, 1964); and Fred Warshofsky, *Doomsday: The Science of Catastrophe* (New York: Reader's Digest Press, 1977).

6. Uuras Saarnivaara, *Can the Bible Be Trusted?: Old and New Testament Introduction and Interpretation* (Minneapolis: Osterhus, 1983). Velikovsky's chronology was quite popular in books of arcana and occult knowledge. One coffee-table book on mysteries of science and history from 1978 proudly trumpeted both Velikovsky's cosmology and his history, but laid its stronger endorsement on the latter. Francis Hitching, *The Mysterious World: An Atlas of the Unexplained* (New York: Holt, Rinehart and Winston, 1978), 26–32, 168–75.

7. See comments by Martin Gardner, *The New Age: Notes of a Fringe Watcher* (Buffalo, NY: Prometheus Books, 1988), 70.

8. E-mail from C. Leroy Ellenberger to author, April 22, 2011; Ronald H. Fritze, *Invented Knowledge: False History, Fake Science, and Pseudo-Religions* (London: Reaktion Books, 2009), 184; and James P. Hogan, *Kicking the Sacred Cow: Questioning the Unquestionable and Thinking the Impermissible* (Riverdale, NY: Baen, 2004), 236.

9. Charles Ginenthal, *Carl Sagan & Immanuel Velikovsky* (Tempe, AZ: New Falcon Publications, 1995); and Dale Ann Pearlman, ed., *Stephen J. Gould and Immanuel Velikovsky: Essays in the Continuing Velikovsky Affair* (Forest Hills, NY: Ivy Press, 1996). In his book, Ginenthal thanks Earl Milton, Lynn Rose, Frederic Jueneman, and David Talbott, all veterans of the 1970s circle.

10. David Morrison, "Velikovsky at Fifty: Cultures in Collision on the Fringes of Science," *Skeptic* 9, no. 1 (2001): 62–76; Philip C. Plait, *Bad Astronomy: Misconceptions and Misuses Revealed, from Astrology to the Moon Landing "Hoax"* (New York: John Wiley and Sons, 2002), chap. 18; Henrietta W. Lo, "Velikovsky's Interpretation of the Evidence Offered by China in *Worlds in Collision*," *Skeptical Inquirer* 11 (Spring 1987): 282–91; Nigel Calder, *The Comet Is Coming!: The Feverish Legacy of Mr. Halley* (New York: Viking, 1980), 122–23.

11. Vera Kerkhof, "Minutes of the Meeting at the Post House Hotel, Heathrow, April 7, 1979," IVP 85:12.

12. Duane Leroy Vorhees, "The 'Jewish Science' of Immanuel Velikovsky: Culture and Biography as Ideational Determinants" (PhD diss., Bowling Green State University, 1990), 929; Alfred De Grazia, *Cosmic Heretics: A Personal History of Attempts to Establish and Resist Theories of Quantavolution and Catastrophe in the Natural and Human Sciences, 1963 to 1983* (Princeton, NJ: Metron, 1984), 191. On the British Society for Inter-

disciplinary Studies, see Harold Tresman's keynote address at the 1993 SIS Cambridge Conference "Evidence That the Earth Has Suffered Catastrophes of Cosmic Origin in Historical Times," entitled "The SIS, Its History and Achievements: A Personal Perspective," available at http://www.sis-group.org.uk/sis-history.htm.

13. There was a separate problem with Christoph Marx, a German acolyte who, in the 1970s, was accused by Velikovsky and later his family of profiting from Velikovsky's doctrines through his German translations of the books, as well as offering a Velikovskian interpretation of the Holocaust that Velikovsky himself found objectionable. The matter went into litigation in the early 1980s. Frederic B. Jueneman, *Velikovsky: A Personal View* (Glassboro, NJ: Kronos Press, 1980 [1975]), 66–67; Vorhees, "The 'Jewish Science' of Immanuel Velikovsky," 936–39.

14. Henry H. Bauer, *Beyond Velikovsky: The History of a Public Controversy* (Urbana: University of Illinois Press, 1984), 82.

15. Anthony E. Larson, *And the Moon Shall Turn to Blood: Velikovsky Applied to Prophecy* (Newport Beach, CA: Crown Summit Books, 1982).

16. Frederic Jueneman to Carl Sagan, April 30, 1974, IVP 82:17; emphasis in original.

17. Frederic Jueneman, *Limits of Uncertainty: Essays in Scientific Speculation* (Chicago: Industrial Research, 1975).

18. Frederic Jueneman, *Raptures of the Deep: Essays in Speculative Science* (Des Plaines, IL: Cahners, 1995).

19. See, for example, Alfred De Grazia, *Chaos and Creation: An Introduction to Quantavolution in Human and Natural History* (Princeton, NJ: Metron, 1981); and De Grazia, *The Lately Tortured Earth: Exoterrestrial Forces and Quantavolutions in the Earth Sciences* (Princeton, NJ: Metron, 1983). There are several other volumes.

20. David N. Talbott, *The Saturn Myth* (Garden City, NY: Doubleday, 1980).

21. Wallace Thornhill and David Talbott, *The Electric Universe* (Portland, OR: Mikamar, 2007 [2002]), ii.

22. Ibid., i, 56, 59, 100.

23. C. Leroy Ellenberger to Alfred De Grazia, October 6, 1982, IVP 74:3. See also C. Leroy Ellenberger, "Still Facing Many Problems (Part I)," *Kronos* 10, no. 1 (Fall 1984): 87–102; and "Still Facing Many Problems (Part II)," *Kronos* 10, no. 3 (Summer 1985): 1–24; quote on 8.

24. Ellenberger, "Still Facing Many Problems (Part I)," 87–88; emphasis in original.

25. Ellenberger, "Falsifying Velikovsky," *Nature* 316 (August 1, 1985): 386.

26. Ellenberger, "A Lesson from Velikovsky," *Skeptical Inquirer* (Summer 1986): 380–81; emphasis in original.

27. Ellenberger, "Hysterical Velikovskians Flee Own Frankenstein-Mongoose!" *DIO* 7 (February 1997): 31–32.

28. Victor Clube and Bill Napier, *The Cosmic Serpent: A Catastrophist View of Earth History* (New York: Universe Books, 1982), 257. A similar point was made in Derek Ager's summary of neo-catastrophism within establishment geology, where he fumed in passing at Velikovsky, whom he held responsible for damaging the integrity of the

science. "I will not encourage such pseudo-science by giving a reference," he wrote. "I do not want to be associated in any way with such nonsense." Derek Ager, *The New Catastrophism: The Importance of the Rare Event in Geological History* (Cambridge: Cambridge University Press, 1993), 179–80. He also included a disclaimer against creationists in his preface, xi.

29. S. V. M. Clube and W. M. Napier, "Catastrophism Now," *Astronomy Now* (April 1991): 49.

30. Mike Baillie, *Exodus to Arthur: Catastrophic Encounters with Comets* (London: B. T. Batsford, 1999), 170, 172. For a related interpretation of ancient myth as historical evidence, see Elizabeth Wayland Barber and Paul T. Barber, *When They Severed Earth from Sky: How the Human Mind Shapes Myth* (Princeton, NJ: Princeton University, 2005). The Barbers do not cite Velikovsky.

31. L. Piccardi and W. B. Masse, *Myth and Geology* (London: Geological Society, 2007). Contrast this with the earlier Richard Huggett, *Cataclysms and Earth History: The Development of Diluvialism* (Oxford: Clarendon Press, 1989), where Velikovsky is invoked (to be debunked) on 165–69.

32. Erich Auerbach, *Mimesis: The Representation of Reality in Western Literature*, trans. Willard R. Trask (Princeton, NJ: Princeton University Press, 2003 [1953]); Michael Taussig, *Mimesis and Alterity: A Particular History of the Senses* (New York: Routledge, 1993); and Homi K. Bhabha, *The Location of Culture* (New York: Routledge, 2010 [1994]), chap. 4. In his classic 1974 commencement address at Caltech, physicist Richard P. Feynman drew similar conclusions about imitation. See his "Cargo Cult Science," *Engineering and Science* (June 1974): 10–13. I thank Bhavani Raman for bringing this point to my attention.

33. Velikovsky, "My Challenge to Conventional Views in Science," *Kronos* 3, no. 2 (1977): 9.

34. Velikovsky to George Sarton, December 21, 1950, IVP 94:16.

35. Velikovsky to Claude Schaeffer, April 17, 1961; reproduced in Velikovsky, "Ash," *Pensée*, 4, no. 1 (Winter 1973–74): 13.

36. Richard Hofstadter, *The Paranoid Style in American Politics and Other Essays* (New York: Vintage, 2008 [1965]), 32.

37. There is resonance here with Bart Simon's notion of "undead science," that is, science that has ceased to garner mainstream discussion but that is still pursued in all seriousness by a certain community detached from the mainstream. It is not alive, but not quite dead either. Simon's example is cold fusion research. Bart Simon, "Undead Science: Making Sense of Cold Fusion after the (Arti)Fact," *Social Studies of Science* 29 (1999): 61–85; David Goodstein, "Pariah Science: Whatever Happened to Cold Fusion?" *American Scholar* 63 (1994): 527–41.

38. Hogan, *Kicking the Sacred Cow*, chap. 3; and David Marriott, *The Velikovsky Inheritance: An Essay in the History of Ideas* (Cambridge: Vanguard Press, 2006).

39. "The Velikovsky Debate," available at http://www.coasttocoastam.com/shows/2005/03/30. Wal Thornhill and David Talbott appeared on the show on May 18, 2011, to discuss their "Electric Universe" theory, and Velikovsky was again invoked. See "Comets & Electric Universe," available at http://www.coasttocoastam.com/show/2011/05/18.

Coast to Coast regularly features discussion of fringe histories, and Velikovsky is easily classed in that camp. See Fritze, *Invented Knowledge*, 8.

40. Kenneth Chang, "Quakes, Tectonic and Theoretical," *New York Times*, January 15, 2011, WK4.

41. Henry H. Bauer, "Velikovsky's Place in the History of Science: A Lesson on the Strengths and Limitations of Science," *Skeptic* 3, no. 4 (1994): 52.

42. Charles Coulston Gillispie, *Science and Polity in France: The End of the Old Regime* (Princeton, NJ: Princeton University Press, 1980), 257.

43. For example, Keith Parsons, ed., *The Science Wars: Debating Scientific Knowledge and Technology* (Amherst, NY: Prometheus Books, 2003); and Keith M. Ashman and Philip S. Barringer, eds., *After the Science Wars* (London: Routledge, 2001).

44. Paul R. Gross, introduction to *The Flight from Science and Reason*, ed. Paul R. Gross, Norman Levitt, and Martin W. Lewis (New York: New York Academy of Sciences, 1996), 1–7; Gerald Holton, *Science and Anti-Science* (Cambridge, MA: Harvard University Press, 1993), 148; and Francis B. Harrold, Raymond A. Eve, and Geertruida C. de Goede, "Cult Archaeology and Creationism in the 1990s and Beyond," in *Cult Archaeology and Creationism: Understanding Pseudoscientific Beliefs about the Past*, exp. ed., ed. Francis B. Harrold and Raymond A. Eve (Iowa City: University of Iowa Press, 1995), 168.

45. Paul R. Gross and Norman Levitt, *Higher Superstition: The Academic Left and Its Quarrels with Science* (Baltimore: Johns Hopkins University Press, 1994), 222; emphasis in original.

46. Ibid., 268n32.

47. Mario Bunge, "In Praise of Intolerance to Charlatanism in Academia," in *The Flight from Science and Reason*, ed. Gross et al., 105–6; and Keith Soehnk, "Velikovsky Checkmated," *Skeptic* 9, no. 3 (2002): 16.

48. Bauer, "Velikovsky's Place in the History of Science," 55.

49. Fredric Jameson, foreword to *The Postmodern Condition: A Report on Knowledge*, by Jean-François Lyotard, trans. Geoff Bennington and Brian Massumi (Minneapolis: University of Minnesota Press, 1984), viii. My thanks to David Kaiser, who kindly brought this reference to my attention.

50. Michael Mulkay, "Some Aspects of Cultural Growth in the Natural Sciences," *Social Research* 36 (1969): 22–52.

51. See Naomi Oreskes and Erik M. Conway, *Merchants of Doubt: How a Handful of Scientists Obscured the Truth on Issues from Tobacco Smoke to Global Warming* (New York: Bloomsbury, 2010). On the global-warming debates, which are the most politically charged at present, see the survey in Mike Hulme, *Why We Disagree about Climate Change: Understanding Controversy, Inaction and Opportunity* (Cambridge: Cambridge University Press, 2009). On the origins of global warming denial, see Spencer Weart, "Global Warming: How Skepticism Became Denial," *Bulletin of the Atomic Scientists* 67 (January 2011): 41–50.

52. Allan Brandt, *The Cigarette Century: The Rise, Fall, and Deadly Persistence of the Product That Defined America* (New York: Basic Books, 2007), 160 ("scientific skepticism"), 171 ("Frank Statement"). See also Richard Kluger, *Ashes to Ashes: America's*

Hundred-Year Cigarette War, the Public Health, and the Unabashed Triumph of Philip Morris (New York: Vintage, 1997 [1996]), 205–12 (data on money from 206); and Pamela E. Pennock, *Advertising Sin and Sickness: The Politics of Alcohol and Tobacco Marketing, 1950–1990* (DeKalb: Northern Illinois University Press, 2007), 97–100.

53. Chris Mooney, *The Republican War on Science*, rev. ed (New York: Basic Books, 2006 [2005]), 71–74; David Michaels, *Doubt Is Their Product: How Industry's Assault on Science Threatens Your Health* (New York: Oxford University Press, 2008), xi–xii; Oreskes and Conway, *Merchants of Doubt*, 236–37; and Paul N. Edwards, *A Vast Machine: Computer Models, Climate Data, and the Politics of Global Warming* (Cambridge, MA: MIT Press, 2010), 411–12. See also Michael Specter, *Denialism: How Irrational Thinking Hinders Scientific Progress, Harms the Planet, and Threatens Our Lives* (New York: Penguin, 2009), on the public's acceptance of such claims.

54. Oreskes and Conway, *Merchants of Doubt*.

55. Robert Jastrow, "Velikovsky, a Star-Crossed Theoretician of the Cosmos," *New York Times*, December 2, 1979, E22; Jastrow, "Hero or Heretic?" *Science Digest* Special (September/October 1980): 92–96; and S. Fred Singer, ed., *The Universe and Its Origins: From Ancient Myth to Present Reality and Fantasy* (New York: Paragon House, 1990), which contains a whole section of essays on the Velikovsky affair based on a 1985 conference, yet appearing in print a whole decade after his death and years after he had faded from public consciousness.

56. Oreskes and Conway, *Merchants of Doubt*, 32, 154; Mooney, *The Republican War on Science*, 118–19; Gregory J. Feist, *The Psychology of Science and the Origins of the Scientific Mind* (New Haven, CT: Yale University Press, 2006), 221; Michael Polanyi, *The Tacit Dimension* (Chicago: University of Chicago Press, 2009 [1966]), 64; and Harry Collins and Robert Evans, *Rethinking Expertise* (Chicago: University of Chicago Press, 2007), 60.

57. Robert C. Cowan, "Scientists Rally to Quell Anti-Science Political Movements," *Christian Science Monitor*, June 20, 1995, 13.

58. Anthony Aveni, *The End of Time: The Maya Mystery of 2012* (Boulder: University Press of Colorado, 2009), 160.

59. John Major Jenkins, *The 2012 Story: The Myths, Fallacies, and Truth Behind the Most Intriguing Date in History* (New York: Jeremy P. Tarcher/Penguin, 2009), 6.

60. Ibid., 353.

61. Ibid., 162.

62. Erich von Däniken, *Twilight of the Gods: The Mayan Calendar and the Return of the Extraterrestrials*, trans. Nicholas Quintmere (Pompton Plains, NJ: New Page Books, 2010), 112.

63. Geoff Stray, *Beyond 2012: Catastrophe or Awakening?* (Rochester, VT: Bear & Co., 2009 [2005]), 107–10.

64. Randolph Weldon, *Doomsday 2012: A Survival Manual* (Indianapolis: Cork Hill Press, 2003), 134.

65. Ibid., 142. Weldon attributes some of the anti-Velikovskian hostility of the 1950s to anti-Semitism and anti-Communist Russophobia (144).

66. Ibid., 161, 168.

INDEX